VOLUME SIX HUNDRED AND TWENTY SEVEN

METHODS IN ENZYMOLOGY

Enzymatic Polymerizations

METHODS IN ENZYMOLOGY

Editors-in-Chief

ANNA MARIE PYLE
Departments of Molecular, Cellular and Developmental Biology and Department of Chemistry
Investigator, Howard Hughes Medical Institute
Yale University

DAVID W. CHRISTIANSON
Roy and Diana Vagelos Laboratories
Department of Chemistry
University of Pennsylvania
Philadelphia, PA

Founding Editors

SIDNEY P. COLOWICK and NATHAN O. KAPLAN

VOLUME SIX HUNDRED AND TWENTY SEVEN

METHODS IN ENZYMOLOGY

Enzymatic Polymerizations

Edited by

NICO BRUNS
Adolphe Merkle Institute, University of Fribourg, Fribourg, Switzerland;
Department of Pure and Applied Chemistry, University of Strathclyde, Glasgow, United Kingdom

KATJA LOOS
Macromolecular Chemistry and New Polymeric Materials, Zernike Institute for Advanced Materials, University of Groningen, Groningen, The Netherlands

ELSEVIER

ACADEMIC PRESS
An imprint of Elsevier

Academic Press is an imprint of Elsevier
50 Hampshire Street, 5th Floor, Cambridge, MA 02139, United States
525 B Street, Suite 1650, San Diego, CA 92101, United States
The Boulevard, Langford Lane, Kidlington, Oxford OX5 1GB, United Kingdom
125 London Wall, London, EC2Y 5AS, United Kingdom

First edition 2019

ISBN: 978-0-12-817095-3
ISSN: 0076-6879

For information on all Academic Press publications
visit our website at https://www.elsevier.com/books-and-journals

Publisher: Zoe Kruze
Acquisition Editor: Ashlie M. Jackman
Editorial Project Manager: Shellie Bryant
Production Project Manager: James Selvam
Cover Designer: Mark Rogers

Typeset by SPi Global, India

Contents

Contributors

Azis Adharis
Macromolecular Chemistry and New Polymeric Materials, Zernike Institute for Advanced Materials, University of Groningen, Groningen, The Netherlands

Razan Alaneed
Department of Chemistry; Department of Pharmacy, Martin Luther University Halle-Wittenberg, Halle (Saale), Germany

Carmen Alvarado-Canché
Centro de Investigación en Química Aplicada, Saltillo, Coah., México

Véronique Amstutz
HES-SO Valais-Wallis, Sion, Switzerland

Zesheng An
State Key Laboratory of Supramolecular Structure and Materials, College of Chemistry, Jilin University, Changchun, China

Edmondo M. Benetti
Laboratory for Surface Science and Technology, Department of Materials, ETH Zürich, Zürich, Switzerland

Muhammad Humayun Bilal
Department of Chemistry, Martin Luther University Halle-Wittenberg, Halle (Saale), Germany

Nico Bruns
Adolphe Merkle Institute, University of Fribourg, Fribourg, Switzerland; Department of Pure and Applied Chemistry, University of Strathclyde, Glasgow, United Kingdom

Robert J. Chadwick
Department of Pure and Applied Chemistry, University of Strathclyde, Glasgow, United Kingdom

Ashutosh Chilkoti
Department of Biomedical Engineering; Department of Mechanical Engineering and Materials Science, Duke University, Durham, NC, United States

In Jin Cho
Department of Chemical and Biomolecular Engineering (BK21 Plus Program), Korea Advanced Institute of Science and Technology (KAIST); Metabolic and Biomolecular Engineering National Research Laboratory and Institute for the BioCentury; Systems Metabolic Engineering and Systems Healthcare Cross-Generation Collaborative Laboratory, KAIST, Daejeon, Republic of Korea

So Young Choi
Department of Chemical and Biomolecular Engineering (BK21 Plus Program), Korea Advanced Institute of Science and Technology (KAIST); Metabolic and Biomolecular Engineering National Research Laboratory and Institute for the BioCentury; Systems Metabolic Engineering and Systems Healthcare Cross-Generation Collaborative Laboratory; Applied Science Research Institute, KAIST, Daejeon, Republic of Korea

Arxel De-León
Centro de Investigación en Química Aplicada, Saltillo, Coah., México

Sonal Deshpande
Department of Biomedical Engineering, Duke University, Durham, NC, United States

Mohammad Divandari
Laboratory for Surface Science and Technology, Department of Materials, ETH Zürich, Zürich, Switzerland

Carlos Gallardo-Vega
Centro de Investigación en Química Aplicada, Saltillo, Coah., México

Lucia Gardossi
University of Trieste, Department of Chemical and Pharmaceutical Sciences, Trieste, Italy

Georg M. Guebitz
Institute for Environmental Biotechnology, IFA Tulln, University of Natural Resources and Life Sciences, Vienna; ACIB—Austrian Centre of Industrial Biotechnology, Graz, Austria

Chao Guo
Department of Pure and Applied Chemistry, University of Strathclyde, Glasgow, United Kingdom

Nils Hanik
HES-SO Valais-Wallis, Sion, Switzerland

Paola Jiménez-Cárdenas
Centro de Investigación en Química Aplicada, Saltillo, Coah., México

Jun-ichi Kadokawa
Graduate School of Science and Engineering, Kagoshima University, Kagoshima, Japan

Jörg Kressler
Department of Chemistry, Martin Luther University Halle-Wittenberg, Halle (Saale), Germany

Antonio Ledezma-Pérez
Centro de Investigación en Química Aplicada, Saltillo, Coah., México

Sang Yup Lee
Department of Chemical and Biomolecular Engineering (BK21 Plus Program), Korea Advanced Institute of Science and Technology (KAIST); Metabolic and Biomolecular Engineering National Research Laboratory and Institute for the BioCentury; Systems Metabolic Engineering and Systems Healthcare Cross-Generation Collaborative Laboratory; Applied Science Research Institute, KAIST, Daejeon, Republic of Korea

Youngjoon Lee
Department of Chemical and Biomolecular Engineering (BK21 Plus Program), Korea Advanced Institute of Science and Technology (KAIST); Metabolic and Biomolecular Engineering National Research Laboratory and Institute for the BioCentury; Systems Metabolic Engineering and Systems Healthcare Cross-Generation Collaborative Laboratory, KAIST, Daejeon, Republic of Korea

Katja Loos
Macromolecular Chemistry and New Polymeric Materials, Zernike Institute for Advanced Materials, University of Groningen, Groningen, The Netherlands

Karsten Mäder
Department of Pharmacy, Martin Luther University Halle-Wittenberg, Halle (Saale), Germany

Manuel Martínez-Cartagena
Centro de Investigación en Química Aplicada, Saltillo, Coah., México

Seongjin Park
Department of Chemical and Biomolecular Engineering (BK21 Plus Program), Korea Advanced Institute of Science and Technology (KAIST); Metabolic and Biomolecular Engineering National Research Laboratory and Institute for the BioCentury; Systems Metabolic Engineering and Systems Healthcare Cross-Generation Collaborative Laboratory, KAIST, Daejeon, Republic of Korea

Alessandro Pellis
University of York, Department of Chemistry, Green Chemistry Centre of Excellence, York, United Kingdom

Michela M. Pellizzoni
Adolphe Merkle Institute, University of Fribourg, Fribourg, Switzerland

Markus Pietzsch
Department of Pharmacy, Martin Luther University Halle-Wittenberg, Halle (Saale), Germany

Julien Pott
HES-SO Valais-Wallis, Sion, Switzerland

Felice Quartinello
Institute for Environmental Biotechnology, IFA Tulln, University of Natural Resources and Life Sciences, Vienna, Austria

Doris Ribitsch
Institute for Environmental Biotechnology, IFA Tulln, University of Natural Resources and Life Sciences, Vienna; ACIB—Austrian Centre of Industrial Biotechnology, Graz, Austria

Kyle J. Rodriguez
Adolphe Merkle Institute, University of Fribourg, Fribourg, Switzerland

Jorge Romero-García
Centro de Investigación en Química Aplicada, Saltillo, Coah., México

Jonas Steiner
Department of Pharmacy, Martin Luther University Halle-Wittenberg, Halle (Saale), Germany

Camila Utsunomia
HES-SO Valais-Wallis, Sion, Switzerland

Xiao Wang
Institute of Nanochemistry and Nanobiology, College of Environmental and Chemical Engineering, Shanghai University, Shanghai, China

Yunqi Yang
Department of Mechanical Engineering and Materials Science, Duke University, Durham, NC, United States

Stefan Zauscher
Department of Biomedical Engineering; Department of Mechanical Engineering and Materials Science, Duke University, Durham, NC, United States

Hua Zhao
Department of Chemistry and Biochemistry, University of Northern Colorado, Greeley, CO, United States

Manfred Zinn
HES-SO Valais-Wallis, Sion, Switzerland

Preface

When Hermann Staudinger won the 1953 Nobel Prize in Chemistry for introducing the concept of macromolecules, he emphasized that "macromolecular compounds include the most important substances occurring in nature such as proteins, enzymes, the nucleic acids, besides the polysaccharides (…), as well as rubber, and lastly the large number of new, fully synthetic plastics and artificial fibers" (Staudinger, 1953). The naturally occurring polymers are synthesized by the biochemical machinery of living cells. Not surprisingly, a plethora of enzymes can catalyze polymerization reactions in which low molecular weight monomers are linked into high molecular weight macromolecules. These reactions are essential for life, and the resulting polymers are major building blocks of living matter. Importantly, many enzymes have been found to be promiscuous with respect to the reactions they can catalyze so that many non-natural polymerizations can nowadays also be catalyzed by enzymes.

Since Staudinger's time, synthetic macromolecules have become indispensable materials that everyone uses on a daily basis, ranging from plastics in packaging, to light-weight high performance materials in the automotive and the aerospace sector, or as materials for biomedical applications. At the same time, the sheer volume in which plastics are produced and scattered across our planet make synthetic macromolecules a global environmental threat if not properly discarded and recycled. While most synthetic polymers and plastics are produced by chemical means, it is feasible to synthesize polymers via biocatalytic pathways, employing either enzymes or whole cells as polymerization catalysts. There are many reasons to use nature's catalysts for polymerizations, some of the most prominent being the potential environmental benefit that most enzymes are non-toxic and biocompatible, or that they allow to produce polymers with chemo- or regioselectivity. Moreover, some biologically derived polymers such as polyhydroxyalkanoates (PHAs) are biodegradable polymers that can play a role in tackling the plastic waste crisis. Biocatalysis in polymer science is a highly active and burgeoning field, as exemplified by multitude of scientific papers being published each year (Douka, Vouyiouka, Papaspyridi, & Papaspyrides, 2018; Hollmann & Arends, 2012; Jiang & Loos, 2016; Kobayashi, Uyama, & Kadokawa, 2019; Loos, 2010; Rodriguez et al., 2018; Shoda, Uyama, Kadokawa,

Kimura, & Kobayashi, 2016), and by the implementation of biotechnological routes toward monomers and polymers in industry, e.g., for the synthesis of PHAs or the synthesis of lactic acid. We therefore decided to collate important methods for the biocatalytic synthesis of polymers, monomers, and the modification of polymers by enzymatic means in this volume of *Methods in Enzymology*. The chapters are organized into six sections that cover five classes of polymers, that is polyesters, DNA, polysaccharides and glycomonomers, vinyl polymers, and poly(aromatics), as well as the use of enzymes to modify the surface of polymeric materials. The type of polymerizations includes polycondensations and ring-opening polymerizations (ROPs), whole-cell polyester synthesis, free radical and controlled radical chain growth polymerizations, as well as radical addition polymerizations.

The first part of this volume is dedicated to polyesters, which are an important class of technical relevant polymers. In Chapter 1, Zhao reviews the methods of lipase-catalyzed polymerizations to polyesters in a variety of non-aqueous media that include organic solvents, ionic liquids, and supercritical CO_2. The reactions include polycondensations and ROPs. In Chapter 2, Pellis and Gardossi provide protocols for the enzyme catalyzed solvent-free synthesis of polyesters in a thin film method under reduced pressure, using immobilized cutinase and lipase. They highlight the importance of computational methods to design and optimize experimental conditions. These methods are the visualization of protein structures and their active sites in three-dimensional models, as well as design of experiment (DoE). An advantage of enzymatic polymerizations over their chemical counterparts is that enzymes can discriminate between similar functional groups on the monomers, thereby selectively reacting a subset of them in the polymerization reaction. This is demonstrated in Chapter 3 by Kressler and coworkers who prepare polyesters that feature hydroxyl groups on their backbone by lipase-catalyzed polycondensations. The subsequent chemical modification of some of these functional groups with hydrophobic fatty acids yields amphiphilic polymers. They can self-assemble in water into microparticles or stabilize cubosomes. Polyesters are not only readily accessible by enzymatic means, but also occur naturally in many organisms such as in plants or bacteria. Some bacteria use PHAs as carbon and energy storage. These polyesters are biodegradable and possess many desirable properties. Fermentation routes have been established to produce PHAs with tunable mechanical properties, even on the industrial scale. Zinn and coworkers provide state-of-the-art methods for the biotechnological production of PHAs in Chapter 4. In Chapter 5, Lee and coworkers review the possibility to synthesize the technologically important polymer poly(lactate) (PLA) by fermentative approaches. To this end, microbial strains that produce PHAs were hijacked and reengineered to produce PLA, feeding on simple carbon sources such as glucose.

The second part of this volume shows that also nucleic acids can be used as polymers. Many DNA nanotechnology applications require high molecular weight DNA in amounts that are difficult to synthesize by classical nucleic acid synthesis on solid-phase or on DNA templates. In Chapter 6, Chilkoti, Zauscher, and coworkers comprehensively review the field and methods for the synthesis of high molecular weight single-stranded homo- and co-polynucleotides by terminal deoxynucleotidyl transferase. These polymerizations can be conducted in solution and from surfaces.

The third section of this volume covers enzymatic routes toward polysaccharides and glycopolymers, as well as sugar-containing monomers. In Chapter 7, Kadokawa reviews the methods for α-glucan phosphorylase-catalyzed polymerizations of α-D-glucose 1-phosphate into amylose and into amylose-containing graft polymers. Moreover, the polymerization of amylose-analogs into, e.g., amylosamine, is presented. An alternative strategy toward glycopolymer is to prepare sugar-containing monomers that can then be polymerized by other means. In Chapter 8, Adharis and Loos present various methods for the synthesis of carbohydrate-containing vinyl monomers. These monomers are important building blocks for polymers, since they allow to incorporate renewable biomass into polymers that are synthesized by conventional radical methods. Moreover, the resulting carbohydrate-containing polymers are often bioactive and therefore interesting for biomedical applications.

The fourth section of the volume focuses on radical polymerizations of vinyl monomers (e.g., acrylates, methacrylates, acrylamides, styrene, and styrene derivatives). They are an import class of polymerizations because the resulting polymers are widely used in many applications. Moreover, radical polymerizations are less sensitive to the presence of functional groups and protic solvents than many other types of polymerizations. The simplest form of radical chain growth polymerization is free radical polymerizations in which the radicals react in an uncontrolled way. Chapter 9, authored by Bruns and coworkers, describes the use of horseradish peroxidase (HRP) as a catalyst to initiate free radical polymerizations. The enzymatic polymerization techniques is successfully applied to different vinyl monomers, including 4-acryloylmorpholine, 2-hydroxyethyl methacrylate, and poly(ethylene glycol) methyl ether acrylate. The advantage of free radical polymerizations lies in its simplicity. However, control over molecular weight and the synthesis of polymers with narrow molecular weight distribution and complex molecular architecture (e.g., block copolymers) is not possible or very difficult to achieve. In contrast, reversible-deactivation radical polymerizations (also termed controlled radical polymerizations) have emerged as powerful methods to obtain precise macromolecules. Controlled radical polymerizations can also be catalyzed or initiated by enzymes. In Chapter 10, Benetti, Bruns and coworkers present their methods for biocatalytic atom transfer radical polymerizations (bioATRP) in solution and from surfaces, which involve reversible halogen transfer between enzymes such as HRP or hemoglobin, and the growing polymer chains. Reversible addition-fragmentation chain transfer (RAFT) polymerization is another very popular controlled radical polymerization. In Chapter 11, Wang and An review the published literature on peroxidase-initiated RAFT polymerization and provide a wealth of experimental protocols for bio-RAFT, including homopolymer synthesis, multiblock synthesis, synthesis of high molecular weight polymers, and a protocol for bioRAFT-based polymerization-induced self-assembly.

The next section of the volume is related to the previous one, as it also covers radical polymerizations. However, the enzymatic synthesis of poly(aromatics) is not a chain growth reaction, but an addition polymerization of aromatic radicals that are created by oxidoreductases. In Chapter 12, Romero-García and coworkers review and present methods for the enzymatic synthesis of polyaniline as an archetype for a conductive polymer. The polymerization utilizes the native reactivity of peroxidases in the presence of hydrogen peroxide and can be easily adapted to the synthesis of other conjugated polymers.

The final section and chapter of the volume (Chapter 13) does not cover a polymer synthesis, but instead gives intriguing examples of enzymatic methods to modify and functionalize polymeric materials once they have been synthesized (independent if the synthesis was through an enzymatic route or by conventional chemical means). Guebitz and Ribitsch provide an overview over the possibilities to modify the surface of polyesters such as poly(ethylene terephthalate) (PET) and PLA by treatment with cutinases, lipases, and other enzymes. Moreover, the methods to characterize the resulting polymer samples are summarized and described in detail.

We hope that this methods collection will help colleagues to synthesize useful polymers by biocatalytic means, inspire the reader to elucidate open scientific and technological questions of enzymatic polymerizations, and motivate the reader to exploit the vast opportunities that this field has to offer. We would like to thank all authors for their excellent contributions to this volume of *Methods in Enzymology* and for their continuous efforts to advance the field of enzymatic polymerizations. We believe that the lessons learned from enzymatic polymerizations will not only open up synthetic pathways to novel materials, but will also contribute to a deeper understanding of the polymerization reactions that happen in nature. As Herman Studinger concluded his Nobel Lecture: "In the light of this new knowledge of macromolecular chemistry, the wonder of life in its chemical aspect is revealed in the astounding abundance and masterly macromolecular architecture of living matter" (Staudinger, 1953).

Nico Bruns
University of Strathclyde, Glasgow, United Kingdom

Katja Loos
University of Groningen, Groningen, The Netherlands

References

Douka, A., Vouyiouka, S., Papaspyridi, L.-M., & Papaspyrides, C. D. (2018). A review on enzymatic polymerization to produce polycondensation polymers: The case of aliphatic polyesters, polyamides and polyesteramides. *Progress in Polymer Science*, *79*, 1–25. https://doi.org/10.1016/j.progpolymsci.2017.10.001.

Hollmann, F., & Arends, I. W. C. E. (2012). Enzyme initiated radical polymerizations. *Polymers*, *4*(1), 759–793.

Jiang, Y., & Loos, K. (2016). Enzymatic synthesis of biobased polyesters and polyamides. *Polymers*, *8*(7), 243.

Kobayashi, S., Uyama, H., & Kadokawa, J.-i. (Eds.), (2019). *Enzymatic polymerization towards green polymer chemistry*. Singapore: Springer.

Loos, K. (Ed.), (2010). *Biocatalysis in polymer chemistry*. Weinheim, Germany: Wiley-VCH.

Rodriguez, K. J., Gajewska, B., Pollard, J., Pellizzoni, M. M., Fodor, C., & Bruns, N. (2018). Repurposing biocatalysts to control radical polymerizations. *ACS Macro Letters*, 7(9), 1111–1119. https://doi.org/10.1021/acsmacrolett.8b00561.

Shoda, S.-i., Uyama, H., Kadokawa, J.-i., Kimura, S., & Kobayashi, S. (2016). Enzymes as green catalysts for precision macromolecular synthesis. *Chemical Reviews (Washington, DC, United States)*, *116*(4), 2307–2413. https://doi.org/10.1021/acs.chemrev.5b00472.

Staudinger, H. (1953). *Macromolecular chemistry, Nobel lecture 1953*. Retrieved from https://www.nobelprize.org/prizes/chemistry/1953/staudinger/lecture/.

CHAPTER ONE

Enzymatic polymerization to polyesters in nonaqueous solvents

Hua Zhao*
Department of Chemistry and Biochemistry, University of Northern Colorado, Greeley, CO, United States
*Corresponding author: e-mail addresses: hua.zhao@unco.edu; huazhao98@gmail.com

Contents

Abstract

Aliphatic polyesters are thermoplastic and biodegradable polymers with promising potentials to substitute synthetic polymers derived from petrochemicals. In particular, polylactides (PLAs) and other polylactones can be renewable and biocompatible. A more benign approach for polyester synthesis is the enzymatic polycondensation or ring-opening polymerization (ROP) reactions, whose outcomes largely depend on the reaction conditions including solvents, water content and temperature. This chapter illustrates several examples of enzymatic polymerization to polyesters using various solvents (i.e., organic solvents, supercritical fluids, ionic liquids, and aqueous biphasic systems). Hydrophobic solvents containing little water tend to promote the enzymatic polymerization and lead to high molecular masses of polyesters. Since some enzymatic polymerization reactions are performed at high temperatures (such as ring-opening

Methods in Enzymology, Volume 627
ISSN 0076-6879
https://doi.org/10.1016/bs.mie.2019.03.002

polymerization of lactide at >100 °C), these processes demand solvents with high boiling points (such as many ionic liquids). Supercritical fluids (such as supercritical CO_2) can be "green" solvents, but their compatibility with enzymes and their practicability of scaling up remain as challenges. On the other hand, ionic liquids can be tailored to be compatible with enzymes and to have high thermal stability although the studies of their uses in enzymatic polycondensation and ROP reactions are still at an early stage.

1. Introduction

Aliphatic polyesters are biodegradable and thermoplastic polymers that have comparable mechanical properties with polystyrene (PS) or polyethylene terephthalate (known as PET or PETE) (Drumright, Gruber, & Henton, 2000). These polyesters have numerous applications in packaging, textiles, and horticultural industries (Auras, Harte, & Selke, 2004). More attractively, due to their high biocompatibility and low toxicity, polylactides and other polylactones are being widely adopted by biomedical sectors as controlled delivery carriers for drugs (Piotrowska & Sobczak, 2015; Seyednejad, Ghassemi, van Nostrum, Vermonden, & Hennink, 2011), tissue engineering scaffolds, surgical suture, and bone fixation materials (Albertsson & Varma, 2003; Cheng, Deng, Chen, & Ruan, 2009; Nampoothiri, Nair, & John, 2010; Zinck, 2009). Aliphatic polyesters are commonly produced by two methods: polycondensation and ring-opening polymerization (ROP), traditionally achieved by anionic, cationic or metallic catalysts (Kamber et al., 2007; Kobayashi, 2009; Labet & Thielemans, 2009). Due to the rising concern of residual metals in polyesters for medical and food applications, many recent studies focused on organocatalysts (Kamber et al., 2007) and biocatalysts (Kobayashi, 2015; Pretula, Slomkowski, & Penczek, 2016; Zhang et al., 2014). In particular, biocatalysts such as lipases and esterases have been extensively evaluated in the polycondensation of diacid derivatives and diols, or oxyacid derivatives (Kobayashi, Uyama, & Kimura, 2001), as well as in the ROP of lactides and other lactones (Kobayashi, 2015; Kundys, Białecka-Florjańczyk, Fabiszewska, & Małajowicz, 2018; Lecomte & Jérôme, 2012; Matsumura, 2006; Yang, Ge, Zhao, Shi, & Li, 2011; Yang, Yu, et al., 2011).

Polylactide (PLA) or poly(lactic acid) is usually synthesized from renewable lactic acid or lactide. Lactide (LA) has two chiral centers, and thus there are three unique stereoisomers L,L-lactide (L-LA), D,D-lactide (D-LA), and meso-lactide (DL-LA) (see Fig. 1), which correspond to polyesters of PLLA, PDLA, and PDLLA respectively. Some lactones are naturally occurring

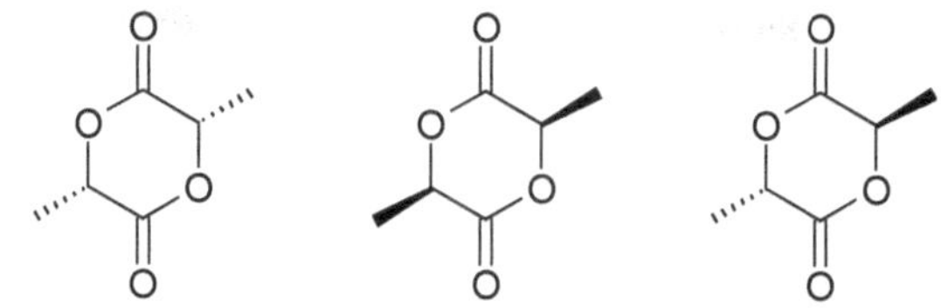

Fig. 1 Structures of lactides.

(Janecki, 2014) while many others are synthesized; for example, ε-caprolactone is synthesized in a large quantity from the oxidation of cyclohexanone by peracetic acid (Labet & Thielemans, 2009). PLA and other polylactones can be prepared by several methods: (i) Direct polycondensation polymerization often produces a small-molecular-mass polymer with less desirable mechanical properties; (ii) Azeotropic condensation polymerization can prepare high-molecular-weight PLA (M_n up to 300,000 Da) (Ajioka, Enomoto, Suzuke, & Yamaguchi, 1995); however, this method suffers from several drawbacks: high temperature, byproduct (primarily water) removal during the reaction, and an extended reaction time; (iii) Solid state polymerization (SSP) is conducted at a temperature between the glass transition temperature and the melting temperature (Moon, Taniguchi, Miyamoto, Kimura, & Lee, 2001) and could yield high molecular masses with minimized side reactions, but this reaction often requires a significantly longer time than those in melt state or solutions; (iv) Ring-opening polymerization (*ROP*) is conventionally conducted as the coordination polymerization using metal derivatives including tin(II) octoate [$Sn(Oct)_2$] or tin(II) butoxide (Kowalski, Libiszowski, Duda, & Penczek, 2000), anionic polymerization, or cationic polymerization in bulk (solventless) (Nampoothiri et al., 2010; Pretula et al., 2016). As pointed out earlier, one concern of this approach is the likely metal residues in polyester products, which potentially could be problems for medical, food and electronic applications although it has been argued that 10 ppm tin(II) octoate residue in PLA is generally not toxic (Pretula et al., 2016).

Many enzymatic polymerization reactions have been carried out under in bulk (solventless) conditions. Although the solvent-free condition eliminates the solvent and could be a "green" approach, some of these reactions could be problematic: (i) Some reactants (especially at high concentrations) are not compatible with the enzyme. For example, lactides at melted state could be inhibitory to the enzymatic propagation due to their polar and hydrophilic natures (Jérôme & Lecomte, 2008; Piotrowska & Sobczak, 2015).

(ii) The polymer product tends to solidify from the reaction mixture, and thus limits the further polymer chain growth and leads to reaction heterogeneity and a large polydispersity index (PDI). Therefore, many groups have focused on the use of various (co-)solvent systems (such as organic solvents, ionic liquids, supercritical fluids, and aqueous biphasic systems) for the enzymatic polycondensation and ROP to polyesters.

2. General methods

2.1 Equipment

- Karl-Fischer titrator
- HPLC/Gel Permeation Chromatography (GPC)/Size Exclusion Chromatography (SEC) equipped with UV–visible detector, Refractive Index detector (RID), or evaporative light-scattering detector (ELSD)
- Nuclear magnetic resonance (NMR) spectrometer
- Fourier-transform infrared spectrometer (FT-IR)
- Rudolph Autopol III polarimeter
- Analytical balance
- Mini centrifuge

2.2 Measurements

The reaction mixture (50 μL) was diluted by 1.0 mL $CDCl_3$. After centrifugation, the clear solution was subject to ^{1}H NMR analysis. For lactide polymerization, the ^{1}H NMR peak areas of methine proton (CH) next to the carbonyl group in the monomer (5.04 ppm) and in the polymer (5.16 ppm) were used to determine the lactide conversion (Yoshizawa-Fujita, Saito, Takeoka, & Rikukawa, 2008). For ε-caprolactone, the peak size of the methylene proton adjacent to the carbonyl group in the monomer (4.23 ppm) and poly(ε-caprolactone)/PCL (4.07 ppm) were used to calculate the monomer conversion (Piotrowska, Sobczak, Oledzka, & Combes, 2016). The mass-average molecular mass (M_w) and polydispersity index (PDI) of the polyester (10 mg/mL in THF) were measured by a GPC/HPLC equipped with UV–visible dual-wavelength, refractive index, and ELSD detectors, two PLgel MIXED-B 10 μm, 300 × 7.5 mm columns (Agilent) eluted with 1.0 mL/min THF at 40 °C (Mena et al., 2013). The calibration curve was generated by using polystyrene standards in the range of M_w 570–62,500 Da (Omay & Guvenilir, 2013). The polyester structure was confirmed by ^{1}H and ^{13}C NMR in $CDCl_3$ and FT-IR (Barrera-Rivera, Flores-Carreón, & Martínez-Richa, 2008; Dong et al., 2012; Nobes, Kazlauskas, & Marchessault, 1996;

Omay & Guvenilir, 2013; Sobczak, 2012; Yoshizawa-Fujita et al., 2008). The specific optical rotation of polylactide in chloroform (10 mg/mL), [α], was determined by a Rudolph Autopol III polarimeter at the wavelength of 589 nm under room temperature.

2.3 Materials

These materials can be obtained from Sigma-Aldrich (St. Louis, MO): L-(−)-lactide, DL-lactide, D-(+)-lactide, ε-caprolactone, pentadecanolide, divinyl adipate, diethyl adipate, 1,4-butanediol, *Candida antarctica* lipase B (CALB) immobilized on acrylic resin (so called Novozym 435) (Catalog # L4777), *Candida antarctica* lipase B (free CALB), lipase PS from *Burkholderia cepacia*, lipase from *Pseudomonas fluorescens* (lipase PF), lipase from porcine pancreas (PPL) type II, *Pseudomonas cepacia* lipase immobilized on Celite (known as lipase PS-C), and Amano lipase A from *Aspergillus niger*. 1-Butyl-3-methylimidazolium hexafluorophosphate ([BMIM][PF_6], high purity) and 1-butyl-3-methylimidazolium bis(trifluoromethylsulfonyl)imide ([BMIM][Tf_2N], synthesis grade), acquired from VWR (Radnor, PA) are products of Merck KGaA (EMD Millipore Corporation, Billerica, MA). Other ionic liquids can be prepared by the literature methods (Tang, Baker, Ravula, Jones, & Zhao, 2012; Wasserscheid & Welton, 2008; Zhao et al., 2008).

3. Enzymatic polycondensation

Polyesters can be produced by lipase-catalyzed polycondensation of dicarboxylic acids or their derivatives (such as esters, activated esters, cyclic and polymeric anhydrides) with diols (Douka, Vouyiouka, Papaspyridi, & Papaspyrides, 2018; Kobayashi, 1999, 2015; Kobayashi et al., 2001; Uyama & Kobayashi, 2006). Some commonly used dicarboxylic acids and their derivatives, nucleophiles (diols, oxiranes, and sugars), and reaction media (organic solvents, ionic liquids, supercritical fluids) are summarized in Table 1. Many dicarboxylic acids and their alky esters show low enzymatic activities in (trans)esterification polymerization reactions. Therefore, activated esters of halogenated alcohols (such as 2-chloroethanol, 2,2,2-trifluoroethanol, and 2,2,2-trichloroethanol), vinyl esters (generating a volatile byproduct acetaldehyde from the tautomerization of vinyl alcohol), and acid anhydrides are more desirable substrates for polycondensation reactions. Many of these reactions have been carried out in solventless conditions; in the following sections, several representative reactions using nonaqueous solvents are discussed.

Table 1 Commonly used acyl donors, nucleophiles and solvents in enzymatic polycondensations.

Substrates or solvents		Examples
Acyl donors	Dicarboxylic acids	Adipic acid, sebacic acid, isophthalic acid, succinic acid, glutaric acid
	Dicarboxylic acid alkyl esters	Dimethyl succinate, dimethyl adipate, diethyl adipate, dimethyl terephthalate, dimethyl isophthalate, diethyl sebacate, dimethyl maleate, dimethyl 5-hydroxyisophthalate
	Dicarboxylic acid-activated esters	Bis(2,2,2-trichloroethyl)glutarate, bis(2,2,2-trichloroethyl)alkanedioates, bis(2-chloroethyl)succinate, bis(2,2,2-trichloroethyl)adipate, bis(2,2,2-trifluoroethyl)sebacate, bis(2-chloroethyl) fumarate, bis(2,2,2-trichloroethyl)-*trans*-3,4-epoxyadipate, bis(2-chloroethyl)2,5-dibromoadipate, divinyl adipate, divinyl sebacate, divinyl esters of isophthalic acid, terephthalic acid, *p*-phenylene diacetic acid
	Cyclic and polymeric anhydrides	Succinic anhydride, glutaric anhydride, diglycolic anhydride, itaconic anhydride, poly(azelaic anhydride)
	Oxyacid derivatives	10-Hydroxydecanoic acid, 11-hydroxyundecanoic acid, 10-hydroxyundecanoic acid, ricinoleic acid, 12-hydroxyoctadecanoic acid, 16-hydroxyhexadecanoic acid, 12-hydroxydodecanoic acid, cholic acid, lactic acid methyl 6-hydroxyhexanoate, dimethyl β-hydroxyglutarate, ethyl esters (of 3- and 4-hydroxybutyric acids, 5- and 6-hydroxyhexanoic acids, 5-hydroxydodecanoic acid, and 15-hydroxypentadecanoic acid)
Nucleophiles	Diols, oxiranes, sugars	1,4-Butanediol, 1,6-hexanediol, 1,8-octanediol, neopentyl glycol, diethylene glycol, 3,3,4,4,5,5,6,6-octafluorooctan-1,8-diol, xylylene glycol (1,4-benzenedimethanol), 1,4-cyclohexanedimethanol, glycerol, 1,2,4-butanetriol, 1,2,6-hexanetriol, oxiranes (glycidyl phenyl ether and benzyl glycidate), sugars (sucrose, sorbitol, mannitol and *meso*-erythritol)
Solvents	Organic solvents, ionic liquids, supercritical fluids	Diisopropyl ether, diphenyl ether, 1,2-dimethoxybenzene (eratrole), toluene, diethyl ether, [BMIM][BF_4], [BMIM][PF_6], supercritical carbon dioxide (scCO_2), supercritical fluoroform

Note: summarized from Refs. Kobayashi (2015); Kobayashi et al. (2001); Uyama & Kobayashi (2006).

3.1 Polycondensation of vinyl ester in an organic solvent

Uyama and Kobayashi reported an effective polycondensation of divinyl adipate with glycols catalyzed by lipase PF in diisopropyl ether (Uyama & Kobayashi, 1994) as shown in Fig. 2. Divinyl adipate (0.40 g, 2.0 mmol) is mixed with 0.18 g (2.0 mmol) 1,4-butanediol, 10 mL diisopropyl ether, and 0.20 g lipase from *Pseudomonas fluorescens*. The mixture was gently agitated at 45 °C for 48 h, followed by the addition of 2.0 mL chloroform to dissolve the polymer and then the insoluble fraction is filtered off. After evaporating chloroform, methanol is added to precipitate the polyester, which is isolated and dried for GPC analysis.

3.2 Polycondensation of diester in an ionic liquid

Ionic liquids are organic salts that melt at temperatures lower than 100 °C. Some common cations and anions are sketched in Fig. 3. As a new type of nonaqueous solvents, ionic liquids have become alternative solvents or (co-)

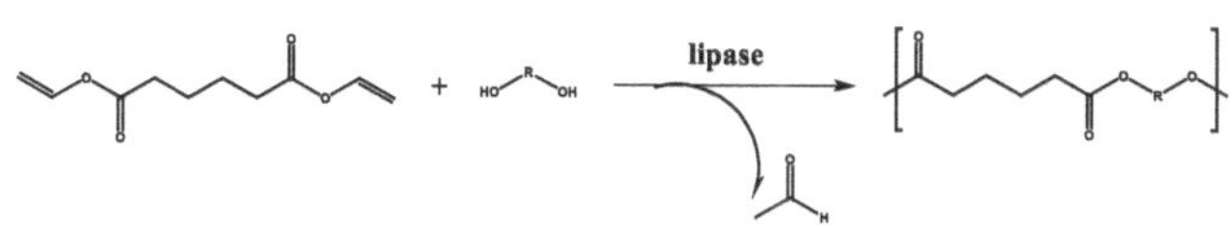

Fig. 2 Enzymatic polymerization of divinyl adipate with glycols.

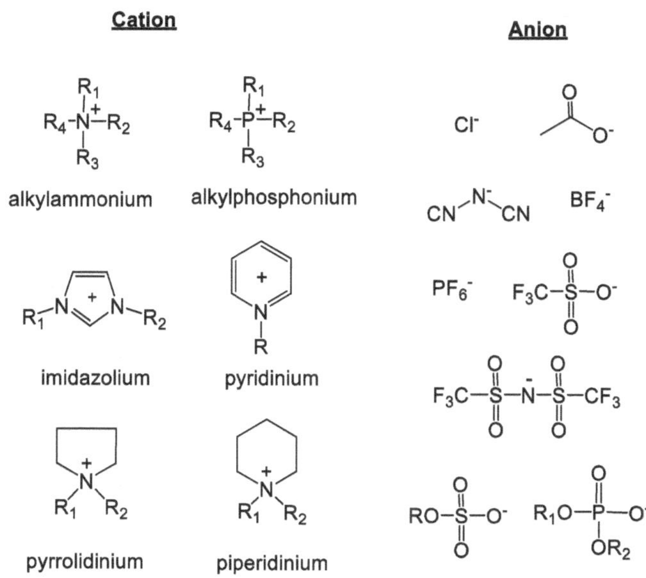

Fig. 3 Representative ionic liquids.

catalysts in many applications including organic catalysis (Jain, Kumar, Chauhan, & Chauhan, 2005; Wasserscheid & Welton, 2008; Zhao & Malhotra, 2002), inorganic synthesis (Endres & Welton, 2008), biocatalysis (Moniruzzaman, Nakashima, Kamiya, & Goto, 2010; van Rantwijk & Sheldon, 2007; Zhao, 2016), polymerization (Kubisa, 2004), and engineering fluids (Brennecke & Maginn, 2003; Zhao, 2006). Ionic liquids can be customized to have different physiochemical properties targeting individual applications (so called "task-specific ionic liquids") (Wasserscheid & Welton, 2008). Due to the tailorable properties of ionic liquids for different biocatalytic needs, a variety of different enzymatic reactions have been studied in ionic liquids (Kragl, Eckstein, & Kaftzik, 2002; Moniruzzaman et al., 2010; Park & Kazlauskas, 2003; van Rantwijk & Sheldon, 2007; Zhao, 2010).

The enzymatic polycondensations of dicarboxylic acid alkyl esters with 1,4-butanediol have been evaluated in some common ionic liquids such as [BMIM][PF_6] and [BMIM][BF_4] (Nara, Harjani, Salunkhe, Mane, & Wadgaonkar, 2003; Uyama, Takamoto, & Kobayashi, 2002). As an example, diethyl adipate (2 mmol, 0.405 g), 1,4-butanediol (2 mmol, 0.180 g), 2.0 mL [BMIM][PF_6], and 50 mg *Pseudomonas cepacia* lipase immobilized on Celite (known as lipase PS-C) are mixed and stirred under reduced pressure (60 mmHg) at 60 °C for 7 days. At the completion of reaction, the reaction mixture is cooled to room temperature first, followed by the addition of 2.0 mL chloroform to dissolve the polyester under stirring. After filtering off the lipase, chloroform is evaporated under vacuum. The polymer is precipitated in cold methanol and then collected by centrifugation (or vacuum filtration). The polyester is dried in air for 24 h before the GPC analysis.

3.3 Polycondensation of divinyl adipate and 1,4-butanediol in $scCO_2$

A literature method is followed (Takamoto, Uyama, & Kobayashi, 2001). Divinyl adipate (2.0 g, 10.0 mmol) is mixed with 0.90 g (10.0 mmol) 1,4-butanediol and 100 mg Novozym 435 in a stainless steel vessel (10 mL), followed by the injection of liquid CO_2 at 10 MPa from a CO_2 delivery pump (JASCO SCF-Get) attached to a CO_2 gas cylinder. The reactor is controlled at 60 °C and 10 MPa using a back pressure regulator to maintain the supercritical state. The reaction is gently stirred for 24 h and then carefully degassed. THF (10 mL) is added to dissolve the polymer for further GPC analysis.

4. Enzymatic ring-opening polymerization (ROP)

4.1 ROP in organic solvents and ionic liquids

The protocol in this section (see Figs. 4 and 5) is based on several literature methods (Henderson, Svirkin, Gross, Kaplan, & Swift, 1996; Kobayashi, Takeya, Sudu, & Uyama, 1998; Nobes et al., 1996; Yoshizawa-Fujita et al., 2008; Zhao, Nathaniel, & Merenini, 2017). The enzymatic ROP reactions of lactones are typically conducted at 50–130 °C, therefore, the choice of organic solvents is limited by their boiling points; in particular, the ROP of lactide is often performed at above 100 °C (Zhao, 2018). A short list of high-boiling solvents includes toluene, xylene, *N,N*-dimethylformamide (DMF), glymes, 1-methyl-2-pyrrolidone, and *N,N*-dimethylacetamide (DMA) although some of them are not compatible with enzymes (Zhao et al., 2017).

Table 2 collects some representative organic solvents, ionic liquids and enzymes used in the enzymatic ROP of lactones. Novozym 435 is a common lipase used, which contains about 10 wt% free *Candida antarctica* lipase B (CALB) attached to a macroporous acrylic resin (Deng & Gross, 1999; Loeker et al., 2004). All organic solvents are anhydrous grades and all ionic liquids are dried in a vacuum oven at 80 °C for 3 days (the water content is usually controlled under 0.01–0.03 wt%). Novozym 435 is dried over P_2O_5 in a closed container for 1 week. The Karl Fischer (KF) Coulometric

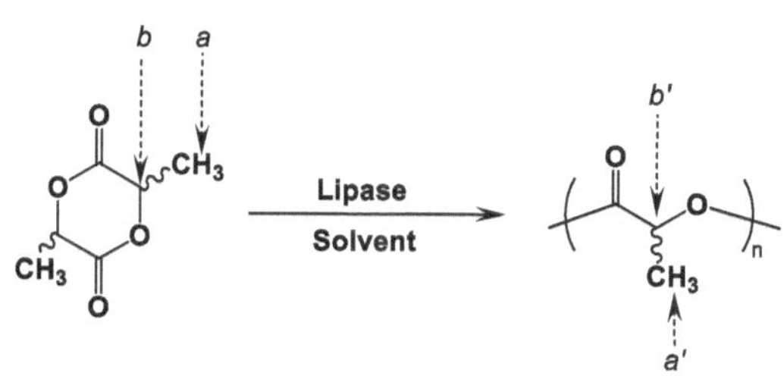

Fig. 4 Enzymatic ROP synthesis of PLA.

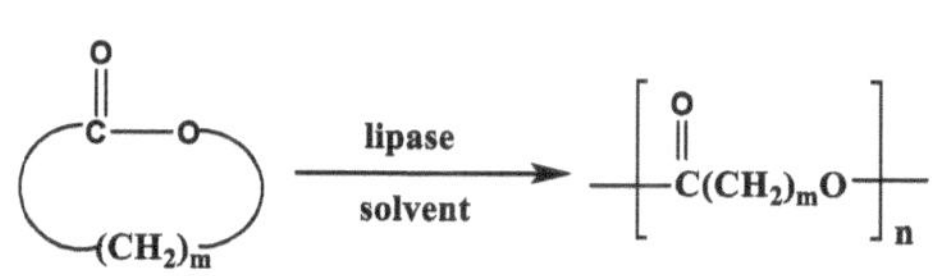

Fig. 5 Enzymatic ROP of common lactones (β-propiolactone when $m=2$; γ-butyrolactone when $m=3$; δ-valerolactone when $m=4$; ε-caprolactone when $m=5$).

Table 2 Examples of solvents and lipases used in ROP reactions.

Monomer	Solvent (temperature)	Lipase	Reference
DL-lactide	Toluene at 60 °C	Novozym 435	Hans, Keul, and Moeller (2009)
L-lactide	Toluene at 100 °C	Free CALB	Yoshizawa-Fujita et al. (2008)
L-lactide, DL-lactide	Toluene at 60–80 °C	Lipase from *Candida antarctica*, *Pseudomonas cepacia*, or *Pseudomonas fluorescens*	Sobczak (2012)
D,L-lactide	Toluene at 80 °C	Novozym 435 and free CALB	Omay and Guvenilir (2013)
L-lactide	Mixture of toluene and triethylamine (2:1 ratio) at 70–90 °C	Lipase from *Burkholderia cepacia*	Duchiron, Pollet, Givry, and Avérous (2015)
D-lactide		Novozym 435	
ε-caprolactone	*n*-Hexane at 40 °C	PPL	Knani, Gutman, and Kohn (1993)
β-butyrolactone, β-propiolactone, γ-butyrolactone and ε-caprolactone	*n*-Hexane or isooctane at r.t. to 69 °C	PPL and lipase from *Pseudomonas cepacia*	Nobes et al. (1996)
β-butyrolactone	Isooctane at 80 °C	Thermophilic lipases from the ESL-001 ClonZyme library	Xie, Li, Chen, and Wang (1997)
ε-caprolactone, δ-valerolactone	Isooctane at 45 °C	*Pseudomonas fluorescens* lipase	Kobayashi et al. (1998)
ε-caprolactone	Heptane at 65 °C	PPL	Henderson et al. (1996)
ε-caprolactone	Isopropyl ether, toluene, butyl ether and isooctane at 70 °C	Novozym 435	Kumar and Gross (2000)
ε-caprolactone	Toluene, cyclohexane and *n*-hexane at 45–80 °C	Novozym 435	Yang, Ge, et al. (2011)
ε-caprolactone	Toluene at 70 °C	Novozym 435	Panova and Kaplan (2003)
ε-caprolactone	*n*-Heptane at 50–70 °C	Lipase from *Yarrowia lipolytica*	Barrera-Rivera et al. (2008)
ε-caprolactone, ω-pentadecalactone	Toluene at 70 °C	Immobilized *Humicola insolens* cutinase	Hunsen, Abul, Xie, and Gross (2008)

Table 2 Examples of solvents and lipases used in ROP reactions.—cont'd

Monomer	Solvent (temperature)	Lipase	Reference
δ-valerolactone	Toluene, cyclohexane and *n*-hexane at 70 °C	Esterase from *Archaeoglobus fulgidus*	Cao et al. (2012)
ε-caprolactone	Toluene, diisopropyl ether, *n*-hexane at 40–80 °C	Novozym 435	Ozsagiroglu, Iyisan, and Avcibasi-Guvenilir (2012)
ε-caprolactone	Toluene, cyclohexane, *n*-hexane at 50–90 °C	Esterase from *Archaeoglobus fulgidus*	Wang et al. (2014)
ε-caprolactone	Toluene at 70 °C	Novozym 435	Poojari, Beemat, and Clarson (2013)
ε-caprolactone	Toluene at 50 °C	Novozym 435	Engel et al. (2016)
Nonadecalactone (C_{19}), tricosalactone (C_{23})	Toluene at 50–110 °C	Novozym 435	Witt, Häußler, and Mecking (2017)
ε-caprolactone	*N*,*N*-dimethylacetamide at 70 °C	Novozym 435	Zhao et al. (2017)
L-lactide	[BMIM][BF_4], [BMIM][Tf_2N], or [BMIM][PF_6] at 110 °C	Free CALB	Yoshizawa-Fujita et al. (2008)
L-lactide	[BMIM][PF_6] at 65 °C	Novozym 435	Mena, Chanfreau, Gimeno, and Bárzana (2010)
L-lactide	[BMIM][PF_6] at 90 °C	Novozym 435	Mena et al. (2013)
L-lactide	[HMIM][PF_6] at 65–90 °C	Novozym 435	Chanfreau et al. (2010)
L-lactide	[BMIM][PF_6] at 130 °C	Novozym 435	Zhao et al. (2017)
L-lactide	[$CH_3OCH_2CH_2$-PBu_3][Tf_2N] at 130 °C	Novozym 435	Zhao, Afriyie, Larm, and Baker (2018)
ε-caprolactone	[BMIM][BF_4] at 60 °C	CALB	Uyama et al. (2002)
ε-caprolactone	[BMIM][PF_6], [BMIM][BF_4], or [BMIM][Tf_2N] at 60 °C	Novozym 435	Marcilla et al. (2006)
β-propiolactone, δ-valerolactone, ε-caprolactone	[BMIM][Tf_2N] at 60–110 °C	Novozym 435	Gorke, Okrasa, Louwagie, Kazlauskas, and Srienc (2007)

Continued

Table 2 Examples of solvents and lipases used in ROP reactions.—cont'd

Monomer	Solvent (temperature)	Lipase	Reference
ε-caprolactone	[BuPy][BF_4] at 60–150 °C	*Yarrowia lipolytica* lipase (YLL), *Candida rugosa* lipase (CRL), porcine pancreatic lipase (PPL)	Barrera-Rivera, Marcos-Fernández, Vera-Graziano, and Martínez-Richa (2009)
ε-caprolactone	[$C_4(C_6IM)_2$][PF_6]$_2$ at 90 °C	Novozym 435	Wu, Zhang, He, and Zhuo (2013)
ε-caprolactone	[BMIM][Tf_2N], [BMIM][PF_6]	Novozym 435	Piotrowska et al. (2016)
ε-caprolactone	[$CH_3OCH_2CH_2$-PBu_3] [Tf_2N], [BMIM][PF_6] at 70 °C	Novozym 435	Zhao et al. (2017, 2018)

titration (Mettler Toledo C20X) is used to determine the water contents of solvents, monomers and enzymes at 20 °C, where the analyte for the titration is Hydranal® Coulomat AG. A typical reaction follows these steps: the substrate (0.5 g of L-lactide, or ε-caprolactone) is mixed/melted with 0.25 mL co-solvent at an elevated temperature, and 100 mg Novozym 435 (containing ~10 mg free CALB) is added. The reaction mixture is firmly sealed and stirred at 130 °C (for lactide) or 70°C (for ε-caprolactone) in an oil bath. When the reaction completes, the mixture is cooled to room temperature, 2.0 mL $CDCl_3$ is added under stirring to dissolve the polyester. A small amount of the mixture (50 μL) is diluted with 1.0 mL $CDCl_3$ and then centrifuged prior to ^{1}H NMR analysis. To collect the final product, chloroform is evaporated under vacuum, followed by the addition of ice-cold methanol to precipitate the polymer; the polyester is separated by centrifugation or vacuum filtration. The polymer is dried in air for 24 h before dissolving in THF (10 mg/mL) for GPC analysis (see GPC conditions in Section 2.2).

4.2 ROP in supercritical fluids

Supercritical fluids, such as carbon dioxide ($scCO_2$) and 1,1,1,2-tetrafluoroethane (known as R-134a), are alternative solvents for the enzymatic preparation of poly(ε-caprolactone) with molecular masses (M_n) up to 50,000–80,000 Da and polylactide with M_w up to 12,900 Da (see reaction conditions in Table 3).

Table 3 Enzymatic ROP of lactones in supercritical fluids.

Supercritical fluid	Monomer	Conditions	Lipase	Reference
$scCO_2$	ε-caprolactone	0 °C and 10 MPa	CALB	Takamoto et al. (2001)
$scCO_2$	ε-caprolactone	35–65 °C and 8.14–22.1 MPa	Novozym 435	Loeker et al. (2004)
$scCO_2$	ε-caprolactone	35 °C and 10.3 MPa	Novozym 435	Thurecht et al. (2006)
$scCO_2$	ε-pentadecalactone	70 °C and 20 MPa	Novozym 435	Polloni et al. (2017)
$scCO_2$	ε-caprolactone	65 °C and 14.5 MPa	Novozym 435	Santos et al. (2012)
$scCO_2$	ε-caprolactone	65 °C and 12–28 MPa	Novozym 435	Comim Rosso, Bianchin, de Oliveira, Oliveira, and Ferreira (2013)
$scCO_2$	ε-caprolactone	65 °C and 12 MPa	Novozym 435	Rosso Comim, Veneral, de Oliveira, Ferreira, and Oliveira (2015)
1,1,1,2-tetrafluoroethane (R-134a)	ε-caprolactone	65 °C and 2.5 MPa	Novozym 435	García-Arrazola, Gimeno, and Bárzana (2007)
$scCO_2$	L-lactide	65 °C and 30 MPa	Novozym 435	García-Arrazola, López-Guerrero, Gimeno, and Bárzana (2009)

4.2.1 Equipment

The supercritical reactor described by the Oliveira group (Esmelindro et al., 2008; Santos et al., 2012; Valério et al., 2009) comprises a jacketed 50 mL stainless steel reactor (Parr Instrument Company, model 4843, Moline, IL, USA) with mechanical stirring, a cylinder of carbon dioxide, and a syringe pump (Isco Inc., model 260D, Lincoln, USA) with the reservoir attached to a thermostatic bath (Fig. 6).

4.2.2 Method

The method is based on literatures (Santos et al., 2012; Takamoto et al., 2001). The solvent/substrate mass ratio is kept roughly at 4:1. The amounts

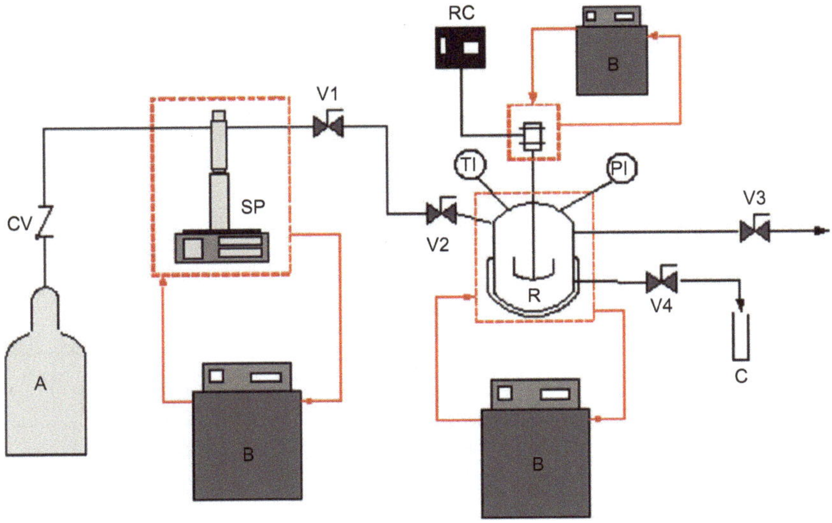

Fig. 6 Supercritical fluid reactor apparatus: A, solvent reservoir; B, thermostatic baths; SP, syringe pump; R, reactor vessel; RC, reactor control system; TI, temperature indicator; PI, pressure indicator; C, collector tube; CV, check valve; V1, V2, V3 and V4, ball valves. *Used with permission from Esmelindro, A. F. A., Fiametti, K. G., Ceni, G., Corazza, M. L., Treichel, H., Oliveira, D., et al. (2008). Lipase-catalyzed production of monoglycerides in compressed propane and AOT surfactant.* The Journal of Supercritical Fluids, 47, *64–69.*

of solvent and substrates are calculated accordingly to fill the volume of reactor based on the densities of pure compounds at the experimental conditions. For example, at 65 °C and 18 MPa, CO_2 density is 0.66 g/cm^3; thus, 7 g of ε-caprolactone and 0.7 g Novozym 435 are loaded first, followed by the charge of about 28 g CO_2 (~42 cm^3) using a syringe pump until the desired pressure is reached. The amount of CO_2 can be estimated from the decreased volume of the syringe pump reservoir (maintained at 5 °C and at the system pressure). The reaction mixture is stirred at 200 rpm for 24 h. At the end of reaction, the reactor is carefully degassed followed by the addition of tetrahydrofuran (30 mL) to the polymer. As a safety precaution, it is known that PCL can be easily plasticized by $scCO_2$, which could potentially lead to the polyester solidifies in the valve upon the depressurization to atmospheric conditions and prevent further release of pressure; therefore, it is suggested to cool the reactor/autoclave by an acetone/dry ice bath (−78 °C) to freeze the CO_2/polymer mixture till the pressure decreases to about atmospheric pressure; the monomer and low-molecular-weight oligomers can be further purged with $scCO_2$ at 4000 psi and 35 °C for 4 h (flow rate: ~5 mL of CO_2 min^{-1}); the polymer is recovered after depressurizing the autoclave

(Loeker et al., 2004). The enzymatic ROP of L-lactide in $scCO_2$ follows a similar procedure where 65 °C and 30 MPa are applied to create melted lactide and form an emulsion with $scCO_2$ (L-lactide solubility in $scCO_2$ is only 6.15%, *w*/*v* at 60 °C and 20 MPa) (García-Arrazola et al., 2009).

4.3 ROP in aqueous biphasic systems

Relatively large and hydrophobic lactones could form "mini-emulsion" with water, therefore, the enzymatic ROP of these lactones could be performed in water as an aqueous biphasic system. These lactones include 11-undecanolide (UDL), 12-dodecanolide (DDL), and 15-pentadecanolide (PDL) (Namekawa, Uyama, & Kobayashi, 1998; Panlawan et al., 2013), and hexadecanolide (HDL) (Målberg, Finne-Wistrand, & Albertsson, 2010).

4.3.1 Equipment

- High Intensity Ultrasonic processor, 400 W, Sweden
- Zetasizer Nano ZS Dynamic light scattering (DLS), Malvern Instruments Ltd., UK

4.3.2 Measurements

The monomer conversion is calculated by comparing the ^{1}H NMR peak areas of monomer and polyester. The GPC method (Målberg et al., 2010) is similar to Section 2.2 with slight modifications: 1.0 mL/min chloroform as the mobile phase, and evaporative light-scattering detector (ELSD). To analyze the polyester particle size, the mini-emulsion sample is diluted 50 times by deionized water first, followed by the measurement using DLS with a 514.5 nm red laser collecting the scattered light at an angle of 90° at room temperature.

4.3.3 Method

At first, the mini-emulsion is prepared following the literature method (Chiaradia et al., 2018; Inprakhon et al., 2014; Målberg et al., 2010; Taden, Antonietti, & Landfester, 2003). Pentadecanolide (2.40 g), 100 mg of hexadecane, and 10.0 g of 1.0 wt% Brij 58 solution in water are mixed under agitation for 1 h at 40 °C. The mixture is subject to ultrasonication for 120 s at 40% amplitude to create the mini-emulsion. Into the mini-emulsion, 50 mg lipase from *Burkholderia cepacia* (lipase PS) in 5.0 g of surfactant solution is added and stirred at 45 °C for 5 h. The reaction is terminated by adding sodium dodecyl sulphate (SDS). The product can be extracted into 20 mL chloroform, followed by a thorough washing with distilled water three times to remove the residual surfactant. After drying the chloroform solution by anhydrous Na_2SO_4, the polymer is obtained after filtering off the drying salt and evaporating the solvent. Alternatively, the product is taken into a small amount of chloroform (1–2 mL), and then poured in 100 mL of cold methanol to precipitate high-molar-mass polymer while low-molar-mass oligomers remain in supernatant (Inprakhon et al., 2014).

5. Conclusions

This chapter discusses some representative examples of different solvents (such as organic solvents, supercritical fluids, ionic liquids, and aqueous biphasic systems) for the polycondensation of diacids (or their derivatives) and the ROP of lactones. Different solvent conditions have exhibited different pros and cons, and it remains a challenge to design suitable solvents that have high thermal-stability as well as high enzyme-compatibility especially for high-temperature enzymatic applications. Nevertheless, the enzymatic polymerization to polyesters in nonaqueous solvents remains as the academic curiosity with high industrial potentials.

Acknowledgments

The author acknowledges the support from the Henry Dreyfus Teacher-Scholar Award (2012–2018).

References

Ajioka, M., Enomoto, K., Suzuke, K., & Yamaguchi, A. (1995). Basic properties of polylactic acid produced by the direct polycondensation polymerization of lactic acid. *Bulletin of the Chemical Society of Japan*, *68*, 2125–2131.

Albertsson, A.-C., & Varma, I. K. (2003). Recent developments in ring opening polymerization of lactones for biomedical applications. *Biomacromolecules*, *4*, 1466–1486.

Auras, R., Harte, B., & Selke, S. (2004). An overview of polylactides as packaging materials. *Macromolecular Bioscience*, *4*, 835–864.

Barrera-Rivera, K. A., Flores-Carreón, A., & Martínez-Richa, A. (2008). Enzymatic ring-opening polymerization of ε-caprolactone by a new lipase from *Yarrowia lipolytica*. *Journal of Applied Polymer Science*, *109*, 708–719.

Barrera-Rivera, K. A., Marcos-Fernández, Á., Vera-Graziano, R., & Martínez-Richa, A. (2009). Enzymatic ring-opening polymerization of ε-caprolactone by *Yarrowia lipolytica* lipase in ionic liquids. *Journal of Polymer Science, Part A: Polymer Chemistry*, 47(21), 5792–5805.

Brennecke, J. F., & Maginn, E. J. (2003). Purification of gas with liquid ionic compounds, US 6,579,343.

Cao, H., Han, H., Li, G., Yang, J., Zhang, L., Yang, Y., et al. (2012). Biocatalytic synthesis of poly(δ-valerolactone) using a thermophilic esterase from *Archaeoglobus fulgidus* as catalyst. *International Journal of Molecular Sciences*, *13*, 12232–12241.

Chanfreau, S., Mena, M., Porras-Domínguez, J. R., Ramírez-Gilly, M., Gimeno, M., Roquero, P., et al. (2010). Enzymatic synthesis of poly-L-lactide and poly-L-lactide-co-glycolide in an ionic liquid. *Bioprocess and Biosystems Engineering*, *33*, 629–638.

Cheng, Y., Deng, S., Chen, P., & Ruan, R. (2009). Polylactic acid (PLA) synthesis and modifications: A review. *Frontiers of Chemistry in China*, *4*(3), 259–264.

Chiaradia, V., Polloni, A. E., de Oliveira, D., de Oliveira, J. V., Araújo, P. H. H., & Sayer, C. (2018). Polyester nanoparticles from macrolactones via miniemulsion enzymatic ring-opening polymerization. *Colloid and Polymer Science*, *296*(5), 861–869.

Comim Rosso, S. R., Bianchin, E., de Oliveira, D., Oliveira, J. V., & Ferreira, S. R. S. (2013). Enzymatic synthesis of poly(ε-caprolactone) in supercritical carbon dioxide medium by means of a variable-volume view reactor. *Journal of Supercritical Fluids*, *79*, 133–141.

Deng, F., & Gross, R. A. (1999). Ring-opening bulk polymerization of ε-caprolactone and trimethylene carbonate catalyzed by lipase novozym 435. *International Journal of Biological Macromolecules*, *25*, 153–159.

Dong, F.-X., Zhang, L., Tong, X.-Z., Chen, H.-B., Wang, X.-L., & Wang, Y.-Z. (2012). Ionic liquid coated lipase: Green synthesis of high molecular weight poly(1,4-dioxan-2-one). *Journal of Molecular Catalysis B: Enzymatic*, *77*, 46–52.

Douka, A., Vouyiouka, S., Papaspyridi, L.-M., & Papaspyrides, C. D. (2018). A review on enzymatic polymerization to produce polycondensation polymers: The case of aliphatic polyesters, polyamides and polyesteramides. *Progress in Polymer Science*, *79*, 1–25.

Drumright, R. E., Gruber, P. R., & Henton, D. E. (2000). Polylactic acid technology. *Advanced Materials*, *12*, 1841–1846.

Duchiron, S. W., Pollet, E., Givry, S., & Avérous, L. (2015). Mixed systems to assist enzymatic ring opening polymerization of lactide stereoisomers. *RSC Advances*, *5*, 84627–84635.

Endres, F., & Welton, T. (2008). Inorganic synthesis. In P. Wasserscheid & T. Welton (Eds.), *Ionic liquids in synthesis* (pp. 289–318). Weinheim: Wiley-VCH Verlag.

Engel, S., Höck, H., Bocola, M., Keul, H., Schwaneberg, U., & Möller, M. (2016). CaLB catalyzed conversion of ε-caprolactone in aqueous medium. Part 1: Immobilization of CaLB to microgels. *Polymers*, *8*(10), 372.

Esmelindro, A. F. A., Fiametti, K. G., Ceni, G., Corazza, M. L., Treichel, H., Oliveira, D., et al. (2008). Lipase-catalyzed production of monoglycerides in compressed propane and AOT surfactant. *Journal of Supercritical Fluids*, *47*, 64–69.

García-Arrazola, R., Gimeno, M., & Bárzana, E. (2007). Use of liquid 1,1,1,2-tetrafluoroethane as solvent media for enzyme-catalyzed ring-opening polymerization of lactones. *Macromolecules*, *40*, 4119–4120.

García-Arrazola, R., López-Guerrero, D. A., Gimeno, M., & Bárzana, E. (2009). Lipase-catalyzed synthesis of poly-L-lactide using supercritical carbon dioxide. *Journal of Supercritical Fluids*, *51*, 197–201.

Gorke, J. T., Okrasa, K., Louwagie, A., Kazlauskas, R. J., & Srienc, F. (2007). Enzymatic synthesis of poly(hydroxyalkanoates) in ionic liquids. *Journal of Biotechnology*, *132*, 306–313.

Hans, M., Keul, H., & Moeller, M. (2009). Ring-opening polymerization of DD-lactide catalyzed by novozyme 435. *Macromolecular Bioscience*, *9*(3), 239–247.

Henderson, L. A., Svirkin, Y. Y., Gross, R. A., Kaplan, D. L., & Swift, G. (1996). Enzyme-catalyzed polymerizations of ε-caprolactone: Effects of initiator on product structure, propagation kinetics, and mechanism. *Macromolecules*, *29*, 7759–7766.

Hunsen, M., Abul, A., Xie, W., & Gross, R. (2008). Humicola insolens cutinase-catalyzed lactone ring-opening polymerizations: Kinetic and mechanistic studies. *Biomacromolecules*, *9*, 518–522.

Inprakhon, P., Panlawan, P., Pongtharankul, T., Marie, E., Wiemann, L. O., Durand, A., et al. (2014). Toward one-pot lipase-catalyzed synthesis of poly(ε-caprolactone) particles in aqueous dispersion. *Colloids and Surfaces, B: Biointerfaces, 113*, 254–260.

Jain, N., Kumar, A., Chauhan, S., & Chauhan, S. M. S. (2005). Chemical and biochemical transformations in ionic liquids. *Tetrahedron, 61*(5), 1015–1060.

Janecki, T. (Ed.), (2014). *Natural lactones and lactams: Synthesis, occurrence and biological activity.* Weinheim: Wiley-VCH.

Jérôme, C., & Lecomte, P. (2008). Recent advances in the synthesis of aliphatic polyesters by ring-opening polymerization. *Advanced Drug Delivery Reviews, 60*(9), 1056–1076.

Kamber, N. E., Jeong, W., Waymouth, R. M., Pratt, R. C., Lohmeijer, B. G. G., & Hedrick, J. L. (2007). Organocatalytic ring-opening polymerization. *Chemical Reviews, 107*, 5813–5840.

Knani, D., Gutman, A. L., & Kohn, D. H. (1993). Enzymatic polyesterification in organic media. Enzyme-catalyzed synthesis of linear polyesters. I. Condensation polymerization of linear hydroxyesters. II. Ring-opening polymerization of ε-caprolactone. *Journal of Polymer Science, Part A: Polymer Chemistry, 31*, 1221–1232.

Kobayashi, S. (1999). Enzymatic polymerization: A new method of polymer synthesis. *Journal of Polymer Science, Part A: Polymer Chemistry, 37*, 3041–3056.

Kobayashi, S. (2009). Recent developments in lipase-catalyzed synthesis of polyesters. *Macromolecular Rapid Communications, 30*, 237–266.

Kobayashi, S. (2015). Enzymatic ring-opening polymerization and polycondensation for the green synthesis of polyesters. *Polymers for Advanced Technologies, 26*, 677–686.

Kobayashi, S., Takeya, K., Sudu, S., & Uyama, H. (1998). Lipase-catalyzed ring-opening polymerization of medium-size lactones to polyesters. *Macromolecular Chemistry and Physics, 199*, 1729–1736.

Kobayashi, S., Uyama, H., & Kimura, S. (2001). Enzymatic polymerization. *Chemical Reviews, 101*, 3793–3818.

Kowalski, A., Libiszowski, J., Duda, A., & Penczek, S. (2000). Polymerization of L,L-dilactide initiated by tin(II) butoxide. *Macromolecules, 33*(6), 1964–1971.

Kragl, U., Eckstein, M., & Kaftzik, N. (2002). Enzyme catalysis in ionic liquids. *Current Opinion in Biotechnology, 13*, 565–571.

Kubisa, P. (2004). Application of ionic liquids as solvents for polymerization processes. *Progress in Polymer Science, 29*(1), 3–12.

Kumar, A., & Gross, R. A. (2000). *Candida antarctica* lipase B catalyzed polycaprolactone synthesis: Effects of organic media and temperature. *Biomacromolecules, 1*, 133–138.

Kundys, A., Białecka-Florjańczyk, E., Fabiszewska, A., & Małajowicz, J. (2018). Candida antarctica lipase B as catalyst for cyclic esters synthesis, their polymerization and degradation of aliphatic polyesters. *Journal of Polymers and the Environment, 26*, 396–407.

Labet, M., & Thielemans, W. (2009). Synthesis of polycaprolactone: A review. *Chemical Society Reviews, 38*, 3484–3504.

Lecomte, P., & Jérôme, C. (2012). Recent developments in ring-opening polymerization of lactones. *Advances in Polymer Science, 245*, 173–218.

Loeker, F. C., Duxbury, C. J., Kumar, R., Gao, W., Gross, R. A., & Howdle, S. M. (2004). Enzyme-catalyzed ring-opening polymerization of ε-caprolactone in supercritical carbon dioxide. *Macromolecules, 37*, 2450–2453.

Målberg, S., Finne-Wistrand, A., & Albertsson, A.-C. (2010). The environmental influence in enzymatic polymerization of aliphatic polyesters in bulk and aqueous mini-emulsion. *Polymer, 51*, 5318–5322.

Marcilla, R., de Geus, M., Mecerreyes, D., Duxbury, C. J., Koning, C. E., & Heise, A. (2006). Enzymatic polyester synthesis in ionic liquids. *European Polymer Journal, 42*(6), 1215–1221.

Matsumura, S. (2006). Enzymatic synthesis of polyesters via ring-opening polymerization. *Advances in Polymer Science, 194*, 95–132.
Mena, M., Chanfreau, S., Gimeno, M., & Bárzana, E. (2010). Enzymatic synthesis of poly-L-lactide-co-glycolide in the ionic liquid 1-butyl-3-methylimidazolium hexafluorophosphate. *Bioprocess and Biosystems Engineering, 33*, 1095–1101.
Mena, M., López-Luna, A., Shirai, K., Tecante, A., Gimeno, M., & Bárzana, E. (2013). Lipase-catalyzed synthesis of hyperbranched poly-L-lactide in an ionic liquid. *Bioprocess and Biosystems Engineering, 36*, 383–387.
Moniruzzaman, M., Nakashima, K., Kamiya, N., & Goto, M. (2010). Recent advances of enzymatic reactions in ionic liquids. *Biochemical Engineering Journal, 48*(3), 295–314.
Moon, S. I., Taniguchi, I., Miyamoto, M., Kimura, Y., & Lee, C. W. (2001). Synthesis and properties of high molecular weight poly(L-lactic acid) by melt/solid polycondensation under different reaction conditions. *High Performance Polymers, 13*, 189–196.
Namekawa, S., Uyama, H., & Kobayashi, S. (1998). Lipase-catalyzed ring-opening polymerization of lactones in water. *Polymer Journal, 30*, 269–271.
Nampoothiri, K. M., Nair, N. R., & John, R. P. (2010). An overview of the recent developments in polylactide (PLA) research. *Bioresource Technology, 101*, 8493–8501.
Nara, S. J., Harjani, J. R., Salunkhe, M. M., Mane, A. T., & Wadgaonkar, P. P. (2003). Lipase-catalysed polyester synthesis in 1-butyl-3-methylimidazolium hexafluorophosphate ionic liquid. *Tetrahedron Letters, 44*(7), 1371–1373.
Nobes, G. A. R., Kazlauskas, R. J., & Marchessault, R. H. (1996). Lipase-catalyzed ring-opening polymerization of lactones: A novel route to poly(hydroxyalkanoate)s. *Macromolecules, 29*, 4829–4833.
Omay, D., & Guvenilir, Y. (2013). Synthesis and characterization of poly(D,L-lactic acid) via enzymatic ring opening polymerization by using free and immobilized lipase. *Biocatalysis and Biotransformation, 31*(3), 132–140.
Ozsagiroglu, E., Iyisan, B., & Avcibasi-Guvenilir, Y. (2012). Effects of different reaction mediums on ring opening polymerization of poly(ε-caprolactone) by lipase. *African Journal of Biotechnology, 11*(63), 12688–12696.
Panlawan, P., Luangthongkam, P., Wiemann, L. O., Sieber, V., Marie, E., Durand, A., et al. (2013). Lipase-catalyzed interfacial polymerization of ω-pentadecalactone in aqueous biphasic medium: A mechanistic study. *Journal of Molecular Catalysis B: Enzymatic, 88*, 69–76.
Panova, A. A., & Kaplan, D. L. (2003). Mechanistic limitations in the synthesis of polyesters by lipase-catalyzed ring-opening polymerization. *Biotechnology and Bioengineering, 84*, 103–113.
Park, S., & Kazlauskas, R. J. (2003). Biocatalysis in ionic liquids—Advantages beyond green technology. *Current Opinion in Biotechnology, 14*(4), 432–437.
Piotrowska, U., & Sobczak, M. (2015). Enzymatic polymerization of cyclic monomers in ionic liquids as a prospective synthesis method for polyesters used in drug delivery systems. *Molecules, 20*, 1–23.
Piotrowska, U., Sobczak, M., Oledzka, E., & Combes, C. (2016). Effect of ionic liquids on the structural, thermal, and in vitro degradation properties of poly(ε-caprolactone) synthesized in the presence of *Candida antarctica* lipase B. *Journal of Applied Polymer Science, 133*(31), 43728.
Polloni, A. E., Veneral, J. G., Rebelatto, E. A., de Oliveira, D., Oliveira, J. V., Araújo, P. H. H., et al. (2017). Enzymatic ring opening polymerization of ω-pentadecalactone using supercritical carbon dioxide. *Journal of Supercritical Fluids, 119*, 221–228.
Poojari, Y., Beemat, J. S., & Clarson, S. J. (2013). Enzymatic synthesis of poly(ε-caprolactone): Thermal properties, recovery, and reuse of lipase B from *Candida antarctica* immobilized on macroporous acrylic resin particles. *Polymer Bulletin, 70*, 1543–1552.

Pretula, J., Slomkowski, S., & Penczek, S. (2016). Polylactides—Methods of synthesis and characterization. *Advanced Drug Delivery Reviews*, *107*, 3–16.

Rosso Comim, S. R., Veneral, J. G., de Oliveira, D., Ferreira, S. R. S., & Oliveira, J. V. (2015). Enzymatic synthesis of poly(ε-caprolactone) in liquified petroleum gas and carbon dioxide. *Journal of Supercritical Fluids*, *96*, 334–348.

Santos, R. D., Rosso Comim, S. R., de Oliveira, D., Treichel, H., Di Luccio, M., Ferreira, S. R. S., et al. (2012). Lipase-catalyzed synthesis of poly(ε-caprolactone) in supercritical carbon dioxide. *Biocatalysis and Agricultural Biotechnology*, *1*(4), 280–283.

Seyednejad, H., Ghassemi, A. H., van Nostrum, C. F., Vermonden, T., & Hennink, W. E. (2011). Functional aliphatic polyesters for biomedical and pharmaceutical applications. *Journal of Controlled Release*, *152*, 168–176.

Sobczak, M. (2012). Enzyme-catalyzed ring-opening polymerization of cyclic esters in the presence of poly(ethylene glycol). *Journal of Applied Polymer Science*, *125*(5), 3602–3609.

Taden, A., Antonietti, M., & Landfester, K. (2003). Enzymatic polymerization towards biodegradable polyester nanoparticles. *Macromolecular Rapid Communications*, *24*(8), 512–516.

Takamoto, T., Uyama, H., & Kobayashi, S. (2001). Lipase-catalyzed synthesis of aliphatic polyesters in supercritical carbon dioxide. *e-Polymers*, *1*(1). https://doi.org/10.1515/epoly.2001.1.1.19. no 4.

Tang, S., Baker, G. A., Ravula, S., Jones, J. E., & Zhao, H. (2012). PEG-functionalized ionic liquids for cellulose dissolution and saccharification. *Green Chemistry*, *14*(10), 2922–2932.

Thurecht, K. J., Heise, A., deGeus, M., Villarroya, S., Zhou, J., Wyatt, M. F., et al. (2006). Kinetics of enzymatic ring-opening polymerization of ε-caprolactone in supercritical carbon dioxide. *Macromolecules*, *39*, 7967–7972.

Uyama, H., & Kobayashi, S. (1994). Lipase-catalyzed polymerization of divinyl adipate with glycols to polyesters. *Chemistry Letters*, *23*(9), 1687–1690.

Uyama, H., & Kobayashi, S. (2006). Enzymatic synthesis of polyesters via polycondensation. *Advances in Polymer Science*, *194*, 133–158.

Uyama, H., Takamoto, T., & Kobayashi, S. (2002). Enzymatic synthesis of polyesters in ionic liquids. *Polymer Journal*, *34*(2), 94–96.

Valério, A., Fiametti, K. G., Rovani, S., Franceschi, E., Corazza, M. L., Treichel, H., et al. (2009). Enzymatic production of mono- and diglycerides in compressed n-butane and AOT surfactant. *Journal of Supercritical Fluids*, *49*, 216–220.

van Rantwijk, F., & Sheldon, R. A. (2007). Biocatalysis in ionic liquids. *Chemical Reviews*, *107*(6), 2757–2785.

Wang, M., Shi, H., Wu, D., Han, H., Zhang, J., Xing, Z., et al. (2014). Glutaraldehyde cross-linking of immobilized thermophilic esterase on hydrophobic macroporous resin for application in poly(ε-caprolactone) synthesis. *Molecules*, *19*, 9838–9849.

Wasserscheid, P., & Welton, T. (2008). *Ionic liquids in synthesis* (2nd ed.). Weinheim: Wiley-VCH.

Witt, T., Häußler, M., & Mecking, S. (2017). No strain, no gain? Enzymatic ring-opening polymerization of strainless aliphatic macrolactones. *Macromolecular Rapid Communications*, *38*, 1600638.

Wu, C., Zhang, Z., He, F., & Zhuo, R. (2013). Enzymatic synthesis of poly(ε-caprolactone) in monocationic and dicationic ionic liquids. *Biotechnology Letters*, *35*, 879–885.

Xie, W., Li, J., Chen, D., & Wang, P. G. (1997). Ring-opening polymerization of β-butyrolactone by thermophilic lipases. *Macromolecules*, *30*, 6997–6998.

Yang, Y., Ge, Y., Zhao, H., Shi, W., & Li, Q. (2011). Lipase-catalyzed synthesis of poly (ε-caprolactone) and characterization of its solid-state properties. *Biocatalysis and Biotransformation*, *29*(6), 337–343.

Yang, Y., Yu, Y., Zhang, Y., Liu, C., Shi, W., & Li, Q. (2011). Lipase/esterase-catalyzed ring-opening polymerization: A green polyester synthesis technique. *Process Biochemistry, 46*, 1900–1908.

Yoshizawa-Fujita, M., Saito, C., Takeoka, Y., & Rikukawa, M. (2008). Lipase-catalyzed polymerization of L-lactide in ionic liquids. *Polymers for Advanced Technologies, 19*, 1396–1400.

Zhang, J., Shi, H., Wu, D., Xing, Z., Zhang, A., Yang, Y., et al. (2014). Recent developments in lipase-catalyzed synthesis of polymeric materials. *Process Biochemistry, 49*(5), 797–806.

Zhao, H. (2006). Innovative applications of ionic liquids as 'green' engineering liquids. *Chemical Engineering Communications, 193*(12), 1660–1677.

Zhao, H. (2010). Methods for stabilizing and activating enzymes in ionic liquids—A review. *Journal of Chemical Technology and Biotechnology, 85*(7), 891–907.

Zhao, H. (2016). Protein stabilization and enzyme activation in ionic liquids: Specific ion effects. *Journal of Chemical Technology and Biotechnology, 91*, 25–50.

Zhao, H. (2018). Enzymatic ring-opening polymerization (ROP) of polylactones: Roles of non-aqueous solvents. *Journal of Chemical Technology and Biotechnology, 93*(1), 9–19.

Zhao, H., Afriyie, L. O., Larm, N. E., & Baker, G. A. (2018). Glycol-functionalized ionic liquids for high-temperature enzymatic ring-opening polymerization. *RSC Advances, 8*, 36025–36033.

Zhao, H., Baker, G. A., Song, Z., Olubajo, O., Crittle, T., & Peters, D. (2008). Designing enzyme-compatible ionic liquids that can dissolve carbohydrates. *Green Chemistry, 10*(6), 696–705.

Zhao, H., & Malhotra, S. V. (2002). Applications of ionic liquids in organic synthesis. *Aldrichimica Acta, 35*(3), 75–83.

Zhao, H., Nathaniel, G. A., & Merenini, P. C. (2017). Enzymatic ring-opening polymerization (ROP) of lactides and lactone in ionic liquids and organic solvents: Digging the controlling factors. *RSC Advances*, 7(77), 48639–48648.

Zinck, P. (2009). One-step synthesis of polyesters specialties for biomedical applications. *Reviews in Environmental Science and Bio/Technology, 8*, 231–234.

CHAPTER TWO

Integrating computational and experimental methods for efficient biocatalytic synthesis of polyesters

Alessandro Pellis[a], Lucia Gardossi[b,*]
[a]University of York, Department of Chemistry, Green Chemistry Centre of Excellence, York, United Kingdom
[b]University of Trieste, Department of Chemical and Pharmaceutical Sciences, Trieste, Italy
*Corresponding author: e-mail address: gardossi@units.it

Contents

ISSN 0076-6879
https://doi.org/10.1016/bs.mie.2019.07.040

Abstract

The research on biocatalyzed polycondensation has delivered an array of polyesters having molecular weights below 20,000 g mol^{-1} but characterized by controlled structures and desired functionalities. Their unique catalytic efficiency under mild conditions enables enzymes to catalyze the polycondensation of monomers bearing labile lateral moieties that can be easily accessed *via* post-polymerization modifications. Despite this great potential, nowadays biocatalysts are not employed for polycondensation on industrial scale due to some bottlenecks related to the formulation of biocatalysts and the process configuration, which make the enzymatic technology non-economic. Recycling the enzymatic catalysts is not only a matter of producing an active and robust formulation, but it also requires the optimal integration of such biocatalyst within a specific reactor and process configuration that must enable efficient mass-transfer while preserving the integrity of the enzymatic preparation. In this chapter, we describe examples of integrated experimental-computational approaches for the rational planning and implementation of enzymatic polycondensation using lipase B from *Candida antarctica* and cutinase 1 from *Thermobifida cellulosilytica*. They rely on molecular visualization, molecular modeling and chemometrics, which are methods requiring very modest computational power and approachable by operators who do not have specific computational background. The examples also address the sustainability issue, by describing solvent-free processes involving bio-based monomers and biocatalysts immobilized on renewable carriers.

1. Introduction

Nowadays chemistry, and the polymer sector in particular, is facing the challenge of moving beyond the simple target of process sustainability by integrating renewable feedstock streams in their productive routes, thus realizing the potential of the new sustainable and circular bioeconomy (Pellis, Herrero Acero, Gardossi, et al., 2016).

Biocatalysis for the synthesis of polyesters represents an opportunity for increasing the value and the performance of bio-based polymers, renewable products that are expected to mitigate the environmental impact of fossil-based polymers and plastics. Structurally different bio-based monomers, produced *via* fermentation or chemical transformation, are already produced at industrial scale and the bio-based plastic market is expected to increase up to $ 5.2 billion by 2030 (Pellis, Cantone, Ebert, & Gardossi, 2018). Hydrolytic enzymes, such as lipase B from *Candida antarctica*, are able to catalyze the polycondensation of structurally different bio-based diacids and polyols *via* polycondensation. Biocatalysts are efficient and sustainable alternatives to toxic metal catalysts used in polycondensation, as tin, titanium and antimony. Biocatalysts are demonstrated to work also in solvent-free systems and catalyze highly selective processes at mild reaction temperatures ranging between 40 and 90 °C, whereas conventional chemo-catalytic polycondensations are carried out at $T > 150$ °C. These features enable the synthesis of functionalized as well as biodegradable polyesters with controlled architectures. Biocatalyzed polycondensation generally yields polymers with moderate molecular weights if compared to polymers obtained *via* conventional chemical processes. However, this drawback can be overcome by using two-steps procedures, where an initial biocatalyzed polycondensation is followed by a second reaction step, after the removal of the biocatalyst, working at higher temperature and/or lower pressure (Cerea, Gardossi, Sinigoi, & Fattor, 2013).

The increase of molecular weight is achievable also by the synthesis of oligomers and short telechelics prepolymers with functional ends (Bassanini, Hult, & Riva, 2013). It must be noted that polyesters with highly regular structures and molecular weights below 2500 Da have technological and commercial relevance in coating and adhesive applications and that biodegradable oligomers can find applications as film forming in cosmetic formulations (Ansorge-Schumacher & Thum, 2013). Research efforts in this field aim not only at substituting the well-established fossil-based polymers (drop-in products) but also at the design of a new generation of highly performing and competitive polymers and materials (Fig. 1).

However, these new challenges require a stringent integration between process engineering, enzyme technology and organic chemistry. On that respect, computational methods and tools for data analysis aid the rational planning of protocols for the formulation and application of biocatalysts under extreme and hostile environments. The enzymatic polycondensations reported in the following paragraphs have largely benefited from different computational methods, embracing molecular simulations but also multivariate statistical analysis. The main objective is to deliver efficient enzymatic tools and methods for the benefit of polymer chemistry.

Actually, despite the wide number of works on enzymatic polyester synthesis present in the scientific literature, this wealth of knowledge and catalytic potential currently is not exploited on the industrial scale. During the 90s, the enzymatic polycondensation of adipic acid with polyols had been implemented industrially by Baxenden Chemicals (UK) for the commercial production of polyesters for coating and adhesive applications but the process was later discontinued (Binns, Harffey, Roberts, & Taylor, 1999). Richard Gross analyzed the bottlenecks for the take-off of enzymatic

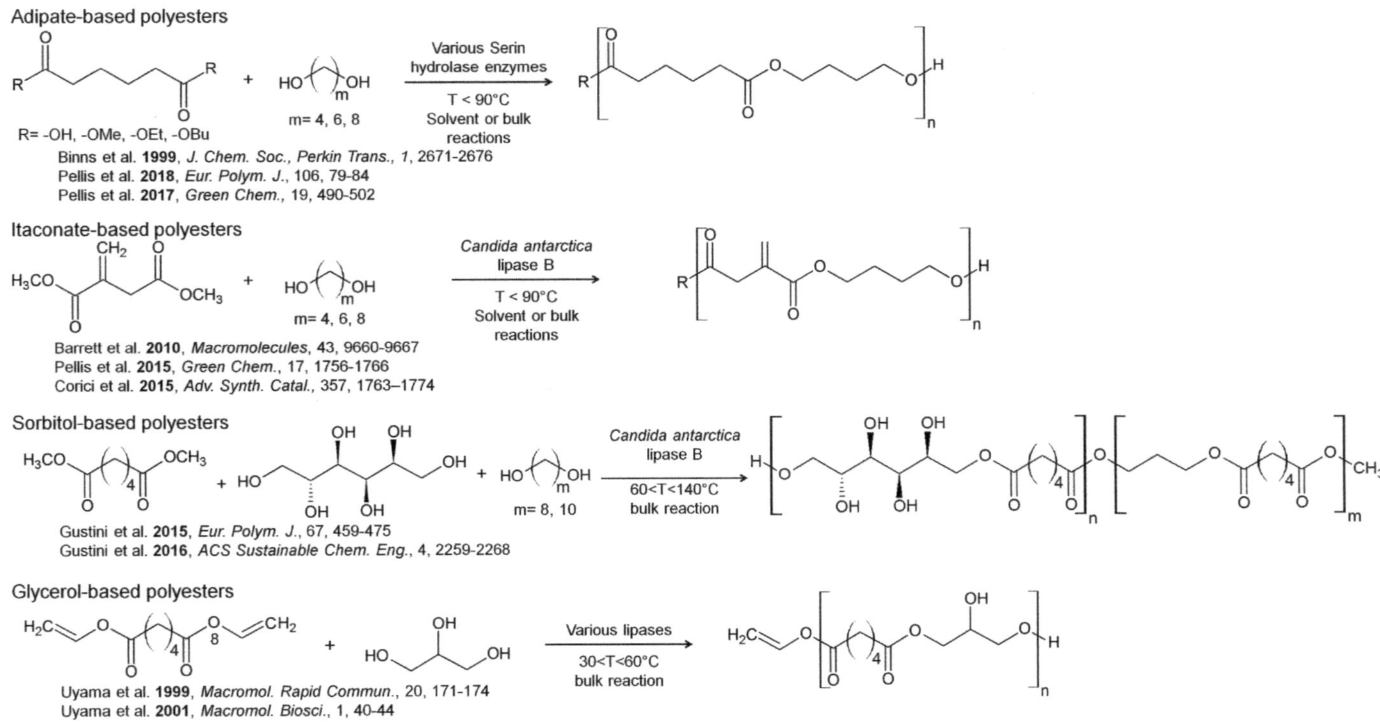

Fig. 1 Aliphatic and functional polymers synthesized *via* enzymatic catalysis.

polycondensation (Gross, Ganesh, & Lu, 2010) and he identified the lack of robustness of the biocatalyst as one of the major limitations. These biocatalyzed processes require the immobilization of the enzymes: firstly, to prevent protein contamination and secondly to enable the recycling the most expensive reaction component, namely the enzymatic immobilized preparation (Cantone et al., 2013).

As in the case of the Baxenden process, most reactions reported in the literature employ of the commercial biocatalyst Novozym® 435, consisting of lipase B from *Candida antarctica* (CaLB) physically adsorbed on a methacrylic resin. Indeed, F. Binns, and co-workers reported the detachment of about 10% of the protein from the carrier during one single synthetic cycle of polycondensation (Binns et al., 1999). The presence of the native and active enzyme in the reaction mixture after the removal of the Novozym® 435 formulation was confirmed by later studies, which also demonstrated that CaLB needs to be immobilized *via* covalent bonds to prevent the enzyme leaking and product contamination (Pellis et al., 2015). Indeed, polycondensations, especially when carried out in solvent-free systems, require vigorous mixing that causes the disruption of the biocatalyst (Hilterhaus et al., 2008). Temperatures between 60 °C and 90 °C are generally exploited to improve mass transfer while reducing the system's viscosity, thus causing additional stress to the biocatalyst, with the immobilization that is regarded as a strategy for improving the enzyme's stability.

Here we report the protocols developed with the intent of formulating robust biocatalysts and, furthermore, the chemometric procedure for the identification and control of experimental variables that affect a biocatalyzed polycondensation. The protocols were developed to meet the guidelines proposed by the European Federation of Biotechnology Section on Applied Biocatalysis, which delivered schematic instructions for reporting experimental data effectively while enabling experiments' reproducibility (Gardossi et al., 2010).

2. Covalent immobilization of hydrolases and their application in the solvent-free synthesis of biobased polyesters

2.1 Rational development of protocols for the immobilization of enzymes: Understanding enzyme and process constraints

CaLB and cutinase 1 from *Thermobifida cellulosilytica* (Thc_Cut1) were immobilized on epoxy methacrylic resins and also on functionalized rice husk (RH), a largely available biomass that is proposed as a renewable alternative to fossil-based acrylic resins (Cespugli et al., 2018; Corici et al., 2016). The immobilization protocols were developed by starting from the analysis of the structural features of the two hydrolases.

Simple molecular simulations can be carried out on conventional personal computers using very modest computational resources, thus acquiring useful information for maximizing the efficiency of the enzyme upon immobilization. More specifically, when covalent bonds take place between the enzyme and the carrier, the following factors can severely affect the final efficiency of the biocatalyst:

(a) The occlusion of the enzyme active site because of the formation of covalent bonds in its proximity. On that respect the visual inspection of the enzyme surface advises whether the lysine residues, which are generally exploited for the covalent anchoring, are nearby the opening of the active site.

(b) A poor affinity of the surface of the protein for the carrier. The analysis of the hydrophilic and hydrophobic areas of the protein surface indicate whether the enzyme is prone to adsorb on hydrophobic carriers or rather will interact preferentially with hydrophilic supports. These factors will also determine the ionic strength of the buffer solution employed for the immobilization, since a higher buffer concentration will force the protein towards the solid support.

(c) The presence of glycan moieties. The localization of the glycosylation sites on the surface of enzymes produced by eukaryotic microorganisms indicate the presence of wide hydrophilic areas that can mask superficial hydrophobic regions, thus modifying the protein-carrier interactions.

The analysis of the hydrophilic-hydrophobic balance of the surface of the proteins provides an indication of their affinity for carriers of different nature (Pellis, Ferrario, Zartl, et al., 2016).

The natural substrates of lipases and cutinases are hydrophobic insoluble molecules, namely triglycerides and cutin. Conversely, their surface display wide hydrophobic areas, especially in the correspondence of the opening of the active site. However, CaLB is expressed in *Aspergillus* sp. and the analysis of the primary sequence of CaLB indicates the presence of an N-glycosylation site at Asn 74 that masks a hydrophobic region on CaLB surface (Basso et al., 2007).

On the contrary, Thc_cut1, which is expressed in *E. coli,* is not glycosylated and that makes all the hydrophobic regions prone to hydrophobic interactions with organic hydrophobic resins (Basso et al., 2007).

Overall, these features justify the high percentage of Thc_cut1 immobilized on the hydrophobic methacrylate resin EC-EP/M (Sepabeads, Resindion) as well as the retention of a large percentage of the original activity upon covalent immobilization. Notably the loading of CaLB and Thc_cut1 were unacceptably low even after prolonged incubation in the case of the methacrylic resin EC-HFA, which displays hetero-functional polar chains (a secondary amine together with a hydroxyl group), which increase the hydrophilic nature of the carrier surface (Table 1).

Finally, the localization of the catalytic Ser 105 in CaLB and Ser 131 in Thc_cut1 indicates the higher accessibility of the superficial active site of the cutinase (Fig. 2A and C), whereas CaLB has a buried catalytic triad (Fig. 2B and D). These simple structural evidences provide useful rational guidelines for the selection of the immobilization carrier, among the large variety available on the market (Cantone et al., 2013). Overall, Thc_cut1 appears as a very promising enzymatic scaffold for polycondensations.

2.2 Computational protocols for the construction of the structural models of enzymes

2.2.1 CaLB structure

The 1TCA (Uppenberg et al. 1994) crystallographic structure retrieved from the Protein Data Bank (https://www.rcsb.org/) (Berman et al., 2000) was used for computing CaLB. It refers to a crystal obtained in 20 mM acetate buffer pH 3.6, in the presence of polyethylene glycol 4000

Table 1 Immobilization yields and recovered activities of Thc_cut1 and CaLB immobilized on epoxy methacrylic resin.

Carrier functional group	Enzyme	Bound enzyme (%)[a]	Recovered activity (%)[b]
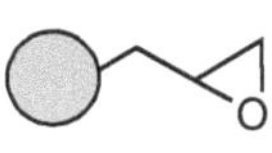	Thc_cut1	>99	37
	CaLB	87	8
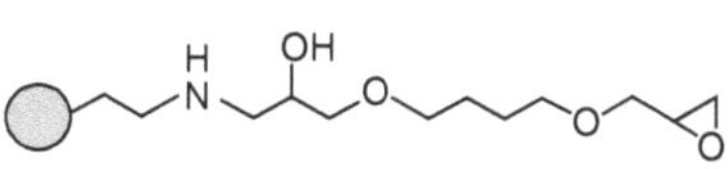	Thc_cut1	<2	n.d.
	CaLB	24	n.d.

[a]Percentage of enzyme activity expressed by the immobilized preparation when compared to the soluble form.
[b]Calculated by evaluating the residual activity and protein concentration in the supernatant.
n.d. = not determined.
All data are the average of two independent immobilization procedures.
The immobilization was performed in 0.1 M Tris-HCl buffer pH 7 at 21 °C for 24 h, using 10 mg of protein per g of dry resin in 10 mL buffer (Pellis, Ferrario, Zartl, et al., 2016).

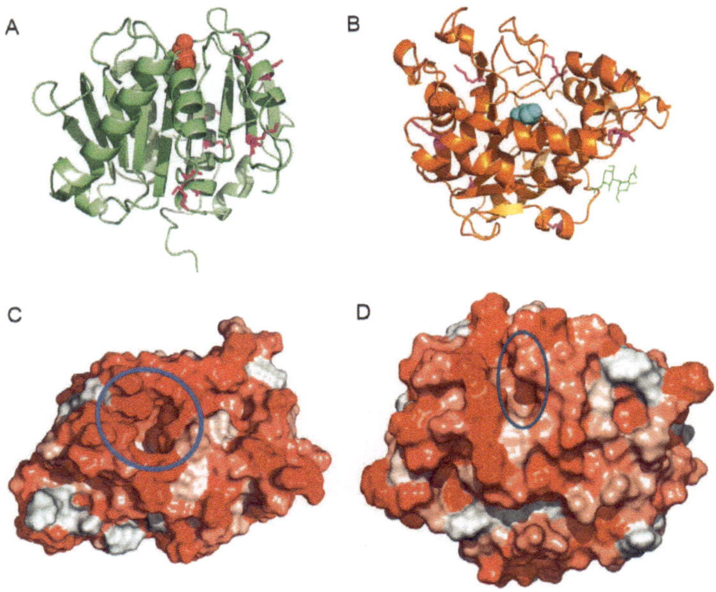

Fig. 2 Three-dimensional models of Thc_cut1 (left) and CaLB (right). Top: Lysine residues are evident on the surface of Thc_cut1 (A) and CaLB (B), highlighted in sticks mode (magenta). The catalytic Ser residues are represented in sphere mode. The CaLB lipase expressed in *Aspergillus* sp. is glycosylated and it displays the *N*-acetyl glucosamine units in the correspondence of the insertion of the glycan (stick mode and green color). Bottom: the models illustrate the distribution of hydrophobic areas on the surface of the two enzymes, Thc_cut1 (C) and CaLB (D), especially in the proximity of the active sites.

(20%) and isopropanol (10%). The CaLB 1TCA structure includes one protein together with the crystal water molecules, which were removed from the model. The sequence corresponding to the N-glycosylation site was identified by means of the open source software NetNGly (http://www.cbs.dtu.dk/services/NetNGlyc/).

2.2.2 Thc_cut1 structure

The homology model of *Thermobifida cellulosilytica* cutinase (Thc_Cut1) was used, which was constructed using the protein sequence from the NCBI GenBank nucleotide sequence HQ147785 (Herrero Acero et al., 2011) and the three-dimensional structure of the cutinase from *Thermobifida fusca* (PDB36 code 4CG1) as a template (99.23% of sequence identity). The file in pdb format of the 3D structure of Thc_cut1 constructed by homology modeling is public (Pellis, Ferrario, Zartl, et al., 2016) and can be download from http://www.rsc.org/suppdata/c5/cy/c5cy01746g/c5cy01746g1.pdf.

2.2.3 Construction of the models

Protein structures (Fig. 2) were recorded and visualized by means of the Open Source PyMOL software (The PyMOL Molecular Graphic System, Version 1.5.0.3, Schrodinger LLC version 1.4.1) (Berman et al., 2000). It allows the 3D visualization of small molecules, proteins, surfaces, density and trajectories. The catalytic serines were highlighted manually in sphere mode to visualize and compare the depth of the two active sites. The lysine residues were manually selected and highlighted in stick mode (magenta) to visualize the location of the potential sites for the formation of the covalent bonds with the carriers.

2.3 From theory to practice: Protocols for the covalent immobilization of cutinase 1 from *Thermobifida cellulosilytica* and lipase B from *Candida antarctica* on epoxy methacrylic resins

2.3.1 Materials: Enzymes and carriers

The recombinant *Thermobifida cellulosilytica* cutinase 1 (Thc_cut1) was expressed in *E. Coli* and purified at the University of Natural Resources and Life Sciences of Vienna (Herrero Acero et al., 2011). The enzyme was stored in 100 mM Tris-HCl buffer at pH 7 and immobilized in the same buffer.

Lipozyme CaLB was from Novozymes (DK). The preparation is a light brown solution composed, as stated by the producer, by 44% (w/w) water, 25% (w/w) sorbitol, 25% (w/w) glycerol, enzyme 6% (w/w dry matter), 0.20% (w/w) sodium benzoate, 0.10% (w/w) potassium sorbate. The protein concentration, determined independently using the protocol 2.3.13 was of 8 mg mL^{-1}.

Sepabeads resins EC-EP/M and EC-HFA/M are from Resindion S.r.l. (Mitsubishi Chemical Corporation, Binasco, Italy). They are highly porous methacrylic spherical beads with average pore diameter of 10–20 nm, particle size in the range of 200–600 μm and a mean diameter of 300–360 μm. They are provided in wet form, having a water retention around 55–65%. The minimum concentration of epoxy functional groups, as stated by the producer, is 100 μmol/g wet for EC-EP/M and 75 μmol/g wet for EC-HFA/M.

The beads, when used in high molar solutions and in common organic solvents have low swelling tendency and mechano-osmotic stability given by the intense crosslinking. The resins must be stored at 5–6 °C and for no longer than 6 months due to the spontaneous reactivity of epoxy groups in slightly acidic aqueous medium.

2.3.2 *Activity assay for the soluble CaLB and Thc_cut1: Hydrolysis of p-nitrophenyl butyrate (PNPB)*

The enzymatic activity was measured at 21 °C using *p*-nitrophenyl butyrate (PNPB) as a substrate. This general hydrolysis test for esterase activity was preferred to the tributyrin hydrolysis titration (see protocol 3.2.5) because lipases and cutinases display different substrate specificities. The assay mixture consists of 200 μL of solution B and 20 μL of the enzyme solution at known protein concentration, where

Solution B = of 40 μL of solution A and 1 mL of 100 mM Tris-HCl buffer at pH 7;

Solution A = 86 μL of PNPB and 1000 μL of 2-methyl-2-butanol.

The hydrolytic release of *p*-nitrophenol (ε 405 nm = 9.36 mL $(\mu mol\, cm)^{-1}$) leads to the increase of the absorbance at 405 nm, which was measured over time using a Tecan plate reader and plastic 96 well plates. The blank sample was prepared by replacing the 20 μL of enzymatic solution with an equal amount of buffer. The specific activity was expressed in units (U), where 1 unit is defined as the amount of enzyme required to hydrolyze 1 μmol of substrate per minute under the given assay conditions.

2.3.3 *Immobilization of Thc_cut1 and CaLB on epoxy-activated beads*

The calculation of the carrier/enzyme ratio was always referred to the weight of the dry resin, which was calculated according the protocol 2.3.4. An amount of wet carrier equal to 1.0 g of dry beads was washed with ethanol (20 mL, 2 ×) and double distilled H_2O (20 mL, 2 ×) prior to use. The resin was suspended in 10 mL of 1 mg mL^{-1} enzyme solution in 0.1 M Tris-HCl buffer pH 7. The immobilization was performed for 24 h at 21 °C using a rotary wheel as a mixing system. The immobilization was monitored by withdrawing samples and measuring the protein concentration and residual activity in the supernatant. At the end of the immobilization, the biocatalysts were rinsed with 50 mL of 0.1 M Tris-HCl buffer pH 7 for three times to remove the protein not covalently bond on the support. To protect the unreacted epoxy groups, the enzymatic preparations were incubated for 24 h in 45 mL of 3 M glycine at 21 °C. Afterward, the enzyme preparation was washed (50 mL × 3) with 0.1 M Tris-HCl buffer pH 7. Finally, the immobilized biocatalyst was dried at 30 °C for 48 h under reduced pressure (13.3 kPa) in a desiccator. That was done because the water content must be minimized in polycondensation reactions to prevent competing hydrolytic reactions that would hamper the elongation of the polyesters.

All immobilized preparations had a final water content lower than 0.05% (w/w). The water present in the immobilization resins and in the immobilized biocatalysts was evaluated according to the protocol 2.3.4.

2.3.4 Determination of the water content of the resins and the immobilized biocatalysts

0.2 g samples, either of resin or of immobilized enzymes, were weighted in a tarred weighting bottle (A). The bottle was then dried for 6 h at 120 ± 5 °C and cooled down in a desiccator until the weight was constant. Afterward the bottle was weighted again (B). The moisture content was calculated as follows:

$$\text{Moisture content}\,(\%) = [(A - B)/A] \times 100$$

2.3.5 Activity assay for the immobilized enzymes

About 20 mg of immobilized enzyme were suspended in an assay solution at 21 °C consisting of 0.1 mL of the substrate solution (86 µL of PNPB and 1000 µL of 2-methyl-2-butanol) and 11 mL of 100 mM Tris-HCl buffer at pH 7. The hydrolytic release of *p*-nitrophenol causes the increase of the absorbance at 405 nm, which was measured over time with a benchtop spectrophotometer (HACH Lange) using plastic cuvettes. The *p*-nitrophenol has a ε 405 nm = 9.36 mL (µmol cm)$^{-1}$. A blank test was carried out using resin beads where glycine was used instead of the enzyme as the reactant for the epoxy groups. The activity was calculated in units (U), where 1 unit is defined as the amount of enzyme required to hydrolyze 1 µmol of substrate per minute under the given assay conditions.

2.3.6 Evaluation of the detachment of the enzyme from the support

Immobilized enzyme preparations were incubated as described in the activity assay protocol 2.3.5 but without adding the PNPB solution. Samples were taken after 5, 10, 15 and 30 min and the biocatalyst was filtered off. The substrate solution was added to the supernatant and the residual esterase activity was measured as previously described (see Section 2.3.2 for details).

2.3.7 Equipment for the solvent-free enzymatic polycondensation using the thin-film method

The thin-film enzymatic reaction systems for solvent-free polycondensation were conducted using a Rotavapor R-114 (BÜCHI) connected to a Vac®

V-513 (BÜCHI) vacuum pump and a V-800 (BÜCHI) pressure controller. Temperature was controlled by means of a B-480 (BÜCHI) water bath.

2.3.8 Enzymatic polycondensation of dimethyl adipate (DMA) and 1,4-butanediol (BDO) using a thin-film reaction system in solvent-free conditions

Dimethyl adipate (DMA) (1.0451 g, 6.0 mmol) and 1,4-butanediol (BDO) (0.5407 g, 6.0 mmol, molar ratio 1.0:1.0) were mixed in a 50-mL round-bottom flask. After heating, the two monomers are in the liquid state and completely miscible in each other. The addition of the biocatalysts (0.16 g of immobilized Thc_cut1 or 0.16 g of immobilized CaLB, 10% (w/w) respect to the total amount of monomers) started the reaction. The flask was connected to a rotary evaporator and immersed in a water bath at 70 °C. The pressure was set at 100 kPa and the flask rotated at 200 rpm for 24 h. The final product was a colorless viscous liquid, which was easily solubilized in DCM. The crude product was analyzed, after solvent evaporation, by ^{1}H NMR, ESI-MS and GPC without any further purification. All reactions were performed in duplicates and compared to a control reaction mixture in the absence of enzyme. The immobilized CaLB led to 76% of conversion of BDO (calculated according the protocol 2.3.9), $M_w = 888$; $M_n = 528$; PD = 1.68 (see protocol 2.3.10). The immobilized Thc_cut1 gave 86% of conversion, $M_w = 1923$; $M_n = 985$; PD = 1.95.

2.3.9 NMR spectroscopy and determination of reaction conversion

The monomer conversion was estimated by ^{1}H NMR according to the following procedure: samples (50 mg) of crude reaction mixture were withdrawn and the biocatalyst was removed by filtration on a borosilicate glass filter under vacuum. The recovered product was dissolved in chloroform-d_1 and analyzed at ^{1}H NMR. The ratio between the signal at δ 1.26 attributed to methyl group of ethyl adipate ($C\underline{H}_3$-CH_2-O) and the signal at δ 2.33 (-CH_2-$C\underline{H}_2$-C(O)O-) assumed as constant, was exploited to evaluate the conversion as previously reported (Corici et al., 2015; Pellis et al., 2015).

NMR spectra were recorded on a Varian (400 MHz) or on an Oxford JEOL (270 MHz) spectrometer.

2.3.10 Gel permeation chromatography (GPC)

Polymer samples in a concentration of 2 mg mL^{-1} were dissolved in the mobile phase (HPLC-grade THF with 250 ppm of BHT as peroxides inhibitor). Samples were filtered through 0.45 μm PTFE filters into glass HPLC vials.

GPC was carried out on an Agilent Technologies HPLC System (Agilent Technologies 1260 Infinity) connected to a 17,369 6.0 mm ID × 40 mm L HHR-H, 5 μm Guard column and a 18,055 7.8 mm ID × 300 mm L GMHHR-N, 5 μm TSKgel liquid chromatography column (Tosoh Bioscience). Operational temperature was of 40 °C and the mobile phase flow rate was set to 1 mL min^{-1}. An Agilent Technologies G1362A refractive index detector was used for detection. The molecular weights of the polymers were calculated using polystyrene calibration standards (250–70,000 g mol^{-1}, Sigma-Aldrich).

2.3.11 Evaluation of the recyclability of the Thc_cut1 covalently immobilized on epoxy methacrylic carriers

The recyclability of the two immobilized *Thc_cut1* was tested in a polycondensation reactions carried out with 1.0451 g of dimethyl adipate (DMA) (6.0 mmol) and 0.5407 g of 1,4-butanediol (BDO) (6.0 mmol, molar ratio 1.0:1.0). The reaction was started by the addition of the biocatalyst (0.1586 g of Thc_cut1, 10% (w/w) respect to the total amount of monomers) and conducted in a 50-mL round-bottom flask connected to a rotary evaporator for 4 h at 50 °C under atmospheric pressure (100 kPa). Withdrawals of the fluid crude reaction mixture at 1, 2, 3 and 4 h (about 50 μL then dissolved in $CDCl_3$) allowed the monitoring of the conversion by ^1H NMR (Pellis et al., 2015; Pellis, Ferrario, Zartl, et al., 2016). The reaction mixture was filtered under reduced pressure without any addition of organic media because it was sufficiently fluid. At the end of the reaction, the immobilized enzyme was recovered with the aid of a sintered glass filter, equipped with a cellulose filter. No organic solvent was used for rinsing the biocatalyst in order to prevent any solvent-derived detrimental effect. The recovered biocatalyst was applied in the subsequent synthetic cycles under the same experimental conditions described above and by using the same amount of fresh monomers. For each cycle the conversion was calculated over time. No reaction occurred in the absence of enzyme. The percentage of conversion decreased by 40% after 10 cycles. The recyclability test was carried out at atmospheric pressure since it was observed that Thc_cut1 is negatively affected by reduced pressure (see Table 2).

2.3.12 Evaluation of the recyclability of CaLB covalently immobilized on epoxy methacrylic carrier

6472 mg of diethyl adipate (DEA) (32 mmol) and 3244 mg of 1,4-butanediol (BDO, 36 mmol) (scale 9.7 g; molar ratio 8:9) were mixed in a round reaction flask (50 mL capacity) connected to a rotary evaporator. The addition of

Table 2 Scheme of the experimental design developed for evaluating the effect of experimental conditions on the polycondensation (24 h) of DMA with BDO or ODO catalyzed by immobilized Thc_cut1.

T (°C)	P (mbar)	H_2O (μL)	Conversion (%)	M_n (Da)
BDO				
30	70	0	85	901
70	1000	0	66	514
70	70	10	22	359
30	1000	10	50	417
50	**535**	**5**	**96**	**1893**
50	**535**	**5**	**90**	**1266**
ODO				
70	70	0	6	271
30	1000	0	54	423
30	70	10	66	716
70	1000	10	53	434
50	**535**	**5**	**85**	**1315**
50	**535**	**5**	**85**	**1221**

The reaction conditions yielding to the best results are highlighted in bold.
Table reproduced from Pellis, A., Ferrario, V., Cespugli, M., Corici, L., Zartl, B., Herrero Acero, E., et al. (2017). Fully renewable polyesters via polycondensation catalyzed by Thermobifida cellulosilytica cutinase 1: An integrated approach. *Green Chemistry, 19*, 490–502. https://doi.org/10.1039/C6GC02142E by permission of The Royal Society of Chemistry.

the biocatalysts (equal amount of enzymatic units) started the reaction that run at 40 °C and 70 mbar for 5 h. The conversion of diethyl adipate into the oligomers was monitored at 10, 20, 40 and 300 min. After the synthetic procedure (300 min) the biocatalyst was recovered by filtration (see protocol 2.3.11) and the reaction was then repeated under the same conditions with the same enzymatic preparation for 8 times to evaluate the recyclability of the enzymatic preparation. The percentage conversion decreases of about 10% during the first three recycles and then remains constant.

2.3.13 Determination of protein concentration

The BioRad protein assay (Bio-Rad Laboratories GmbH, Vienna, Cat. No: 500-0006) was used by applying the following protocol. Aliquots of 10 μL were directly added into the wells of a 96-well micro-titer plate (Greiner 96 Flat Bottom Transparent Polystyrene). Afterward, 200 μL of the BioRad reaction solution (prepared diluting 1:5 with deionized water the stock solution received from the company) were added in each well. The plate was incubated at 21 °C for 5 min with a 400 rpm stirring. The Tris-HCl buffer used for protein dilution (0.1 M pH 7) was employed as blank. The calibration curve was constructed using BSA (bovine serum albumin) as a standard. The concentration was calculated from the average of triplicate measurements of adsorption at λ = 595 nm taken after 5 min.

3. Covalent immobilization of CaLB on functionalized rice husk for the fully renewable synthesis of poly(itaconate)s

3.1 The challenges hampering the scalability of enzymatic synthesis of poly(itaconate)s

Itaconic acid (IA) is a renewable monomer obtainable through fermentation of *Aspergillus terreus* (Willke & Vorlop, 2001). It presents structural similarities with acrylic and methacrylic acids because of the lateral vinyl moiety (Giacobazzi, Gioia, Colonna, & Celli, 2019). The IA offers opportunities for introducing molecular complexity in polymers through post polymerization functionalization since its *exo* double bond is an aza-Michael acceptor (Pellis, Hanson, et al., 2019). Nevertheless, the poly(itaconate)s have been little investigated so far because the chemical polycondensation of IA occurs at temperature above 150 °C and causes Ordelt saturation, isomerization of the C=C bond and cross- linking (Fig. 3) (Pellis, Hanson, et al., 2019). Radical reactions can be avoided using inhibitors (Satoh, Lee, Nagai, & Kamigaito, 2014), while there is no effective solution to limit the Ordelt saturation at such temperatures (Farmer, Castle, Clark, & Macquarrie, 2015). Therefore, the biocatalyzed polycondensation under mild conditions and below 90 °C has been proposed as a route for the synthesis of poly(itaconate)s having fully preserved vinyl groups that are readily available for further post-polymerization functionalizations (Lv, Cui, Du, & Li, 2016). However, the molecular weights reported for poly(itaconate)s synthetized by means of covalently immobilized CaLB are modest, because the conjugation between the C_1-ester group and the C=C bond of IA determines a decrease of the acylic carbon (Corici et al., 2015; Jiang, Woortman, van Ekenstein, & Loos, 2015). It must be noted that the main objective of enzymatic polycondensation of IA is not the synthesis of high

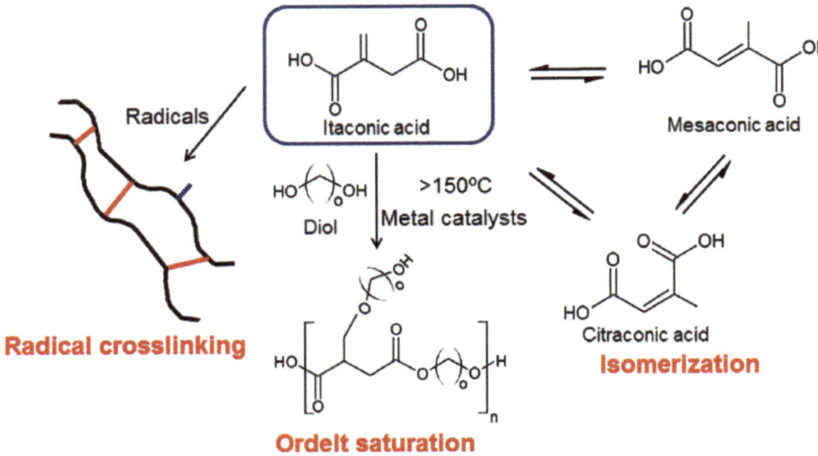

Fig. 3 Undesired side-reactions occurring during the polycondensation of itaconates with polyols. *Scheme reproduced from Pellis, A., Hanson, P. A., Comerford, J. W., Clark, J. H., & Farmer, T. J. (2019). Enzymatic synthesis of unsaturated polyesters: functionalization and reversibility of the aza-Michael addition of pendants.* Polymer Chemistry, 10, *843–851. https://doi.org/10.1039/C8PY01655K by permission of The Royal Society of Chemistry.*

molecular weight poly(itaconate)s but rather the insertion of discrete units of IA within different polyesters. That would allow controlling the structures and the final properties of new bio-based functional materials, which are not achievable trough traditional chemo-catalytic strategies (Jiang et al., 2015; Jiang, Woortman, et al., 2014; Jiang, Woortman, van Ekenstein, & Loos, 2013).

The low reactivity of the C_1 was boosted by promoting the enzyme-substrate interactions and the formation of the acyl-enzyme intermediate. When working in solvent-free conditions that results particularly challenging because of the severe mass transfer limitations caused by the viscosity. At the same time, the use of large amounts of enzymes would be non-economic. However, by using a low amount of enzyme immobilized on a wide surface of carrier it is possible to distribute homogeneously the biocatalyst in the reaction system, thus promoting the chain elongation (Corici et al., 2015). The porosity of the immobilization carriers is of little aid in case of very viscous systems because of the mass transfer limitations. Therefore, the solution was found in the immobilization of CaLB on large volumes of carriers that were used in 40–50% (w/w) on monomers basis.

As already mentioned above (see introduction), most studies utilize the commercial biocatalyst Novozym® 435, an immobilized formulation prepared by physically adsorbing the lipase on acrylic resins

(Barrett, Merkel, Luft, & Yousaf, 2010). It is known that the enzyme can partially detach from the acrylic support when used in viscous systems or in the presence of water (Binns et al., 1999; Jiang et al., 2015; Jiang, van Ekenstein, et al., 2014). Actually, the producer states for Novozym® 435 a specific activity of >5000 U/g referred to the synthesis of propyl laurate in non-aqueous media. In fact, when placed in aqueous buffer some of the adsorbed enzyme detaches from the carrier. Consequently, the evaluation of the hydrolytic activity of Novozym® 435 in buffer is not informative. In order to prevent the leakage of the enzyme from the support, a low amount of lipase was covalently immobilized on large volumes of carriers (Cespugli et al., 2018; Corici et al., 2016; Pellis et al., 2015). The modest specific hydrolytic activity of the biocatalysts (between 309 and 526 U/g_{dry}) is counterbalanced by the wide surface available, which facilitates the access of the substrates to the enzyme active site. These biocatalysts were employed for the polycondensation of dimethyl itaconate (DMI) with BDO in the presence of 158–263 U of CaLB per g of monomers. However, the use of such large amount of carriers sets the problem of the economic and environmental sustainability of the biocatalyzed process. Stringent regulations and environmental concerns are increasing their relevance in determining the viability of biocatalytic processes and Life Cycle Analysis (LCA) studies of biocatalyzed processes disclosed that the fossil-based raw materials of epoxy methacrylic resins for the immobilization of enzymes represent the primary source of greenhouse gas emission for the considered process (Kim, Jiménez-González, & Dale, 2009; Pellis et al., 2018).

In response to these challenges, new renewable and inexpensive carriers made by rice husk were developed. This natural composite material has the further advantage of combining exceptional mechanical stability, conferred by silica and lignin, with a density of about 0.4 $g\,mL^{-1}$. On a weight basis it means that the volume of the rice husk carrier is at least double as compared to methacrylic resins (density ~1.1 $g\,mL^{-1}$), with a consequent increase of the accessibility of the enzyme. This feature allowed to synthetize oligomers with $M_n = 1600\,g\,mol^{-1}$ and $M_w = 2900\,g\,mol^{-1}$ ($Đ = 1.76$) using only 158 U per g of monomers, whereas CaLB covalently immobilized on commercial resins gave products with $M_n = 1100\,g\,mol^{-1}$ and $M_w = 1800\,g\,mol^{-1}$ ($Đ = 1.61$). The rice husk was functionalized by oxidizing the cellulosic component with $NaIO_4$ and then by inserting a diamine as spacer that was activated with glutaraldehyde before lipase immobilization (Fig. 4) (Cespugli et al., 2018). The polycondensation was carried out in a solvent less systems and the high

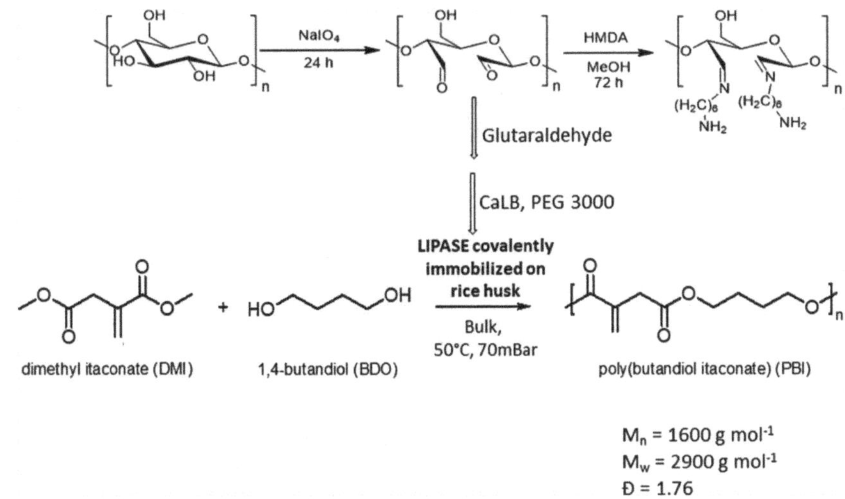

Fig. 4 Rice husk-based carriers. The rice husk was functionalized by oxidizing the cellulosic component with $NaIO_4$ and then by inserting a diamine as spacer that was activated with glutaraldehyde before lipase immobilization.

viscosity was overcome by applying the thin-film method (Cerea et al., 2013; Pellis et al., 2015), which maintains the integrity of the immobilized biocatalysts while assuring optimal mass and heat transfer.

The possibility of combining inexpensive and biodegradable immobilization carriers with sustainable biocatalyzed polycondensations sets the basis for a new generation of fully renewable polyesters, meeting both environmental and economic standards (Fig. 5).

3.2 Protocols for the functionalization of rice husk and immobilization of CaLB

3.2.1 Materials for the preparation of functionalized rice husk (RH)

Samples of RH (*Carnaroli* type) were from Riseria Cusaro S.r.L. (Binasco, Italy). The organic composition, determined according to a procedure previously described (Corici et al., 2016), is the following: 46.5% of cellulose, 22.1% of pentosanes (hemicellulose) and 31.9% of lignin. SiO_2 constitute 20% of the total weight. After milling, the RH has a water retention capacity of 42.6% (w/w), determined as reported here below. The solvents were standard laboratory grade. Alcohol, organic acid, and other reagents were purchased from either Aldrich Chemical Co. (Milwaukee, WI) or Sigma-Aldrich and used as received if not otherwise specified.

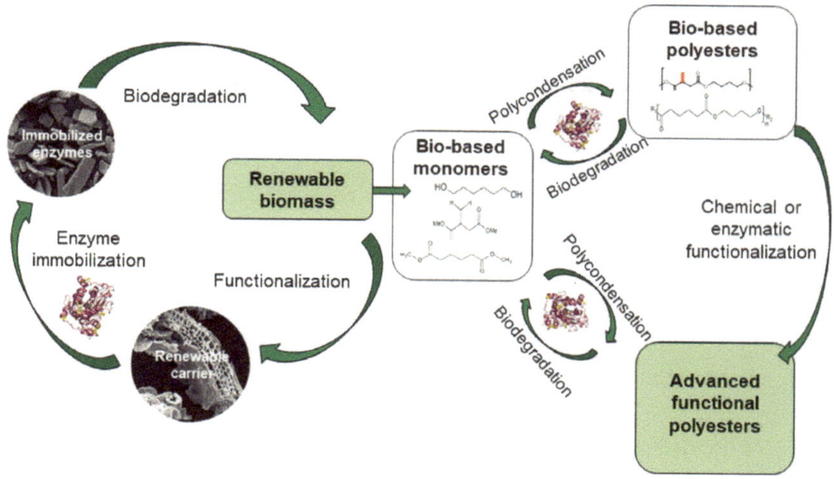

Fig. 5 Schematic illustration of the concept of fully renewable polyesters and their functionalization. *Figure adapted from Pellis, A., Herrero Acero, E., Ferrario, V., Ribitsch, D., Guebitz, G. M., & Gardossi, L. (2016). The closure of the cycle: Enzymatic synthesis and functionalization of bio-based polyesters.* Trends in Biotechnology, 34, *316–328.*

3.2.2 Methods for the characterization of the RH

- Grinding and sieving

RH was milled using a Rotor Mill ZM 200 (Retsch S.r.l., Bergamo, Italy) and the obtained material was sorted by size using four different sieves (450, 250, 200 and 100 μm). The weighted wet particles were dried at 120 °C in an oven for 6 h. Before milling the density of RH was 0.153 g mL^{-1} whereas the milled RH had a density of 0.437 g mL^{-1} (diameter 0.2–0.4 mm).

- Determination of water content

1.0 g of rice husk (sample A) was weighted in a tarred weighing bottle and dried for 2 h in an oven at 105 ± 5 °C. After cooling in a desiccator, the bottle containing the RH was opened to equalize the air pressure and then weighted. The cycles involving drying, cooling and weighting were repeated until a constant weight (sample B) was reached. The moisture content was calculated according to the following equation:

$$\text{Moisture content}\% = [(A - B)/A] \times 100$$

- Determination of the water retention capacity of the milled RH

0.45 g of milled RH were placed in a plastic syringe and treated with a series of seven soaking steps consisting in the addition of 2 mL of H_2O/EtOH solutions. The seven different solutions had the following content of water:

0%; 10%; 30%; 50%; 70%; 90%; 100%. The increasing H_2O/EtOH ratios promotes the full wettability of the matrix and the adsorption of water in the pores. The particles were filtered after each soaking treatment (10 min at 25 °C in a rotary wheel at 200 rpm) and a new solution was added in the syringe. At the end the particles were washed three times using distilled water (10 mL each) to completely remove the EtOH, then filtered under reduced pressure and weighted. Afterwards the RH particles were dried for 6 h at 120 °C in an oven and finally weighted. The weight difference provided the adsorption capacity (42.6%). As a control, a sample of RH was weighted before and after drying but without applying the soaking treatments. It resulted to contain 8.8% (w/w) of water.

3.2.3 Methods for the functionalization of RH

- Oxidation of RH using sodium periodate

2 g of RH (particle size 200–400 μm) were rinsed with a mixture of H_2O: ethanol 50:50 (3 × 3 mL) and then placed in a syringe with septum. A solution of a 0.20 M $NaIO_4$ (50 mL) was added and the reaction was carried out for 22 h in the dark at 25 °C under stirring on a rotary wheel. At the end of the reaction, the RH turned to dark brown and it was filtered and rinsed with deionized water (3 × 10 mL). The oxidized material was analyzed for determining the presence of carbonyl groups.

- Determination of the carbonyl groups content in the oxidized RH using hydroxylamine hydrochloride

The reaction between hydroxylamine hydrochloride and the carbonyl groups forms an oxime, with the concomitant release of HCl that is titrated using a sodium hydroxide solution. The assay was carried out using 25 mL of a solution of hydroxylamine hydrochloride 0.25 M, pH 3.20. About 200 mg of the oxidized RH were added to the solution and reacted for 2 h under stirring. The released HCl was titrated with 0.1 M NaOH, maintaining the pH at 3.20. At the end of the titration, the RH was filtered and dried for 6 h in an oven at 120 °C to determine the weight of the dried analyzed sample. As a reference, a sample of non-oxidized RH was washed six times with a 50:50 solution of ethanol and water, and then treated with the hydroxylamine hydrochloride. Each measurement was performed twice. The carbonyl groups content was calculated using the following formula:

$$\text{mmol aldehyde}/g_{\text{carrier}} = (V_{\text{NaOH}} \times C_{\text{NaOH}})/m_{\text{dry carrier}}$$

where V_{NaOH} is the volume in mL required to maintain the pH at 3.20, C_{NaOH} is the concentration of NaOH (0.1 mmol/mL) and $m_{\text{dry carrier}}$ is the mass in grams of the dry samples. The value of the carbonyl groups content resulted to be 1.00 mmol $^{g-1}$, as an average of two measures.

- Functionalization of oxidized rice husk with a diamine spacer

40 mL of a 0.9 M solution of hexamethylenediamine (HMDA) in MeOH were added to the oxidized rice husk (200 mg) and the mixture was reacted for 72 h at 25 °C on a rotary wheel (200 rpm). Afterwards the RH was filtered and rinsed with methanol (2 × 40 mL).

- Activation of amine functionalized rice husk with glutaraldehyde prior to immobilization

200 mg of amine-functionalized RH were treated at 25 °C with 50 mL of a 1.25% (v/v) solution of glutaraldehyde in 0.05 M phosphate buffer at pH 8. After 5 h, the RH was filtered and rinsed twice with 50 mL of 0.05 M phosphate buffer at pH 8.

3.2.4 Materials for the immobilization of CaLB on functionalized rice husk and epoxy methacrylic carriers

Native lipase B from *Candida antarctica* (batch number: LCN02115), was purchased from Novozymes (DK). The preparation is a light brown solution composed, as stated by the producer, by 44% (w/w) water, 25% (w/w) sorbitol, 25% (w/w) glycerol, enzyme 6% (w/w dry matter), 0.20% (w/w) sodium benzoate, 0.10% (w/w) potassium sorbate. The protein concentration, determined independently using the protein quantification test reported here below, was 8 mg mL^{-1}. Alcohol, organic acid, and other reagents were acquired from Aldrich Chemical Co. (Milwaukee, WI) or Sigma-Aldrich and used as received if not otherwise specified.

Sepabeads carriers were obtained from Resindion (Binasco, Italy).

3.2.5 Method for the determination of the activity of CaLB: Hydrolysis of tributyrin

The hydrolytic activity of the immobilized CaLB towards the tributyrin was assayed by titrating the released butyric acid with 0.1 M NaOH solution. The substrate, in the form of an emulsion, was composed by 1.5 mL tributyrin, 5.1 mL of Arabic gum that acts as emulsifier (0.6%, w/v) and 23.4 mL of distilled water. Afterward, 2 mL of 0.1 M potassium phosphate buffer pH 7.0 were added to the tributyrin emulsion (0.17 M, 30 mL total volume) and the mixture was incubated in a vessel maintained at 30 °C under constant mechanical stirring. Upon pH stabilization, about 100 mg of immobilized biocatalyst or 0.005 mL of a solution of native CaLB were added. The consumption of the NaOH was followed for 14 min.

One unit of activity was defined as the amount of immobilized enzyme required to produce 1 μmol of butyric acid per min at 30 °C. One unit (U) of lipase activity was defined as the amount of enzyme that liberates 1 μmol of butyric acid per minute under the reported assay conditions. Titrations and pH measurements were carried out employing a Graphic DL50 Mettler Toledo automatic titrator.

3.2.6 Protocol for the covalent immobilization of CaLB on RH oxidized with $NaIO_4$ and functionalized with a diamine spacer

In order to have a theoretical enzyme loading of 10,000 U $g_{carrier}^{-1}$, the commercial preparation of Lipozyme CaLB (4005 U/mL) was diluted with phosphate buffer (0.5 M, pH 8), which corresponds to a final enzymatic activity of 900 U/mL. A volume of 11.4 of this solution was added to 1 g of RH functionalized according to the procedure reported above and previously placed in a syringe with septum. Afterwards, 0.5 mL of a solution of PEG-3000 (2 mg mL^{-1} in phosphate buffer 0.5 M at pH 8) were added to stabilize the enzyme throughout the immobilization procedure. The mixture reacted for 48 h at 25 °C on a rotary wheel and 72% of the protein was loaded on the carrier, as verified by determining the residual protein content in the aqueous solutions at the end of the immobilization procedure (see for the protocol of protein quantification assay above). At the end of the immobilization, the immobilized enzyme was filtered and rinsed with 0.02 M phosphate buffer at pH 8 (6 × 1.2 mL). The water in excess was removed by washing with acetone (4 × 1 mL) and by drying the biocatalyst on the septum of the syringe under vacuum (740 mbar). The dried biocatalyst had a final water content of 12% (w/w), determined by weight difference using the method reported above. The specific activity of the immobilized CaLB was 316 U g_{dry}^{-1} (tributyrin assay, determined as the average of two measurements) and it was stored in a vial at 4 °C.

3.2.7 Methods for the characterization of the CaLB immobilized on rice husk

- Determination of protein leaking.

At the end of the tributyrin activity test, the immobilized biocatalyst was removed by filtration and 500 μL of the remaining liquid emulsion were withdrawn and subjected to the tributyrin hydrolysis assay in order to determine the free active enzyme released during the activity assay.

- Determination of the recyclability of the immobilized CaLB.

In order to evaluate the recyclability of the immobilized enzyme, after the activity assay the biocatalyst was filtered and exposed to the activity assay for 9 more cycles, observing a retention of activity of 90%.

3.2.8 Protocol for the preparation of the low-loading formulation of CaLB immobilized covalently on epoxy-methacrylic resins EC-EP

1.0 g of EC-EP/S epoxy resins (having an average particle size between 100 μm and300 μm) and washed once with 5 mL of deionized water were placed in a syringe with septum. To have a theoretical enzyme loading of 10,000 U $g_{carrier}^{-1}$, the commercial preparation of Lipozyme CaLB (4005 U/mL) was diluted with 0.5 M phosphate buffer at pH 8 (final enzymatic activity of 900 U/mL). A volume of 11.4 of this solution was added to the resin. After 24 h of reaction at 25 °C the quantification of the protein in the aqueous solution (see for the protocol of protein quantification assay above) indicated that >95% of the lipase was already immobilized and therefore the reaction was stopped filtering off the immobilized biocatalyst preparation from the supernatant solution. Then the preparation was incubated for 24 h with 15 mL of a saturated solution of glycine to quench any residual unreacted epoxy group. Afterwards, the immobilized biocatalyst was filtered and washed six times using 1 mL of 0.02 M phosphate buffer at pH 8, then rinsed with acetone (3 × 3 mL) and dried on the septum of the syringe under vacuum (740 mbar for 30 min). The biocatalyst was stored at 4 °C in a vial. The dried preparation had a water content of 9% (w/w), determined by weight difference using the method described above. The specific activity of the immobilized CaLB was 526 U g_{dry}^{-1} (tributyrin assay, determined as the average of two measurements) and it was stored in a vial at 4 °C.

3.3 Protocols for the synthesis of poly(itaconate)s catalyzed by low-loading covalently immobilized CaLB

3.3.1 Protocol for the enzymatic synthesis of poly(1,4-butylene itaconate) catalyzed by CaLB covalently immobilized on epoxy methacrylic carriers

The reaction was performed using the thin-film equipment described in protocol 2.3.7. Equimolar amounts of dimethyl itaconate (DMI, 16 mmol, 2.4950 g) and 1,4-butanediol (BDO, 16 mmol, 1.462 g) were mixed with 50% (w/w) of CaLB immobilized on the Resindion Sepabeads EC-EP carrier. The polycondensation was conducted at 50 °C and 70 mbar in a

100-mL round-bottom flask connected to a rotary evaporator. The overall amount of polyol was added in two aliquots, at the start of the reaction and after 6 h, to mitigate the negative effect of the diol on the chain elongation, as previously reported (Corici et al., 2016). After 72 h, the work up of the reaction mixture was performed using DCM and the polymers were characterized by NMR spectroscopy and ESI-MS after filtration of the biocatalyst and evaporation of the solvent, without any prior purification. The final product appears as a transparent viscous liquid.

3.3.2 Protocol for the enzymatic synthesis of poly(1,4-butylene itaconate) using CaLB covalently immobilized on RH

Equimolar amounts of BDO (8 mmol, 0.72 g) and DMI (8 mmol, 1.2485 g) were reacted in the presence of 50% (w/w) of CaLB immobilized on RH (158 U/g monomers). The diol was previously dried using molecular sieves in order to eliminate the moisture. The thin-film biocatalyzed polycondensation[9] was performed at 50 °C and 70 mbar in a 25-mL round-bottom flask connected to a rotary evaporator. The diol was added in two aliquots, as described in Section 3.3.1. The reaction conversion was monitored by thin layer chromatography (method 2.3.7) and NMR spectroscopy (method 2.3.9). After 72 h, the product was dissolved in DCM and the biocatalyst was removed by filtration. After the removal of the solvent, the final product appeared as a transparent viscous liquid that was characterized by NMR spectroscopy (see methods 3.3.3), GPC (protocol 2.3.10) and mass spectrometry (methods 3.3.3) without any further treatment.

3.3.3 Methods for the monitoring of the polycondensation and the characterization of the poly(itaconate)s

- NMR spectroscopy

NMR spectra were recorded in $CDCl_3$ on an Oxford JEOL (270 MHz) or a Varian (400 MHz) spectrometers. The conversion was calculated with the aid of ^{1}H NMR, by taking into account the signal of the α methylene that was compared to the signal of the hydroxyl of 1,4-butanediol and to the signal of the same methylene in the esterification product.

- Thin Layer Chromatography (TLC)

The polycondensation of the monomers was monitored on silica gel glass plates 20 × 20 cm (Macherey-Nagel) using ethyl acetate as the mobile phase and a $KMnO_4$/KOH (1.25/0.5%) staining solution for developing the plate.

- Electrospray Ionization Mass Spectrometry (ESI-MS)

The crude reaction mixtures were analyzed by electrospray positive ionization using an Esquire 4000 ESI-MS ion trap Bruker instrument

(Karlsruhe, Germany) generating the ions in an acidic environment. Approximately 10 mg of sample were dissolved in 1 mL of methanol containing 0.1% (v/v) formic acid. The generated positive ions had m/z ratio falling in the range of 100–1000. A subsequent deconvolution procedure allowed reconstructing the mass peaks of the chemical species starting from the analysis of the generated peaks.

4. Design of experiments (DoE) applied to the analysis of the effect of experimental factors in an enzymatic polycondensation

4.1 Factorial design of experiments

In a classical experimental approach, 'one factor at a time' is varied, while all the other factors are maintained constant. The outcome is an estimation of the effect of the single variable at selected fixed conditions. Therefore, if the variables do no act additively but rather they interact, the method provides incomplete information and the real optimum can be identified by chance or with difficulty. Chemometrics gathers mathematical and statistical techniques that are able to overcome these limitations by collecting and analyzing experimental observations in a multivariate, namely by studying all variables (even discontinuous ones) simultaneously and at multiple levels (Box, Hunter, & Hunter, 2005). This type of analysis is possible with the employment of statistical experimental designs (*i.e.*, factorial design), where sets of experimental conditions are chosen systematically and with the objective of collecting meaningful information. According to such strategies, the statistical analysis of data provides information about the effect of the factors (X variables) but also about the relevance of the interactions on the response considered (Y variable).

Chemometrics applies mathematical or statistical methods for creating relationships between measurements done on a chemical system and the state of the same system. It embraces multivariate statistical analysis and statistical experimental designs, which allow for the study of many variables simultaneously (Boccú et al., 1990; Braiuca, Knapic, Ferrario, Ebert, & Gardossi, 2009; Ferrario, Ebert, Svendsen, Besenmatter, & Gardossi, 2014). Multivariate statistical methods stem from different mathematical basis but they share the ability to make possible the interpretation of complex systems by extracting the relevant information. These methods are of great utility in the study of complex biocatalyzed processes, which cannot be modeled by applying the fundamental enzymology knowledge and are influenced by multiple physical-chemical parameters (Braiuca, Ebert, Basso, Linda, & Gardossi, 2006).

A chemometric approach was applied to the study of the effect of temperature, pressure and water activity (a_w) on the polycondensation catalyzed by immobilized *Thermobifida cellulosilytica* cutinase 1 immobilized on an epoxy-activated carrier (Pellis et al., 2017). By using a fractional factorial design scheme (Box et al., 2005), sets of experimental conditions were selected for carrying out the polycondensation of dimethyl adipate catalyzed by Thc_cut1 in solvent-free conditions. The analysis considered also the effect of two different diols: BDO and 1,8-octanediol (ODO). The objective of the study was to identify the best experimental conditions leading to the highest conversion of the diols and maximum chain elongation (M_n). Table 2 reports the plan of the experiments, consisting of 12 polycondensation reactions, namely a factorial design, where each variable was studied at two extreme values. Additionally, a central point was run for each diol twice and it corresponds to the intermediate levels of the variables.

Special attention was given to the effect of water and a_w more specifically, which was measured at the starting of each reaction. Defined volumes of water were added to the reaction mixtures in closed vessels, which were equilibrated for the time necessary to allow the water to distribute among the phases of the system. By measuring the vapor pressure (or relative humidity, RH) in the gas phase by means of a humidity sensor, it is possible to evaluate the a_w, namely the "free" water available for promoting undesired hydrolytic reactions or for increasing the hydration state of the enzyme (Ulijn et al., 2002). It is assumed that the a_w is the same in all phases when the system reaches the equilibrium (Spizzo et al., 2007). The water activity is the actual informative parameter affecting the polycondensation and it is expected to vary as a function of the hydrophilic nature of the diol and also throughout the course of the reaction as the ratio of the substrates and the products changes.

Table 2 reports the data obtained in the polycondensations using BDO and ODO and they indicate that the best results, in terms of both M_n and conversion, are obtainable at average values of temperature, pressure and humidity (Pellis et al., 2017). The replicates using BDO and ODO provide useful information on the effect of the diol structure as they indicate a moderate preference of Thc_cut1 for BDO. The effects of the variables on the two responses here considered were calculated by multiple linear regression analysis and they are illustrated by the variable importance plots (VIP plots), which also account for the effect of the variable interactions (Fig. 6). The interactions are indicate by the symbol "*" between the two interacting variables. The values correspond to the difference between the averages

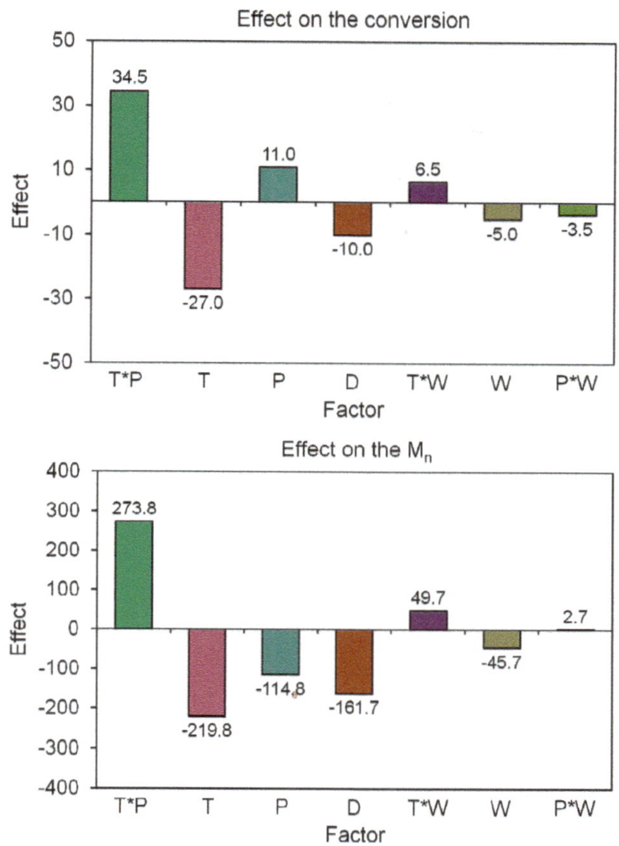

Fig. 6 Variables effect and impact of their interactions on reaction conversion and M_n. The enzymatic preparation used was Thc_cut1 immobilized on an epoxy-activated carrier. Reaction substrates were dimethyl adipate and the two aliphatic diols 1,4-butanediol and 1,8-octanediol. *Table reproduced from Pellis, A., Ferrario, V., Cespugli, M., Corici, L., Zartl, B., Herrero Acero, E., et al. (2017). Fully renewable polyesters via polycondensation catalyzed by Thermobifida cellulosilytica cutinase 1: An integrated approach.* Green Chemistry, 19, *490–502. https://doi.org/10.1039/C6GC02142E by permission of The Royal Society of Chemistry.*

of responses obtained at a high and low level of each experimental factor while the other variables are maintained at the intermediate level. In Fig. 6, the terms having values close to zero influence the reaction at a negligible extent. It must be remembered, however, that the algorithm, as well as the definition of effect, takes into account only the minimum and maximum levels set for the factors, so the net effect is calculated as a linear variation of the response between the two high and low levels of factors, not taking into account the intermediate level.

The analysis of the data collected according to the factorial design allowed to understand how the reaction factors can be tuned in order to optimize the system. The temperature exerts a major detrimental effect on the Thc_cut1, which prefers mild temperatures (50 °C). Notably, the chemometric approch disclosed the severe negative effect of the interaction between low pressure and high temperature. Unexpectedly, adding small quantities of water to the system exerts only a negligible effect on the M_n, since the polycondensation reaction proceeds even in the presence of such small volumes of water (5 μL for 0.006 mol per monomer) and under very mild temperatures (50 °C) and vacuum (535 mbar).

These results indicate that the operative conditions for Thc_cut1 must be considerably different from those generally applied to polycondensations catalyzed by CaLB, which require a strict control of humidity and application of low pressure for the removal of the leaving nucleophile, either water or alcohols (Pellis, Comeford, et al., 2019; Weinberger, Pellis, Comerford, Farmer, & Guebitz, 2018). Apparently, Thc_cut1 is less sensitive to traces of water and promotes the chain elongation under milder conditions if compared to CaLB.

4.2 Protocols for the chemometric analysis

4.2.1 Material

The factorial design was planned using the software MODDE 8.0.2 (MSK Umetrics).

4.2.2 Planning the factorial design

Thc_cut1 immobilized on epoxy methacrylic resin was employed for the polycondensations that was conducted on a thin film system and under solvent-free conditions. The considered independent variables were studied at the following levels (+1 and −1): temperature = 70 °C and 30 °C; Pressure = 1000 mbar and 70 mbar H_2O = 10 μL and 0 μL. The factorial design consists in the combination of the different levels of the variables. Moreover, the effect of the two selected diols (BDO and ODO) was evaluated by dividing the factorial design into two blocks having a central point, which corresponds to the combination of each independent variable considered at an intermediate level (level 0: $T = 50$ °C; $P = 535$ mbar; H_2O = 5 μL). Experiments were run according to a random sequence and the responses measured for each experiment were the conversion of the diol (calculated by ^{1}H NMR spectroscopy as previously described) and the Mn of the product (calculated by GPC analysis). Finally, the matrix of the experimental design and the corresponding responses were analyzed by means of the multiple linear regression analysis algorithm implemented in the software MODDE. The algorithm calculates the effects of the factors on the dependent variables, namely conversion of the diol and Mn of the products. While simple linear regression is a method that determines the relationship between one factor (X) and a continuous output (Y) (*i.e.*, expressed in terms of a mathematical equation such as $Y = b + mX$) the general purpose of multiple linear regression analysis is to quantitate the relationship between several independent variables and a dependent variable. Multiple linear regression analysis models are built with descriptive variables using the least squares methods to minimize the residuals. A general multiple linear regression analysis model can be expressed by the following equation:

$$y = a + b_1 \cdot x_1 + b_2 \cdot x_2 + \cdots + b_n \cdot x_n$$

where y is the dependent variable (quantitative property to predict), x_n an independent variable, a the intercept, and b_n the regression coefficient for x_n.

4.3 Protocols for the enzymatic polycondensation in solvent free conditions

4.3.1 Chemicals and reagents

1,4-Butanediol (BDO), and 1,8-octanediol (ODO) were purchased from Merck. Dimethyl adipate (DMA), and all other chemicals and solvents were purchased from Sigma-Aldrich at reagent grade and used without further purification if not otherwise specified.

4.3.2 Preparation of the immobilized biocatalyst

The recombinant *Thermobifida cellulosilytica* cutinase 1 (Thc_cut1) was produced and purified as reported in protocol 2.3.1 and immobilized according to protocol 2.3.3.

4.3.3 Enzymatic polycondensation using a thin-film reaction system

The reactions composing the factorial design were carried out using equimolar amounts (0.006 mol) of the diester and the diols in the presence of 10% (w/w) of immobilized iThc_cut1. Reactions were conducted in 50 mL round bottom flasks connected to a rotary evaporator and by applying reduced pressure when requested (see protocol 2.3.7). After 24 h, the biocatalyst was removed by filtration and the products dissolved in THF. The products were characterized without any further purification by NMR spectroscopy (protocol 2.3.9), ESI-MS (protocol 4.3.5) and GPC (protocol 2.3.10).

4.3.4 Water activity measurements

The water activity of all reaction systems composing the factorial design was measured at 30 °C by means of a hygrometer (DARAI-Trieste, Italy). The porous immobilization carrier (without the biocatalyst), the monomers and defined volumes of water were mixed at 30 °C in the same type of flasks used for carrying out the reactions. It was assumed that the small amount of protein linked on the carrier has a negligible effect on the a_w of the system if compared with the large amount of the porous resin. The humidity sensor was exposed to the gas phase of the sealed vessels and the relative humidity (RH) was measured until constant readings (a_w), namely the achievement of the apparent equilibrium of the multiphase system. The a_w is correlated to the relative humidity at the equilibrium by the following equation: $RH = a_w \times 100$.

The reaction mixture containing ODO reached constant RH values after 24 h of equilibration whereas, when BDO was used as the diol, the system took up to 96 h to reach the equilibrium (Table 3). This behavior depends on the higher hydrophilicity of BDO and the presence of some water in the same reactive, which took longer time to partition among the different phases. That was also confirmed by a progressive increase of the RH values from 24 (RH = 34%) to 96 h ($a_w = 0.49$). The measurement of the a_w of the equilibrated systems indicated that at the starting of the reactions all different reactions included in the factorial design had a_w values comprised between 0.34 and 0.63.

Table 3 Equilibration of the systems after addition of water using various aliphatic diols.

Water added to the system (μL)	a_w	Time required for the equilibration of the system
DMA-BDO		
0	0.49	96
5	0.63	96
10	0.61	96
DMA-ODO		
0	0.34	24
5	0.35	24
10	0.51	24

4.3.5 Electrospray ionization mass spectrometry (ESI-MS)

ESI-MS (electrospray positive ionization by generating the ions in an acidic environment) was performed on the crude reaction mixtures using an Esquire 4000 (Bruker). Sample (about 10 mg) were dissolved in 2 mL of MeOH containing 0.1% (v/v) of formic acid. The positively charged ions had m/z ratio falling in the 200–1000 range. For the subsequent deconvolution process, see protocol 3.3.3.

5. Conclusions

The synergic used of user friendly and well established computational methodologies allows for the rational planning of protocols to overcome some major bottlenecks hampering the takeoff of enzymatic polycondensation technology. The formulation of robust biocatalysts, preventing the contamination of the polymeric products, along with the control of the experimental variables that affect the biocatalyst set the basis for the full exploitation of the catalytic potential and selectivity of enzymes.

Acknowledgments

Alessandro Pellis thanks the Austrian Science Fund (FWF) for the Erwin Schrödinger Fellowship (Grant Agreement J4014-N34). Lucia Gardossi acknowledges the University of Trieste for financial support (FRA 2018).

References

Ansorge-Schumacher, M. B., & Thum, O. (2013). Immobilised lipases in the cosmetics industry. *Chemical Society Reviews*, *42*, 6475–6490.

Barrett, D. G., Merkel, T. J., Luft, J. C., & Yousaf, M. N. (2010). One-step syntheses of photocurable polyesters based on a renewable resource. *Macromolecules*, *43*, 9660–9667.

Bassanini, I., Hult, K., & Riva, S. (2013). Dicarboxylic esters: Useful tools for the biocatalyzed synthesis of hybrid compounds and polymers. *Beilstein. The Journal of Organic Chemistry*, *11*, 1583–1595.

Basso, A., Braiuca, P., Cantone, S., Ebert, C., Linda, P., Spizzo, P., et al. (2007). In silico analysis of enzyme surface and glycosylation effect as a tool for efficient covalent immobilization of CalB and PGA on Sepabeads®. *Advanced Synthesis and Catalysis*, *349*, 877–886.

Berman, H. M., Westbrook, J., Feng, Z., Gilliland, G., Bhat, T. N., Weissig, H., et al. (2000). The protein data bank. *Nucleic Acids Research*, *28*(1), 235–242.

Binns, F., Harffey, P., Roberts, S. M., & Taylor, A. (1999). Studies leading to the large scale synthesis of polyesters using enzymes. *Journal of the Chemical Society, Perkin Transactions*, *1*, 2671–2676.

Boccú, E., Ebert, C., Gardossi, L., Gianferrara, T., & Linda, P. (1990). Chemometric optimization of an asymmetric reduction catalyzed by Baker's yeast. *Biotechnology and Bioengineering*, *35*, 928–934.

Box, G. E. P., Hunter, J. S., & Hunter, W. G. (2005). *Statistics for experimenters: Design, innovation and discovery* (2nd ed.). Hoboken-NJ: Wiley and Sons Inc.

Braiuca, P., Ebert, C., Basso, A., Linda, P., & Gardossi, L. (2006). Computational methods to rationalize experimental strategies in biocatalysis. *Trends in Biotechnology, 24*, 419–425.

Braiuca, P., Knapic, L., Ferrario, V., Ebert, C., & Gardossi, L. (2009). A three-dimensional quantitative structure-activity relationship (3D-QSAR) model for predicting the enantioselectivity of Candida antarctica lipase B. *Advanced Synthesis and Catalysis, 351*, 1293–1302.

Cantone, S., Ferrario, V., Corici, L., Ebert, C., Fattor, D., Spizzo, P., et al. (2013). Efficient immobilisation of industrial biocatalysts: Criteria and constraints for the selection of organic polymeric carriers and immobilisation methods. *Chemical Society Reviews, 42*, 6262–6276.

Cerea, G., Gardossi, L., Sinigoi, L., & Fattor, D. (2013). *Process for the production of polyesters through synthesis catalyzed by enzyme.* World Patent WO2013110446A1.

Cespugli, M., Lotteria, S., Navarini, L., Lonzarich, V., Del Terra, L., Vita, F., et al. (2018). Rice husk as an inexpensive renewable immobilization carrier for biocatalysts employed in the food, cosmetic and polymer sectors. *Catalysts, 8*, 471.

Corici, L., Ferrario, V., Pellis, A., Ebert, C., Lotteria, S., Cantone, S., et al. (2016). Large scale applications of immobilized enzymes call for sustainable and inexpensive solutions: Rice husks as renewable alternatives to fossil-based organic resins. *RSC Advances, 6* (68), 63256–63270.

Corici, L., Pellis, A., Ferrario, V., Ebert, C., Cantone, S., & Gardossi, L. (2015). Understanding potentials and restrictions of solvent-free enzymatic polycondensation of itaconic acid: An experimental and computational analysis. *Advanced Synthesis and Catalysis, 357*, 1763–1774.

Farmer, T. J., Castle, R. L., Clark, J. H., & Macquarrie, D. J. (2015). Synthesis of unsaturated polyester resins from various bio-derived platform molecules. *International Journal of Molecular Sciences, 16* (7), 14912–14932.

Ferrario, V., Ebert, C., Svendsen, A., Besenmatter, W., & Gardossi, L. (2014). An integrated platform for automatic design and screening of virtual mutants based on 3D-QSAR analysis. *Journal of Molecular Catalysis B: Enzymatic, 101*, 7–15.

Gardossi, L., Poulsen, P. B., Ballesteros, A., Hult, K., Vytas, K., Švedas, V. K., et al. (2010). Guidelines for reporting of biocatalytic reactions. *Trends in Biotechnology, 28*, 171–180.

Giacobazzi, G., Gioia, C., Colonna, M., & Celli, A. (2019). Thia-Michael reaction for a thermostable Itaconic-based monomer and the synthesis of functionalized biopolyesters. *ACS Sustainable Chemistry & Engineering*, 7(5), 5553–5559.

Gross, R. A., Ganesh, M., & Lu, W. (2010). Enzyme-catalysis breathes new life into polyester condensation polymerizations. *Trends in Biotechnology, 28*(8), 435–443.

Herrero Acero, E., Ribitsch, D., Steinkellner, G., Gruber, K., Greimel, K., Eiteljoerg, I., et al. (2011). Enzymatic surface hydrolysis of PET: Effect of structural diversity on kinetic properties of cutinases from thermobifida. *Macromolecules, 44*, 4632–4640.

Hilterhaus, L., Minow, B., Müller, J., Berheide, M., Quitmann, H., Katzer, M., et al. (2008). Practical application of different enzymes immobilized on sepabeads. *Bioprocess and Biosystems Engineering, 31*, 163–171.

Jiang, Y., van Ekenstein, G. O. R. A., Woortman, A. J. J., & Loos, K. (2014). Fully biobased unsaturated aliphatic polyesters from renewable resources: Enzymatic synthesis, characterization, and properties. *Macromolecular Chemistry and Physics, 215*, 2185–2197.

Jiang, Y., Woortman, A. J. J., van Ekenstein, G. O. R. A., & Loos, K. (2013). Enzyme-catalyzed synthesis of unsaturated aliphatic polyesters based on green monomers from renewable resources. *Biomolecules, 3*, 461–480.

Jiang, Y., Woortman, A. J. J., van Ekenstein, G. O. R. A., & Loos, K. (2015). Environmentally benign synthesis of saturated and unsaturated aliphatic polyesters via enzymatic polymerization of biobased monomers derived from renewable resources. *Polymer Chemistry, 6*, 5451–5463.

Jiang, Y., Woortman, A. J. J., van Ekenstein, G. O. R. A., Petrovic´, D. M., & Loos, K. (2014). Enzymatic synthesis of biobased polyesters using 2,5-Bis(hydroxymethyl)furan as the building block. *Biomacromolecules*, *15*, 2482–2493.

Kim, S., Jiménez-González, C., & Dale, B. E. (2009). Enzymes for pharmaceutical applications—A cradle-to-gate life cycle assessment. *International Journal of Life Cycle Assessment*, *14*, 392–400.

Lv, A., Cui, Y., Du, F.-S., & Li, Z.-C. (2016). Thermally degradable polyesters with tunable degradation temperatures via postpolymerization modification and intramolecular cyclization. *Macromolecules*, *49*, 8449–8458.

Pellis, A., Cantone, S., Ebert, C., & Gardossi, L. (2018). Evolving biocatalysis to meet bioeconomy challenges and opportunities. *New Biotechnology*, *40*, 154–169.

Pellis, A., Comeford, J. W., Weinberger, S., Guebitz, G. M., Clark, J. H., & Farmer, T. J. (2019). Enzymatic synthesis of lignin derivable pyridine based polyesters for the substitution of petroleum derived plastics. *Nature Communications*, *10*(1), 1762.

Pellis, A., Corici, L., Sinigoi, L., D'Amelio, N., Fattor, D., Ferrario, V., et al. (2015). Towards feasible and scalable solvent-free enzymatic polycondensations: Integrating robust biocatalysts with thin film reactions. *Green Chemistry*, *17*, 1756–1766.

Pellis, A., Ferrario, V., Cespugli, M., Corici, L., Zartl, B., Herrero Acero, E., et al. (2017). Fully renewable polyesters via polycondensation catalyzed by Thermobifida cellulosilytica cutinase 1: An integrated approach. *Green Chemistry*, *19*, 490–502.

Pellis, A., Ferrario, V., Zartl, B., Brandauer, M., Gamerith, C., Herrero Acero, E., et al. (2016). Enlarging the tools for efficient enzymatic polycondensation: Structural and catalytic features of cutinase 1 from Thermobifida cellulosilytica. *Catalysis Science & Technology*, *6*, 3430–3442.

Pellis, A., Hanson, P. A., Comerford, J. W., Clark, J. H., & Farmer, T. J. (2019). Enzymatic synthesis of unsaturated polyesters: Functionalization and reversibility of the aza-Michael addition of pendants. *Polymer Chemistry*, *10*, 843–851.

Pellis, A., Herrero Acero, E., Gardossi, L., Ferrario, V., & Guebitz, G. M. (2016). Renewable building blocks for sustainable polyesters: New biotechnological routes for greener plastics. *Polymer International*, *65*(8), 861–871.

Satoh, K., Lee, D.-H., Nagai, K., & Kamigaito, M. (2014). Precision synthesis of bio-based acrylic thermoplastic elastomer by RAFT polymerization of Itaconic acid derivatives. *Macromolecular Rapid Communications*, *35*, 161–167.

Spizzo, P., Basso, A., Ebert, C., Gardossi, L., Ferrario, V., Romano, D., et al. (2007). Resolution of (R,S)-flurbiprofen catalysed by dry mycelia in organic solvent. *Tetrahedron*, *63*, 11005–11010.

Ulijn, R. V., De Martin, L., Gardossi, L., Janssen, A. E. M., Moore, B. D., & Halling, P. J. (2002). Solvent selection for solid-to-solid synthesis. *Biotechnology and Bioengineering*, *80*, 509–515.

Uppenberg, J., Hansen, M. T., Patkar, S., & Jones, T. A. (1994). The sequence, crystal structure determination and refinement of two crystal forms of lipase B from Candida antarctica. *Structure*, *2*(4), 293–308.

Weinberger, S., Pellis, A., Comerford, J. W., Farmer, T. J., & Guebitz, G. M. (2018). Efficient physisorption of Candida antarctica lipase B on polypropylene beads and application for polyester synthesis. *Catalysts*, *8*(9), 369.

Willke, T., & Vorlop, K. D. (2001). Biotechnological production of itaconic acid. *Applied Microbiology and Biotechnology*, *56*, 289–295.

CHAPTER THREE

Multiple grafting to enzymatically synthesized polyesters

Muhammad Humayun Bilal[a,†], Razan Alaneed[a,b,†], Jonas Steiner[b], Karsten Mäder[b], Markus Pietzsch[b], Jörg Kressler[a,*]
[a]Department of Chemistry, Martin Luther University Halle-Wittenberg, Halle (Saale), Germany
[b]Department of Pharmacy, Martin Luther University Halle-Wittenberg, Halle (Saale), Germany
*Corresponding author: e-mail address: joerg.kressler@chemie.uni-halle.de

Contents

Abstract

Enzymatic polymerization is an environmentally benign process for the synthesis of biodegradable and biocompatible polymers. The regioselectivity of lipase B from *Candida Antarctica* (CAL-B) produces linear functional polyesters without protection-deprotection steps. In this work, two different methods for the enzymatic synthesis of functional polyesters based on renewable resources, as, e.g., glycerol, using CAL-B

† These authors contributed equally to this study.

Methods in Enzymology, Volume 627
ISSN 0076-6879
https://doi.org/10.1016/bs.mie.2019.04.031

are outlined. Poly(glycerol adipate) was synthesized by enzymatic transesterification between glycerol and divinyl adipate or dimethyl adipate. Methods are also reported to graft poly(glycerol adipate) with different amounts of hydrophobic side chains (lauric, stearic, behenic, and oleic acids) and hydrophilic poly(ethylene glycol) side chains, respectively. The hydrophilicity or lipophilicity of grafted polyesters is well controlled by changing the degree of grafting of hydrophilic and hydrophobic side chains. The multiple grafted polyesters are characterized by NMR spectroscopy, differential scanning calorimetry, gel permeation chromatography, and X-ray diffraction. Furthermore, the self-assembly of the graft copolymers in water and their use as steric stabilizers for cubosomes are discussed. For this purpose mainly dynamic light scattering and small angle X-ray scattering have been employed.

1. Introduction

Enzymatic polyester synthesis is a green approach to produce functional polyesters from monomers preferentially derived from renewable resources having multiple functionalities. Functional polyesters are polyesters possessing pendant functional groups such as mercapto (Tanaka, Kohri, Takiguchi, Kato, & Matsumura, 2012), hydroxyl (Bilal et al., 2017; Hu, Gao, Kulshrestha, & Gross, 2006; Jbeily, Naolou, Bilal, Amado, & Kressler, 2014; Kallinteri, Higgins, Hutcheon, Pourçain, & Garnett, 2005; Naolou, Jbeily, Scholtysek, & Kressler, 2013), epoxy (Olsson, Lindström, & Iversen, 2007), halides (Jérôme & Lecomte, 2008), solketal (Wu, Al-Azemi, & Bisht, 2008), azide groups (Jérôme & Lecomte, 2008; Naolou, Busse, & Kressler, 2010), and unsaturated entities (Olson & Sheares, 2006; Tsujimoto, Uyama, & Kobayashi, 2001, 2002), at the polymer main chain. These pendant functionalities offer the opportunity to conjugate drugs (Suksiriworapong et al., 2018), proteins or other hydrophilic or hydrophobic side chains to achieve graft copolymers of required properties.

Glycerol, which is a by-product of biofuel production, was extensively used for the synthesis of functional polyesters under several reaction conditions and in the presence of lipases derived from different origins (Shoda, Uyama, Kadokawa, Kimura, & Kobayashi, 2016; Varma, Albertsson, Rajkhowa, & Srivastava, 2005). A linear polymer is obtained as the condensation reaction occurs basically at the primary hydroxyl groups of glycerol, only 5–10 mol% of secondary hydroxyl groups are reacted at 50 °C (Kline, Beckman, & Russell, 1998; Kulshrestha, Gao, & Gross, 2005). Poly(glycerol sebacate) (PGS) and poly(glycerol adipate) (PGA) are typical examples of such glycerol-based polyesters (Jiang & Loos, 2016; Yu et al., 2012). Like glycerol, other multihydroxy alcohols such as erythritol, xylitol, ribitol,

D-glucitol, D-mannitol, and galactitol obtained by the reduction of sugars (Sifontes Herrera et al., 2011), were also subjected to enzymatic polymerization (Hu et al., 2006; Kim & Dordick, 2001; Uyama, Klegraf, Wada, & Kobayashi, 2000). These reduced sugars are difficult to polymerize because they are only soluble in highly polar solvents such as DMF, DMSO, and pyridine. These highly polar solvents reduce the enzyme activity as a result of their impact on the enzyme conformation in solution (Martin, Ampofo, Linhardt, & Dordick, 1992; Therisod & Klibanov, 1986). The pendant functionalities in functional polyesters can also be introduced via the acid part of the repeating unit (Jiang & Loos, 2016; Shoda et al., 2016). The presence of functional pendant groups along the polymer backbone is a highly efficient source of tailoring the properties of polyesters including features such as hydrophilicity, biodegradation rates, and bioadhesion (Williams, 2007). *Comb*-like polyesters can be synthesized by grafting from, grafting onto, and grafting through strategies (Feng et al., 2011). Kallinteri et al. obtained amphiphilic *comb*-like polyesters by direct esterification of fatty acids with the pendant hydroxyl groups of the polyester backbone. Later on, they used them to prepare nanoparticles (NP) for drug delivery systems (DDS) (Kallinteri et al., 2005). In another study, Orafai et al. have used fatty acids (caprylic and stearic acid) and an protected amino acid (tryptophan) grafted PGA and determined the surface free energy of nanoparticles prepared from these polymers (Orafai et al., 2008). Zhang et al. have used a one-pot synthesis approach for the preparation of grafted polyesters with unsaturated fatty acids to form polymeric triglyceride analogues (Zhang et al., 2013). These grafted polyesters could also be cured to produce biodegradable polymeric networks (Tsujimoto et al., 2002; Uyama, Kuwabara, Tsujimoto, & Kobayashi, 2003). Our group has extensively studied poly(glycerol adipate) grafted with saturated and unsaturated fatty acids of various lengths. It was discovered that the degree of substitution is important to attain certain morphologies of nanoparticles obtained in aqueous media. These nanoparticles were further used as vehicles for drug delivery (Weiss et al., 2012). Later on, the synthesis of worm-like aggregates for effective drug carriers was reported (Naolou, Meister, Schöps, Pietzsch, & Kressler, 2013).

Besides PGA, sugar-based polyesters such as poly(xylitol adipate) (PXA) and poly(D-sorbitol adipate) (PDSA) were also grafted with stearoyl side chains that produce unique nanoparticles in aqueous media (Bilal et al., 2016). Grafted polymers could also be synthesized using synthetic pathways other than direct esterification. Thus, Pfefferkorn et al. synthesized *comb*-like polymers containing blocks of two crystallizable side chains, i.e., poly(ε-caprolactone) (PCL) and poly(ethylene glycol) (PEG) (Pfefferkorn et al., 2013). They explained comprehensively the crystallization behavior of the grafted copolymers. The phase behavior of these amphiphilic polymers at the air/water interface was also investigated. In another study, polyesters with azide pendant groups were synthesized from polycondensation reaction of an azide group-containing diol (synthesized from 3-hydroxymethyl-3-methyloxetane) and grafted with azide-alkyne "click" reaction using PEG to produce water-soluble polymers (Naolou et al., 2010). Later on, Wu et al. synthesized azido functionalized polyesters starting from 2-azido glycerol which after click reaction with PEG, produced a comb-like polymer (Wu, Wang, Liu, Deng, & Yu, 2014). This comb-like polymer self-assembles in form of spherical micelles in aqueous media, and it has a relatively low critical micelle concentration (CMC). Jbeily et al. have converted hydroxyl pendant groups of polyesters to macro-initiators for atom transfer radical polymerization (ATRP), that after ATRP of glycerol monomethacrylate produced water-soluble copolymers (Jbeily et al., 2014). In another approach, a methacrylic backbone was synthesized by ATRP followed by enzymatic ring opening polymerization (e-ROP) of ε-caprolactone to produce a grafted copolymer (Naolou, Jbeily, et al., 2013). Grafting of functional polyesters can also be used to prepare biodegradable elastomers (Brosnan, Brown, & Ashby, 2013). Mecerreyes et al. obtained grafted polymers by a combination of ROP and ATRP followed by the formation of polymeric networks using a Michael type addition (Mecerreyes et al., 2000).

In this work, poly(glycerol adipate) (PGA) was synthesized by enzymatic polyesterfication of glycerol and dimethyl adipate or divinyl adipate. In the latter case, the produced polyesters are terminated with different end groups like hydroxyl, vinyl, and carboxyl end groups. PGA, which has also vinyl end groups, was used for grafting with fatty acids and PEG chains. A detailed discussion of polymer synthesis and grafting along with characterization is given in the following sections. Furthermore, some applications of the multiple grafted copolyesters will be discussed.

2. Experimental part

General lab safety rules must be observed and followed carefully. Personal protective equipment (e.g., safety goggles or glasses, lab coat, chemically resistant gloves, and closed-toe shoes) must be used in every laboratory while performing any experiment.

2.1 Equipment

2.1.1 Equipment needed for enzymatic polymerization

- Three-neck round-bottom flask (250 mL)
- Two-neck round-bottom flask (250 mL)
- Glass stopper
- Reflux condenser fitted with $CaCl_2$ drying tube
- Overhead mechanical stirrer fitted with a rubber septum, stirrer rod, and Teflon paddle
- Soxhlet extractor (150 mL)
- Magnetic stirring hot plate
- Magnetic stirrer
- Oil bath
- Thermometer
- Glass funnel
- Whatman filter paper
- Single-neck round-bottom flask (100 mL)
- Rotary evaporator

2.1.2 Equipment needed for polymer modifications and purification

- Three-neck round-bottom flask (100 mL)
- Glass stopper
- Rubber septa
- Source of dry nitrogen
- Syringe needle connected to the nitrogen gas supply
- Magnetic stirrer
- Magnetic stirring plate
- Dry ice bath
- Separatory funnel
- Glass funnel
- Packed glass column fitted with stopcock
- Glass beakers (100 and 200 mL)
- Single-neck round-bottom flask (100 mL)
- Rotary evaporator

2.2 Methods

2.2.1 Nuclear magnetic resonance (NMR) spectroscopy

^{1}H NMR spectra were recorded on a Varian Gemini 2000 spectrometer (400 MHz) at 27 °C using tetramethylsilane (TMS) as an internal calibration reference. Approximately 30–40 mg of polymer was dissolved in 0.8 mL of deuterated solvent ($CDCl_3$). Later, NMR spectral data were processed using MestRec (v.4.9.9.6) software (Mestrelab Research).

2.2.2 Gel permeation chromatography (GPC)

GPC measurements (number-average molar mass (M_n), weight-average molar mass (M_w), and molar mass distribution (M_w/M_n) (*PDI*) for PGA, saturated fatty acid grafted PGAs, and unsaturated fatty acid grafted PGAs were performed in THF on a Viscotek GPC max VE 2002 using HHRH Guard-17360 and GMHH-N-18055 columns and a refractive index detector (VE 3580 RI detector, Viscotek). Polystyrene standards were used for calibration. For PGA (M), which was synthesized from glycerol and dimethyl adipate, and PEG and oleate grafted PGAs, GPC measurements were done with DMF + 0.01 M LiBr as mobile phase in a thermostatically controlled column kept at 25 °C and calibrated with poly(methyl methacrylate) standards. Addition of salt such as LiBr to the mobile phase suppresses the interactions of the polymer chains such as hydrogen bonding, hence aggregation is avoided (Hann, 1977). For all samples, the concentration was 3 $mg\,mL^{-1}$, and the flow rate was 1 $mL\,min^{-1}$. The data were analyzed using Origin 8 software.

2.2.3 Differential scanning calorimetry (DSC)

DSC measurements for PGA and saturated fatty acid grafted PGAs were carried out with a Perkin Elmer DSC 8000 under nitrogen flow at a rate of 20 $mL\,min^{-1}$. About 8–12 mg of the sample was filled in standard aluminum pans. Starting at −65 °C, the samples were heated up to 100 °C at a heating rate of 20 $°C\,min^{-1}$. After holding the temperature at 100 °C for 2 min, the cooling/heating cycle was repeated. Glass transitions and melting temperatures were analyzed with the PyrisTM software (Version 13.3.1.0014, Perkin Elmer). DSC measurements for PGA, unsaturated fatty acid grafted PGAs, and PEG grafted PGAs were performed under nitrogen flow using a Mettler Toledo DSC 822e module. About 8–12 mg of the sample was filled in aluminum pans. For all measurements, the sample was first heated to $T = 125$ °C in order to remove the previous thermal history, and after holding this temperature for 5 min, they were cooled to $T = -50$ °C with the rate of 10 $°C\,min^{-1}$. The samples were heated again to 80 °C at 10 $°C\,min^{-1}$ to record their melting endotherms.

2.2.4 X-ray diffraction (XRD)

For XRD measurements, the cubic gel phase or cubosome dispersions were loaded into a glass capillary of 1.5 mm diameter. The capillary was sealed with epoxy resin on both sides. X-ray scattering experiments were performed in transmission mode using a SAXSLAB laboratory setup (Retro-F) equipped with an AXO micro focus X-ray source with an AXO multilayer X-ray optic (ASTIX) as a monochromator for Cu K_{α} radiation ($\lambda = 0.154$ nm). A DECTRIS PILATUS3 R 300K detector was used to record the two-dimensional scattering patterns. A typical procedure of sample preparation to achieve 1 wt% polymer with respect to glycerol monooleate (GMO) is as follows; to 99 mg molten GMO, 1 mg polymer in 40 μL water was added with strong spatulation at 50 °C to obtain the cubic gel phase also known in literature as the bulk cubic phase (Dong, Larson, Hanley, & Boyd, 2006; Rangelov & Almgren, 2005). Similarly, other compositions were also achieved. Cubosome dispersions were prepared, using a top-down approach (Spicer, 2004), in which the bulk gel phase is diluted with distilled water and homogenized using a SilentCrusher S for 15 min at a speed of 45,000 rpm and a break of 1 min after every 1.5 min mixing to provide 10 wt% (of combined mass of GMO and polymer) dispersions. The bulk polymers were measured in a sample holder which is a 2 mm thick aluminum disc. The measurements were performed at room temperature for a q-range of 0.25–7 nm^{-1} for the cubic gel phase or cubosome dispersions. For the measurements of bulk polymer a q-range of 0.25–7 and 1–29 nm^{-1}, respectively, to cover the small and wide angle scattering range were selected.

2.2.5 Scanning electron microscopy (SEM)

SEM experiments were performed using a Philips ESEM XL 30 FEG (Philips Electron Optics). About 2 mg of the sample was sputtered with a layer of chrome (30 nm) and measured under high vacuum with an acceleration voltage of 2 kV.

2.2.6 Surface tension measurements

The surface tensions γ of the aqueous polymer solution at different concentrations were measured by the Wilhelmy plate method using an automated DCAT tensiometer (Data Physics Instruments). The tensiometer worked automatically by injecting predetermined volumes of micelle solution into milli Q water. The surface tension was measured after 10 min of stirring and 30 min waiting period. Measurements were carried out at 25 °C.

2.2.7 Dynamic light scattering (DLS)

DLS measurements were performed using an ALV/DLS-5000 instrument (ALV GmbH). As a light source, a 20 mW He–Ne gas laser (632.8 nm, 20 mW) was used (Uniphase Laser). The DLS instrument was equipped with a goniometer for automatic measurements between scattering angles Θ of 30 and 90°. The correlation functions were analyzed by the CONTIN method, which gives information on the distribution of decay rates Γ. Apparent diffusion coefficients were obtained from $D_{app} = \Gamma/q^2$ (where $q = (4\pi n/\lambda)\ sin(\Theta/2)$, where λ is the wavelength of the light, n is the refractive index, and Θ is the scattering angle). Finally, apparent hydrodynamic radii R_h were calculated via the Stokes–Einstein equation.

$$R_h = \frac{k_B T}{6\pi\eta D_{app}} \quad (1)$$

where k_B is the Boltzmann constant and η is the viscosity of the solvent at the absolute temperature T. To find out the particle size of the cubosomes, the samples prepared for XRD measurements were further diluted to give a $2\,gL^{-1}$ concentration for DLS measurement to avoid undesired multiple scattering.

2.3 Materials

2.3.1 Materials needed for enzymatic polymerization

Lipase B derived from *Candida antarctica* (CAL-B) immobilized on an acrylic resin (Sigma-Aldrich), commercially known as Novozyme (N435), was dried over phosphorous pentoxide (P_2O_5) at 4 °C for 24 h prior to use. Phosphorous pentoxide (P_2O_5) (≥99%) was purchased from Carl Roth. Divinyl adipate (DVA) (stabilized with 4-methoxyphenol (MEHQ), >99.0%) was purchased from TCI and used as received. Tetrahydrofuran (THF, anhydrous, 99.5%, extra dry over a molecular sieve) was purchased from Acros Organics. Dimethyl adipate (≥99%), glycerol (≥99.5%), molecular sieve (5 Å), and calcium chloride (anhydrous, granular, ≤7.0 mm, ≥93.0%) were purchased from Sigma-Aldrich. Glycerol monooleate (GMO) used for this study was a mixture of glycerol 1-monooleate (92 mol%) and glycerol 2-monooleate (8 mol%) (Bilal et al., 2017).

2.3.2 Materials needed for polymer modifications

Stearic acid (grade I, ≥98.5%, capillary GC), lauric acid (98%), succinic anhydride, poly(ethylene glycol) monomethyl ether-750 (mPEG17), pyridine, oleoyl chloride (>90%), and magnesium sulfate ($MgSO_4$, anhydrous,

≥99.5%) were purchased from Sigma-Aldrich. Behenic acid and poly (ethylene glycol) monomethyl ether-1000 (mPEG23) were purchased from TCI-Europe. Silica gel (SiO_2, 0.03–0.2 mm), 4-(dimethylamino)-pyridine (DMAP), 1-ethyl-3-(3-dimethylaminopropyl) carbodiimide hydrochloride (EDC•HCl), dialysis membranes (having a cut-off molar mass of 10 kDa), sodium chloride, and deuterated chloroform ($CDCl_3$, 99.8%) were purchased from Carl Roth. All chemicals were used as received without further purification. All HPLC grade solvents like *n*-hexane, dichloromethane, and acetone were purchased from Carl Roth.

2.4 Enzymatic syntheses of poly(glycerol adipate) (PGA)

2.4.1 PGA synthesis using divinyl adipate (DVA)

A typical enzymatic polycondensation procedure of the PGA backbone is described by Kallinteri et al. (2005). The synthesis procedure is shown in Scheme 1A. A mixture of glycerol (11 g, 120 mmol), an equimolar amount of divinyl adipate (DVA) (23.8 g, 120 mmol), and 23 mL anhydrous tetrahydrofuran (THF) were charged in a 250 mL three-neck round-bottom flask equipped with a top reflux condenser containing a calcium chloride ($CaCl_2$) drying tube at its outlet. The reaction was started by the addition of Novozyme (N435) (0.72 g, 2 wt% of the total mass of the monomers) and the resulting mixture was stirred at 250 rpm using an overhead mechanical stirrer fitted with a Teflon paddle and a rubber septum for 11 h at 50 °C. At the end of the reaction, 50 mL of THF was added, and the product was passed through Whatman filter paper to remove the enzyme beads. THF was removed by rotary evaporation at 60 °C under reduced pressure to yield the slightly yellow and viscous polymer. Then, the crude product was dried

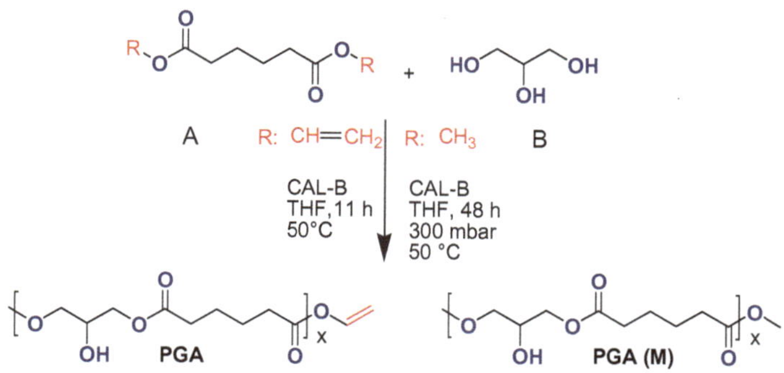

Scheme 1 (A) Synthesis scheme of poly(glycerol adipate) (PGA) using divinyl adipate (DVA) and (B) Synthesis scheme of poly(glycerol adipate) (PGA (M)) using dimethyl adipate (DMA).

overnight under vacuum at 35 °C to remove the residual THF. The total yield was 24 g. The molar masses of the three synthesized PGAs from divinyl adipate were determined by GPC in THF as (PGA_{23}: M_n = 4700 g mol^{-1}, with a polydispersity index (*PDI*) (M_w/M_n) of 2.2, PGA_{33}: M_n = 6600 g mol^{-1}, with a *PDI* of 1.8, and PGA_{26}: M_n = 5300 g mol^{-1}, with a *PDI* of 1.9. The ^{1}H NMR spectrum of PGA_{23} is shown in Fig. 1B. The assignments of all peaks are given according to literature (Kulshrestha et al., 2005). ^{1}H NMR (400 MHz, $CDCl_3$) δ [ppm]: 4.36–3.99 (m, 5H), 2.48–2.20 (m, 4H), 1.75–1.53 (m, 4H).

2.4.2 PGA (M) synthesis using dimethyl adipate (DMA)

A typical enzymatic polycondensation procedure of the PGA (M) backbone using dimethyl adipate (DMA) and glycerol as monomers is shown in Scheme 1B. It was carried out as follows: A mixture of glycerol (11 g, 120 mmol), an equimolar amount of dimethyl adipate (DMA) (23.8 g, 120 mmol), and 13 mL anhydrous tetrahydrofuran (THF) were charged in a 250 mL two-neck round-bottom flask equipped with a soxhlet extractor (150 mL) which is connected to a reflux condenser. The soxhlet extractor was filled with molecular sieve (5 Å, 105 g) and 100 mL anhydrous THF. The reaction mixture was stirred using a magnetic stirrer allowing the

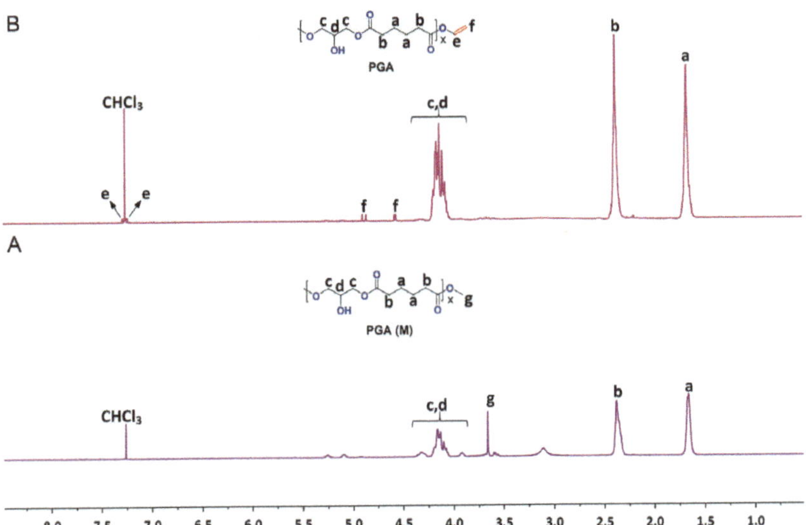

Fig. 1 ^{1}H NMR spectra of (A) PGA (M) synthesized from glycerol and DMA and (B) PGA_{23} synthesized from glycerol and DVA in $CDCl_3$ at 27 °C.

reactants to warm up to the bath's temperature (50 °C) for 30 min. Then, the reaction was started by the addition of Novozyme (N435) (0.72 g, 2 wt% of the total mass of the monomers). The pressure was then reduced gradually to 300 mbar in order to remove the by-product, which is methanol, since the azeotropic mixture of THF and methanol was collected in the soxhlet extractor and methanol was captured by the molecular sieve. The reaction mixture was stirred for 48 h at 50 °C. At the end of the reaction, 50 mL of THF was added, and the product was passed through Whatman filter paper to remove the enzyme beads. THF was removed by rotary evaporation at 60 °C under reduced pressure. Then, the crude product was dried overnight under vacuum at 35 °C to remove the residual THF. The molar mass of PGA (M) was determined by GPC in DMF as $M_n = 2600\,\text{g mol}^{-1}$, with a polydispersity index *PDI* of 1.8. The ^{1}H NMR spectrum of PGA (M) is shown in Fig. 1A. ^{1}H NMR (400 MHz, $CDCl_3$) δ [ppm]: 4.36–3.99 (m, 5H), 3.6 (s, 3H), 2.48–2.20 (m, 4H), 1.75–1.53 (m, 4H).

2.5 Grafting procedures to poly(glycerol adipate)

2.5.1 Grafting with saturated fatty acids

The PGA backbone was modified with various degrees of grafting through the reaction of its pendant free hydroxyl (OH) groups with different fatty acids (lauric acid, stearic acid, behenic acid) according to the acylation reaction procedure described by Kallinteri et al. (2005). The purification was done by precipitation into cold *n*-hexane in the case of acylation with low substitution degrees (<40 mol% of OH-groups). Whereas, column chromatography using (1) acetone and *n*-hexane (5: 95 vol%) and (2) acetone as eluents was applied in the case of higher substitution degrees to ensure the complete removal of unreacted fatty acids. PGA_{23} with molar mass M_n of $4700\,\text{g mol}^{-1}$ was used for the acylation reactions with stearic acid and behenic acid side chains, and PGA_{33} with molar mass M_n of $6600\,\text{g mol}^{-1}$ was used for the acylation reactions with stearic acid and lauric acid side chains. The acylation degrees (given in mol% of converted OH-groups of PGA) were the following:

Stearic acid side chains: 35%, 40%, 70%, and 90% called S_8, S_9, S_{16}, and S_{21}

Behenic acid side chains: 45% called B_{10}

Lauric acid side chains: 20% called L_6

The respective fatty acids are abbreviated as L, S, and B. The index gives the number of acyl chains per polymer chain.

2.5.1.1 Synthesis of poly(glycerol adipate)-graft-stearic acid ($PGA_{23}S_u$)

PGA_{23} with a molar mass M_n of 4700 $g\,mol^{-1}$ and *PDI* of 2.2 was further modified with stearic acid side chains via Steglich esterification reaction in the presence of 1-ethyl-(3-dimethyl aminopropyl)carbodiimide hydrochloride (EDC•HCl) and 4-(dimethylamino)-pyridine (DMAP) to yield $PGA_{23}S_u$ as shown in Scheme 2. In a 100 mL three-neck round-bottom flask equipped with septa and magnetic stirrer, PGA_{23} (with respect to OH-groups, 1 g, 4.9 mmol), was dissolved in anhydrous THF. This was followed by the acylation with stearic acid (intended conversion of 38 mol% of all OH-groups, 0.5 g, 1.7 mmol, intended conversion of 73 mol% of all OH-groups, 1 g, 3.5 mmol, and intended conversion of 93 mol% of all OH-groups, 1.3 g, 4.5 mmol). Afterward, a weighed amount of DMAP (0.06 g, 0.5 mmol) for 38 mol%, (0.13 g, 1.1 mmol) for 73 mol%, and (0.17 g, 1.4 mmol) for 93 mol% was added and the reaction flask was transferred to an ice bath. After

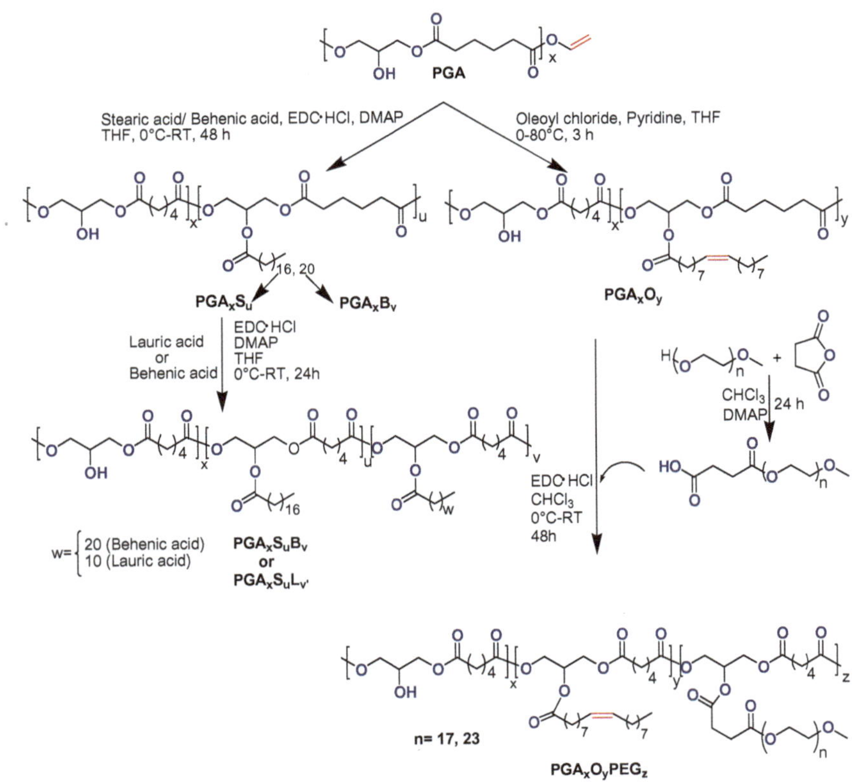

Scheme 2 Grafting of poly(glycerol adipate) with hydrophobic and hydrophilic side chains.

cooling it for 20 min, EDC•HCl (0.8 g, 4.2 mmol) for 38 mol%, (1.7 g, 8.9 mmol) for 73 mol%, and (2.2 g, 11.5 mmol) for 93 mol% was added to the polymer solution. After 1 h, the ice bath was removed and the reaction mixture was stirred at room temperature for 48 h. Then, the solvent was removed by rotary evaporator, and the crude product was dissolved in chloroform and transferred to a separatory funnel and washed 3 times with brine solution. The organic layer was released into a 250 mL beaker, dried over $MgSO_4$ and stirred for 15 min. Inorganic salts were filtered over a funnel through Whatman filter paper, and the filtrate was collected in a 250 mL round-bottom flask and concentrated to a small volume by using a rotary evaporator. The crude product was further purified by precipitation 3 times in ice-cold *n*-hexane for products of <40% of grafting and by column chromatography (1) acetone: *n*-hexane = 5: 95 vol% and (2) acetone, for products of >40% of grafting to remove the unreacted fatty acids. The total yield was 80 wt%. Three types of stearic acid grafted PGAs were synthesized, i.e., $PGA_{23}S_8$ as white waxy product, $PGA_{23}S_{16}$ and $PGA_{23}S_{21}$ as white solid products. The purity of $PGA_{23}S_u$ was confirmed by ^1H NMR spectroscopy as shown in Fig. 2. ^1H NMR (400 MHz, $CDCl_3$) δ [ppm]: 5.28–5.16 (m, 1H), 4.36–3.99 (m, 5H), 2.41–2.25 (m, 6H), 1.73–1.53 (m, 6H), 1.43–1.15 (m, 28H), 0.87 (t, 3H). The degree of substitution and the molar

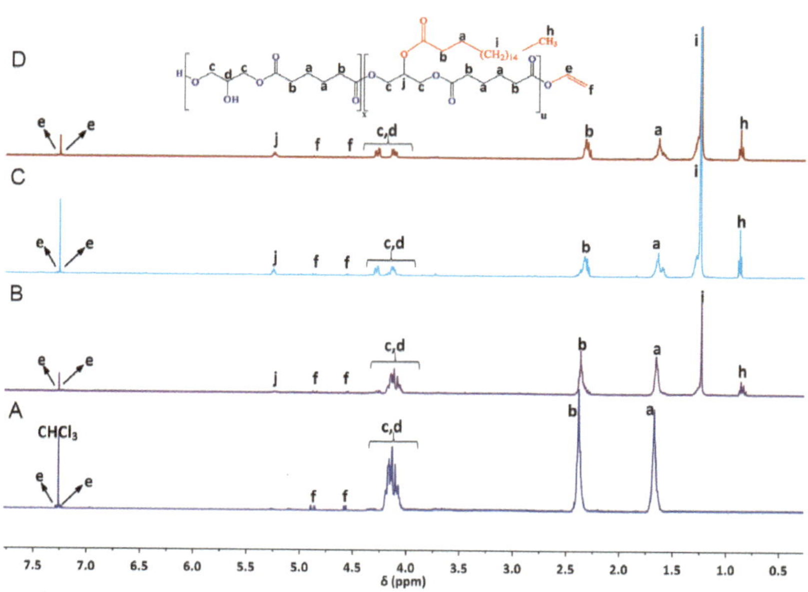

Fig. 2 ^1H NMR spectra of (A) PGA_{23}, (B) $PGA_{23}S_8$, (C) $PGA_{23}S_{16}$, and (D) $PGA_{23}S_{21}$ in $CDCl_3$ at 27 °C.

mass were calculated according to the mol% conversion of all OH-groups from the corresponding ^{1}H NMR spectra (see Fig. 2). Additionally, the purity of PGA_{23} and stearic acid grafted PGAs was confirmed by the unimodal distribution of their GPC traces (see Fig. 6A). The molar masses and polydispersities are given in Table 3.

2.5.1.2 Synthesis of poly(glycerol adipate)-graft-behenic acid ($PGA_{23}B_{10}$)

PGA_{23} with a molar mass M_n of 4700 g mol^{-1} and *PDI* of 2.2 was modified with behenic acid side chains via Steglich esterification reaction in the same synthetic way as $PGA_{23}S_u$ to yield $PGA_{23}B_{10}$ as shown in Scheme 2. In a 100 mL three-neck round-bottom flask equipped with septa and magnetic stirrer, PGA_{23} (with respect to OH-groups, 1 g, 4.9 mmol), was dissolved in anhydrous THF. This was followed by the acylation with behenic acid (intended conversion of 48 mol% of all OH-groups, 0.8 g, 2.4 mmol). Then, a weighed amount of DMAP (0.08 g, 0.7 mmol) and EDC•HCl (1.12 g, 6 mmol) were added to the polymer solution. After 48 h, the purification steps were done exactly as mentioned previously for the purification of $PGA_{23}S_u$. The total yield was 80 wt%. ^{1}H NMR (400 MHz, $CDCl_3$) δ [ppm]: 5.28–5.16 (m, 1H), 4.36–3.99 (m, 5H), 2.41–2.25 (m, 6H), 1.73–1.53 (m, 6H), 1.43–1.15 (m, 36H), 0.87 (t, 3H). The molar mass and polydispersity are given in Table 3.

2.5.1.3 Synthesis of poly(glycerol adipate)-graft-stearic acid-graft-behenic acid ($PGA_{23}S_8B_{10}$)

In the second step of synthesis, $PGA_{23}S_8$ was further modified with behenic acid following the same synthetic route of $PGA_{23}S_u$ as shown in Scheme 2. In a 100 mL three-neck round-bottom flask equipped with septa and magnetic stirrer, $PGA_{23}S_8$ (3 g, 14.8 mmol), was dissolved in anhydrous THF. This was followed by the acylation with behenic acid (intended conversion of 48 mol% of the remaining OH-groups, 2.4 g, 7 mmol). Afterward, a weighed amount of DMAP (0.25 g, 2 mmol) was added and the reaction flask was transferred to an ice bath. After cooling for 20 min, EDC•HCl (3.4 g, 17.7 mmol) was added to the polymer solution. After 1 h, the ice bath was removed and the reaction mixture was stirred at room temperature for 48 h. After that, the purification steps were done exactly as mentioned previously for the purification of $PGA_{23}S_u$ to yield $PGA_{23}S_8B_{10}$ as a white solid product. The product yield was 80 wt%. The ^{1}H NMR spectrum of $PGA_{23}S_8B_{10}$ is shown in Fig. 3. ^{1}H NMR (400 MHz, $CDCl_3$) δ [ppm]: 5.28–5.16 (m, 2H), 4.36–3.99 (m, 10H), 2.41–2.25 (m, 12H), 1.73–1.53

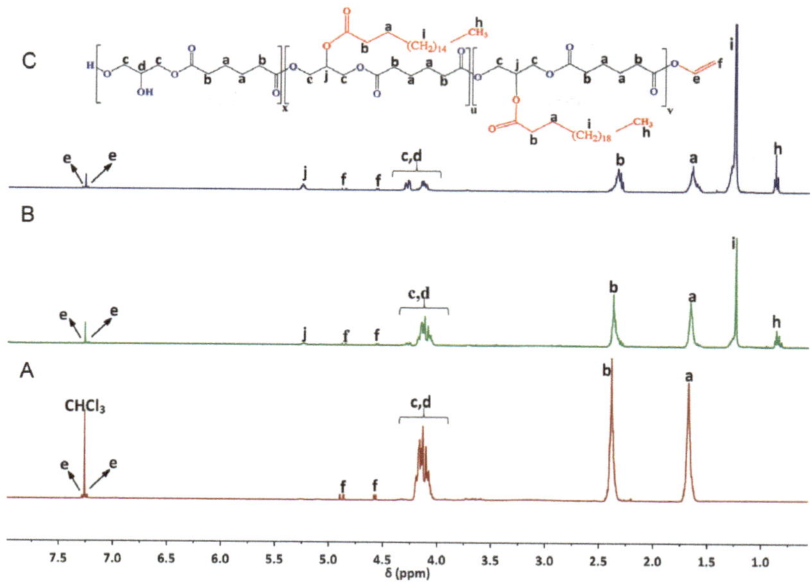

Fig. 3 ^{1}H NMR spectra of (A) PGA_{23}, (B) $PGA_{23}S_8$, and (C) $PGA_{23}S_8B_{10}$ in $CDCl_3$ at 27°C.

(m, 12H), 1.43–1.15 (m, 64H), 0.87 (t, 6H). The degree of substitution and the molar mass were calculated according to the mol% of converted OH-groups, obtained from the corresponding ^{1}H NMR spectra (see Fig. 3). The GPC traces are shown in Fig. 6B. The molar masses and polydispersities are given in Table 3.

2.5.1.4 Synthesis of poly(glycerol adipate)-graft-stearic acid-graft-lauric acid ($PGA_{33}S_9L_6$)

PGA_{33} with a molar mass M_n of 6600 g mol^{-1} and *PDI* of 1.8 was modified with the two different fatty acids, namely, stearic acid and lauric acid. In the first step of synthesis, PGA_{33} was modified with stearic acid side chains via Steglich esterification (see Scheme 2) according to the same procedure described previously to yield $PGA_{33}S_9$. The amounts used for this reaction were as follows: PGA_{33} (with respect to OH-groups, 1 g, 4.9 mmol), stearic acid (intended conversion of 43 mol% of all OH-groups, 0.6 g, 2.1 mmol), DMAP (0.07 g, 0.57 mmol), and EDC•HCl (1 g, 5.2 mmol). In the second step of synthesis, $PGA_{33}S_9$ was further modified with lauric acid as shown in Scheme 2. In a 100 mL three-neck round-bottom flask equipped with septa and magnetic stirrer, $PGA_{33}S_9$ (3 g, 14.8 mmol) was dissolved in anhydrous THF. This was followed by the acylation with lauric acid (intended conversion of 23 mol% of the residual OH-groups, 0.7 g, 3.5 mmol). Afterward, a weighed amount of DMAP (0.12 g, 1 mmol) was added and the reaction flask was transferred to an ice bath. After cooling for 20 min, EDC•HCl (1.7 g, 8.8 mmol) was added to the polymer solution. After 1 h, the ice bath was removed and the reaction mixture was stirred at room temperature for 48 h. After that, the purification steps were done exactly as mentioned previously for the purification of $PGA_{23}S_u$ to yield $PGA_{33}S_9L_6$ as a white waxy product. The product yield was 80 wt%. The formation of $PGA_{33}S_9L_6$ was confirmed by ^1H NMR spectroscopy as shown in Fig. 4. ^1H NMR (400 MHz, $CDCl_3$) δ [ppm]: 5.28–5.16 (m, 2H), 4.36–3.99 (m, 10H), 2.41–2.25 (m, 12H), 1.73–1.53 (m, 12H), 1.43–1.15 (m, 44H), 0.87 (t, 6H). The degree of substitution and the molar mass were calculated according to the mol% of the converted OH-groups from the corresponding ^1H NMR spectra (see Fig. 4). Fig. 6C shows a shift toward the higher molar mass region after grafting. The molar masses and polydispersities are given in Table 3.

2.5.2 Grafting with unsaturated fatty acids

PGA_{26} used for the grafting with the unsaturated fatty acid, namely, oleic acid, has M_n of 5300 g mol^{-1} and *PDI* of 1.9. The grafting of the PGA

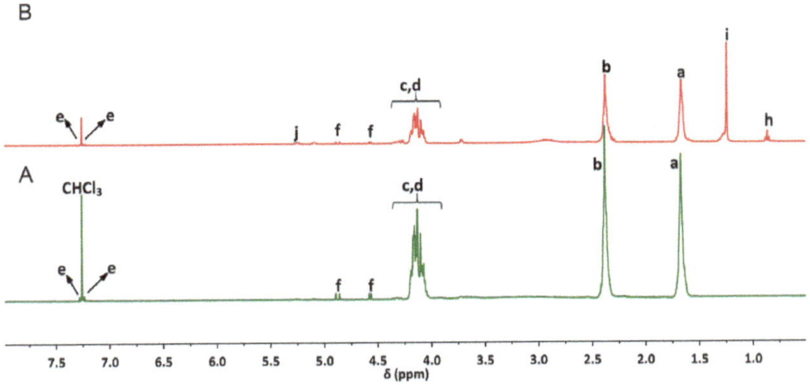

Fig. 4 ^1H NMR spectra of (A) PGA_{33}, (B) $PGA_{33}S_9$, and (C) $PGA_{33}S_9L_6$ in $CDCl_3$ at 27 °C.

backbone was carried out with various degrees of grafting by the reaction between pendant hydroxyl groups of the PGA backbone and oleoyl chloride following the same procedure used in Bilal et al. (2017). A typical procedure to achieve 15 mol% degree of grafting was as follows: PGA_{26} (4.0 g, 19.8 mmol) was dissolved in 50 mL THF in a 100 mL three-neck round-bottom flask equipped with a condenser and magnetic stirrer. Pyridine (3.9 g, 49 mmol) was charged into the reaction flask, and then the reaction flask was transferred to an ice bath. After cooling it for approximately 30 min, the oleoyl chloride (0.88 g, 2.97 mmol) diluted with 5 mL THF was added dropwise at 0 °C. The reaction was allowed to proceed at room temperature for approximately 20 min and then at 80 °C for 3 h. The reaction mixture was filtered to remove the pyridinium salt. The solvent was removed by rotary evaporator, and the crude product was dissolved in dichloromethane (DCM) and extracted 3 times with brine solution. The organic phase was separated and dried over magnesium sulfate. The filtrate was concentrated by rotary evaporator and precipitated 3 times in ice-cold *n*-hexane. The purification of graft polymers with a grafting degree of 50 mol% was carried out via column chromatography using the eluent (1) acetone: *n*-hexane = 5: 95 vol% and (2) acetone. Three different oleic acid grafted PGAs were synthesized, i.e., $PGA_{26}O_4$, $PGA_{26}O_7$, and $PGA_{26}O_{13}$. A spectrum of $PGA_{26}O_4$ is shown in Fig. 5C. ^{1}H NMR (400 MHz, $CDCl_3$) δ [ppm]: 5.36–5.29 (m, 2H), 5.28–5.16 (m, 1H), 4.40–3.99 (m, 5H), 2.48–2.20 (m, 6H), 2.04–1.87 (m, 4H), 1.72–1.56 (m, 6H), 1.36–1.16 (d, J = 14.3 Hz, 20H), 0.84 (t, 3H).

2.5.3 Grafting with poly(ethylene glycol)

2.5.3.1 Synthesis of mPEG17-Succinyl and mPEG23-Succinyl

Carboxylation of poly(ethylene glycol) monomethyl ether (mPEG) was carried out as described by Lu and Zhong (2010), with a small modification. In a typical experiment mPEG23 (10 g, 10 mmol) was added to a 250 mL three-neck round-bottom flask equipped with magnetic stirrer, heating plate and reflux condenser containing a $CaCl_2$ drying tube at its opening. 200 mL chloroform was added. After this, DMAP (0.122 g, 1 mmol) and succinic anhydride (2 g, 20 mmol) were added. The reaction was allowed to run for 24 h at 30 °C. After this time, the reaction mixture was transferred into a separating funnel and the crude product was washed 3 times with brine solution. The organic phase was collected and dried by $MgSO_4$. The crude product was recrystallized using ethyl acetate to have pure

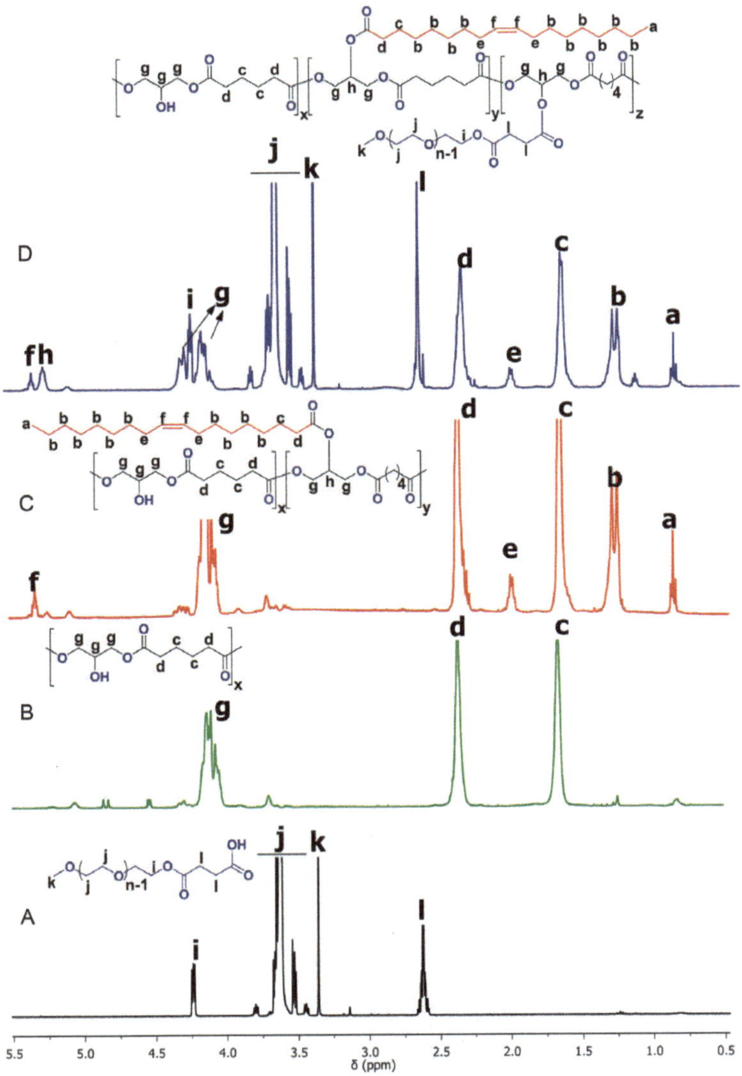

Fig. 5 ^{1}H NMR spectrum of (A) mPEG23-Succinyl, (B) PGA_{26}, (C) $PGA_{26}O_4$, and (D) $PGA_{26}O_4PEG23_{17}$ recorded in $CDCl_3$ at 27 °C.

mPEG23-Succinyl. The number after mPEG gives the number of repeating units. The product was characterized by ^{1}H NMR spectroscopy (see Fig. 5A). ^{1}H NMR (400 MHz, $CDCl_3$) δ [ppm]: 4.27–4.21 (m, 2H), 3.71–3.50 (m, 88H), 3.36 (s, 3H), 2.67–2.56 (m, 4H).

2.5.3.2 Grafting with poly(ethylene glycol) to $PGA_{26}O_y$

Oleic acid grafted polymers were additionally grafted with mPEG17-Succinyl and mPEG23-Succinyl chains via esterification. In a typical experiment $PGA_{26}O_4$ (1 g, 4.08 mmol) was placed in a 100 mL three-neck round-bottom flask equipped with septa and magnetic stirrer. The flask was placed in an ice bath. 60 mL chloroform was then charged into the reaction flask. A weighed amount of mPEG23-Succinyl (0.9 g, 0.8 mmol), EDC•HCl (2.3 g, 12.2 mmol), and DMAP (0.15 g, 1.22 mmol) were then added. The ice bath was removed after 1 h and the reaction was allowed to run for 48 h at room temperature. After this time, the reaction mixture was transferred into a separating funnel and washed 3 times with brine. The organic phase was dried with $MgSO_4$, filtered, and the solvent was evaporated in a rotary evaporator. The crude product was further purified to remove unreacted mPEG23-Succinyl by dialysis in water using regenerated cellulose membrane having a cut-off molar mass of 10 kDa for 7–10 days. After drying the solvent, the slightly yellowish solid product was obtained. Fig. 5D shows the ^{1}H NMR spectrum of $PGA_{26}O_4PEG23_{17}$. ^{1}H NMR (400 MHz, $CDCl_3$) δ [ppm]: 5.37–5.28 (m, 2H), 5.28–5.16 (m, 2H), 4.36–4.24 (m, 5H), 4.23 (m, 2H), 3.84–3.42 (m, 88H), 3.36 (s, 3H), 2.7–2.6 (m, 4H), 2.41–2.25 (m, 6H), 2.02–1.93 (m, 4H), 1.72–1.54 (m, 6H), 1.36–1.17 (m, 20H), 0.86 (m, 3H).

3. Synthesis and polymer characterization

3.1 Poly(glycerol adipate)

Lipase-catalyzed polycondensation or step-growth polymerization is defined as lipase-catalyzed (1) esterification and transesterification of mixtures of diacids or activated diacids with multi-hydroxy alcohols (Bilal et al., 2016; Warwel, Demes, & Steinke, 2001), under anhydrous conditions, (2) ring opening polymerization (ROP) of lactones and other cyclic diesters with various ring sizes (Kikuchi, Uyama, & Kobayashi, 2002). The condensation reaction is reversible since it is associated with continuous formation of low molar mass side products. However, conventional diacids often show low reactivity toward lipase catalyzed condensation with diols which results in polymers with low molar mass. Using activated diacids such as their divinyl esters as an acyl donor helps to improve the molar mass of polyesters as it produces acetaldehyde, an easily removable gaseous side product (Puskas, Seo, & Sen, 2011). Poly(glycerol adipate) is enzymatically synthesized following two strategies. For the first strategy, PGA is enzymatically synthesized using equimolar amounts of glycerol and divinyl adipate (DVA) as shown in Scheme 1A. Since the enzyme has a high affinity toward primary hydroxyl groups rather than secondary hydroxyl groups (Uyama et al., 2000), a linear PGA with free pendant hydroxyl groups can be obtained. DVA plays a significant role in enhancing lipase reactivity as the active site of CAL-B, which is serine, first reacts with the acyl donor of DVA to give an acyl enzyme intermediate. The by-product of this reaction is vinyl alcohol, which tautomerizes to acetaldehyde and evaporates at the reaction temperature making the reaction irreversible (Wang, Lalonde, Momongan, Bergbreiter, & Wong, 1988). The produced polymer has some unsaturated end groups (vinyl end groups) which can react with the nucleophiles such as free amine groups according to aza-Michael addition reaction and forming β-amino carbonyl compounds (Rulev, 2011). According to the PGA synthesis, most of the vinyl end groups will be converted into carboxyl end groups because of hydrolysis (Taresco et al., 2016), but some are remaining at the end of the reaction and can be identified in ^{1}H NMR spectra which exhibited three respective signals at 7.29, 4.89, and 4.58 ppm as shown in Fig. 1B. In addition, ^{13}C NMR spectra exhibited two signals at 141.2 and 97.9 ppm arising from the vinyl end groups. For the second strategy, PGA (M) was enzymatically synthesized using dimethyl adipate (DMA) instead of DVA (see Scheme 1B). Using of DMA has some disadvantages such as the shifting of the equilibrium to the direction of the monomers since methanol is the by-product of this reaction which is more difficult to remove (Naolou, Conrad, Busse, Mäder, & Kressler, 2013). The formed azeotropic mixture of methanol and tetrahydrofuran makes the separation impossible by distillation which leads to low molar mass polymers. For this reason, the polymerization process was carried out in the presence of molecular sieves filled in a soxhlet apparatus attached to the top of the reaction flask. During the enzymatic polymerization, both methanol and THF evaporated, condensed again in the condenser, and were collected into a soxhlet extractor where the mixture interacts with the molecular sieve with a pore size of 5 Å. Only methanol can be captured by the molecular sieve with this size and only THF can reflux to the reaction flask. The GPC results show that the molar masses (M_n) achieved from the polymerization reaction using DVA are higher than that of DMA within shorter reaction time. The produced polymer has no vinyl end groups which can be important for potential polymer-protein conjugations (Besheer, Hertel, Kressler, Mäder, & Pietzsch, 2009), since cross-linking is avoided.

3.2 Modification of PGA backbone with fatty acids

The PGA backbone is acylated to different degrees with different fatty acids (i.e., lauric acid, stearic acid, and behenic acid) via Steglich esterification reaction in the presence of DMAP as a catalyst and EDC•HCl which is used as a coupling promoting agent under mild reaction conditions (see Scheme 2). An excess of EDC•HCl is employed as a drying agent. In all reactions, anhydrous THF is used. The grafted PGAs with stearic acid chains are given names as $PGA_{23}S_8$, $PGA_{23}S_{16}$, $PGA_{23}S_{21}$, and $PGA_{33}S_9$. The grafted PGA_{23} with both, stearic and behenic acid side chains is given name as $PGA_{23}S_8B_{10}$. The grafted PGA_{33} with both, stearic and lauric acid side chains is given name as $PGA_{33}S_9L_6$. The subscript at PGA corresponds to the degree of polymerization, and the other subscripts represent the number of acyl side chains per polymer chain. Furthermore, PGA_{26} was modified with oleic acid side chains in the presence of pyridine, and with mPEG-Succinyl chains in the presence of DMAP and EDC•HCl (see Scheme 2). The grafted PGA_{26} with oleic acid chains are given names as $PGA_{26}O_4$, $PGA_{26}O_7$, and $PGA_{26}O_{13}$. The grafted PGA_{26} with oleic acid and mPEG-Succinyl chains are given names as $PGA_{26}O_4PEG17_{16}$, $PGA_{26}O_4PEG23_{17}$, $PGA_{26}O_7PEG17_{16}$, $PGA_{26}O_7PEG23_{17}$, $PGA_{26}O_{13}PEG17_6$, and PGA_{26} $O_{13}PEG23_8$. 17 and 23 are related to the degree of polymerization of the PEG chains.

3.3 NMR spectroscopy

3.3.1 *Poly(glycerol adipate)*

The formation and purity of the synthesized PGA using DVA and PGA (M) using DMA was confirmed by ^1H NMR spectroscopy (see Fig. 1), which shows the appearance of the methylene and methine signals of the glycerol part in addition to the resonance of the adipic acid part. The resonance due to the secondary proton of the methine group appears at 5.28–5.16 ppm where the grafting is observed. For the synthesized PGA from DVA, most of the vinyl end groups are converted into carboxyl end groups because of hydrolysis, and some are remaining at the end of the reaction and can be identified in the ^1H NMR spectrum as three signals at 7.29, 4.89 and 4.58 ppm, labeled with (e) and (f), as can be seen in Fig. 1B. Since the measurement are done in $CDCl_3$, the signal at 7.29 ppm overlaps with the residual solvent peak. For the PGA (M) which is synthesized with DMA, the resonance due to the methyl protons appears at 3.6 ppm which presents the methoxy end groups, as shown in Fig. 1A.

3.3.2 PGA and saturated fatty acid grafted polyesters

The syntheses of the grafted polymers are confirmed by ^{1}H NMR spectroscopy. All peaks are well assigned which prove the formation and the purity of all grafted polymers. The integral values of all grafted polymers are given in Table 1. The degree of grafting of all fatty acids used is calculated from the integrals of the peaks indicated in the ^{1}H NMR spectra according to Eq. (2).

$$mol\%\ lauric, stearic, behenicacidchaingrafting = \frac{1.33 \times h}{b - 0.67h} \times 100 \qquad (2)$$

Representative spectra of PGA_{23}, $PGA_{23}S_8$, $PGA_{23}S_{16}$, and $PGA_{23}S_{21}$ are shown in Fig. 2. The methine proton of the glycerol part, labeled with (j), appears as multiplet at 5.28–5.16 ppm after grafting with stearic acid. The resonance of the methylene protons of the adipic acid part and the methylene protons of the side chains which are located next to the carbonyl group, labeled with (b) and (a), appears as two multiplets at 2.41–2.25 and 1.73–1.53 ppm, respectively. A triplet at 0.87 ppm, labeled with (h), and a multiplet at 1.43–1.15 ppm, labeled with (i), correspond to the methyl and the methylene protons of stearic acid chains, respectively.

The ^{1}H NMR spectra of PGA_{23}, $PGA_{23}S_8$, and $PGA_{23}S_8B_{10}$ are shown in Fig. 3. The resonance peak, labeled with (j), at 5.28–5.16 ppm corresponds to the methine proton of the glycerol part. In fact, this signal is shifted

Table 1 Integral values from the ^{1}H NMR spectra of all grafted polymers $PGA_xS_uB_vL_{v'}$.

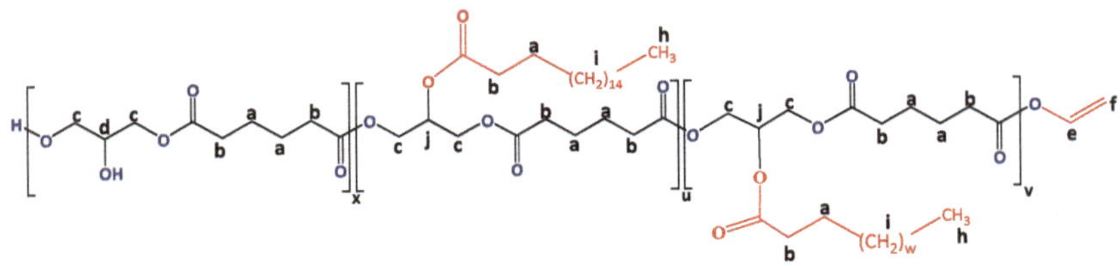

Polymers	a	b	c, d	h	i	J
$PGA_{23}S_8$	12.9	13	14	3	11	1.5
$PGA_{23}S_{16}$	7.9	8.5	7.5	3	24	1.3
$PGA_{23}S_{21}$	6.4	6.5	4.4	3	25.5	1
$PGA_{23}S_8B_{10}$	7.1	7	5.3	3	33	1
$PGA_{33}S_9$	12	12	13	3	11	1.6
$PGA_{33}S_9L_6$	8.8	8.8	7.5	3	16	1

to higher ppm values after grafting with stearic and behenic acids. Moreover, the increase of its intensity indicates the higher degree of grafting as can be observed in $PGA_{23}S_8$ and in the double grafted polymer $PGA_{23}S_8B_{10}$. The integral values of all the grafted polymers are given in Table 1.

Fig. 4 shows representative spectra of PGA_{33}, $PGA_{33}S_9$, and $PGA_{33}S_9L_6$. After acylation with lauric acid, the intensity of methine resonance at 5.28–5.16 ppm was substantially increased due to the higher degree of grafting. The integral values of all the grafted polymers are given in Table 1.

3.3.3 Oleic acid and mPEG-Succinyl grafted PGA

The syntheses of the graft copolymers were confirmed by ^1H NMR spectroscopy. ^1H NMR spectra of mPEG23-Succinyl, PGA_{26}, $PGA_{26}O_4$, and $PGA_{26}O_4PEG23_{17}$ are shown in Fig. 5. The integral values of all the grafted polymers are given in Table 2. The degree of grafting of oleic acid side chains and mPEG-Succinyl side chains is calculated by Eqs. (3) and (4), respectively, according to the integral values of the ^1H NMR spectrum of a particular polymer.

$$mol\%\ oleic\ acid\ chain\ grafting = \frac{1.33 \times a}{d - 0.67a} \times 100 \tag{3}$$

$$mol\%\ mPEGSuccinyl\ grafting = \frac{k}{g+h} \times \frac{5}{3} \times 100 \tag{4}$$

3.4 Gel permeation chromatography

3.4.1 PGA and saturated fatty acid grafted polyesters

The molar mass of the grafted polymers was estimated with gel permeation chromatography (GPC) using THF as eluent and polystyrene as a calibration standards. The GPC traces of PGA and grafted PGAs with stearic acid side chains, both stearic and behenic acid side chains, and both stearic and lauric acid side chains are shown in Fig. 6A–C, respectively. GPC results show molar masses for grafted polymers higher than the polymer backbone. From the analysis of these traces, it results, that the *PDI* values decrease with increasing degrees of grafting. After the modification, the peak shifts to smaller retention times (higher molar masses) which means that the grafting reaction shown in Fig. 6 was successful. Thermal properties are determined by DSC measurements. The GPC results together with thermal properties of all polymers are listed in Table 3.

Table 2 Integral values from the ^{1}H NMR spectra of all grafted polymers, $PGA_xO_yPEG_z$.

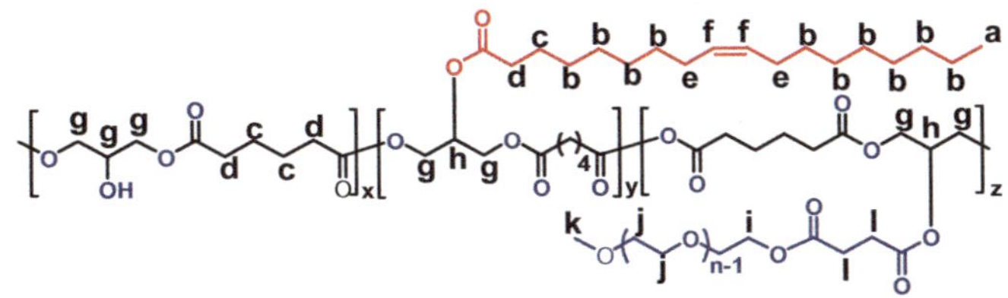

Polymers	a	b	c	d	e	f	g	h	i	j	k	l
$PGA_{26}O_4PEG17_{16}$	3	20	30.1	30.9	3.7	1.5	28	5.1	9.8	335	12.9	20
$PGA_{26}O_4PEG23_{17}$	3	20.4	27.5	26.9	3.6	1.7	25.3	4.2	8.2	380	11.6	15.3
$PGA_{26}O_7PEG17_{16}$	3	20.6	17.9	17.8	4.3	1.5	16	3.5	4.8	178	6.9	11
$PGA_{26}O_7PEG23_{17}$	3	19.4	17.7	17	3.7	1.6	15.2	3	5	240	7.2	10
$PGA_{26}O_{13}PEG17_{16}$	3	20	11.6	11.6	3.2	1.3	9.2	1.6	1.2	35	1.5	2.4
$PGA_{26}O_{13}PEG23_8$	3	20.4	11.3	11.9	3.6	1.6	8.5	1.9	1	59.3	1.9	2.7

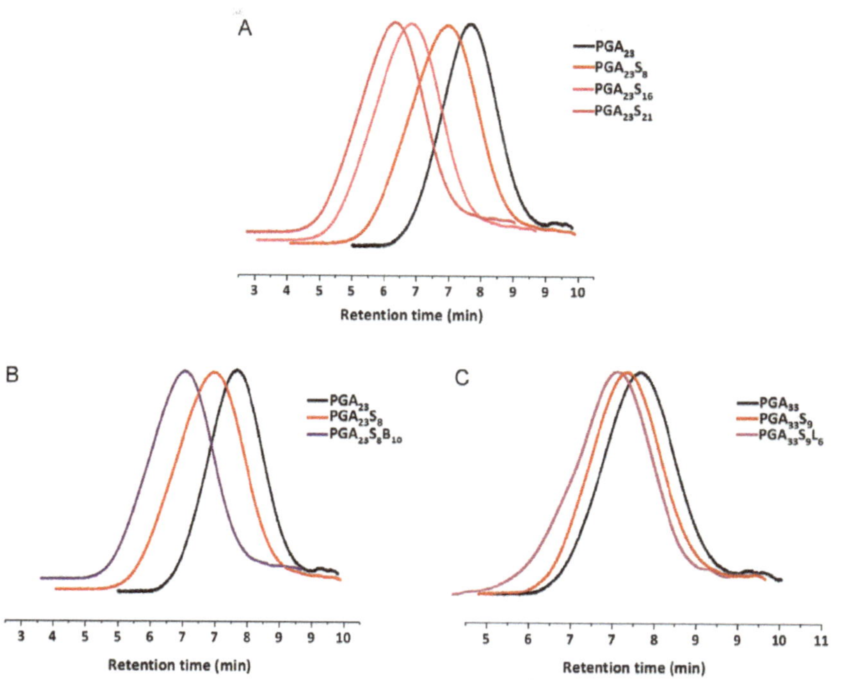

Fig. 6 GPC traces of PGA before and after modification (A) with stearic acid side chains, (B) with both stearic and behenic acid side chains, and (C) with both stearic and lauric acid side chains.

3.4.2 Oleic acid and mPEG-Succinyl grafted PGA

All the synthesized polymers are further characterized by gel permeation chromatography. Because DMF is not a good solvent for oleic acid grafted copolymers, the GPC measurements of PGA_{26} and oleic acid grafted PGAs are carried out in THF. The GPC traces of PGA_{26} and oleic acid grafted PGAs are shown in Fig. 7A. A successive shifting of the peak to higher molar masses confirms the grafting reaction. GPC traces of all graft copolymer samples $PGA_{26}O_yPEG_z$ are shown in Fig. 7B. From the analysis of these chromatograms it is noticed, that the *PDI* of some of the oleic acid grafted polymers increased after grafting with PEG chains instead of decreasing. This could be a sign of small contributions of hydrolytic cleavage of the ester bonds of the polymer backbone. Since the removal of unreacted mPEG-Succinyl chains was difficult, the products were dialyzed against water for 7–10 days. The exposure of the product to water for such a long time might lead to cleavage of ester bonds. However, the degree of this hydrolytic cleavage must be small as all the products have symmetric and unimodal molar mass distributions.

Table 3 Molar masses, polydispersities, and thermal properties of saturated fatty acid grafted polyesters.

Sample	M_n (g mol^{-1})	M_w/M_n (PDI)[a]	T_g (°C)[b]	T_m (°C)[b]
Stearic acid			–	69.7
PGA_{23}	4700[a]	2.2	−24.5	–
$PGA_{23}S_8$	6900[c]	1.9	−20.2	34.5
$PGA_{23}S_{16}$	9200[c]	2	−15.2	35.8
$PGA_{23}S_{21}$	10,500[c]	1.9	Not detectable	36.2
Behenic acid			–	83
$PGA_{23}B_{10}$	8200[c]	1.9	−8.1	55.8
$PGA_{23}S_8B_{10}$	10,400[c]	1.9	−2.3	46.2
Lauric acid			–	45.3
PGA_{33}	6600[a]	1.8	−24.5	–
$PGA_{33}S_9$	10,300[c]	1.8	−19	
$PGA_{33}S_9L_6$	11,600[c]	1.8	−40.1	−10.8

[a]Obtained from GPC using THF as eluent and polystyrene as calibration standard.
[b]Obtained from DSC with a heating rate of 20 °C min^{-1}.
[c]Calculated on the basis of the degree of grafting obtained from ^{1}H NMR spectra.

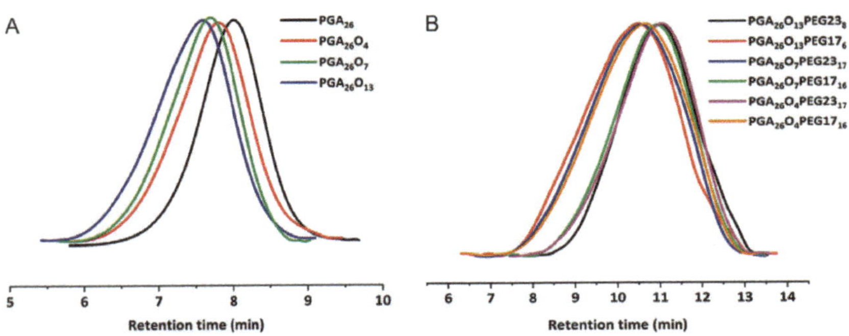

Fig. 7 GPC traces of (A) PGA_{26} and oleic acid grafted PGA_{26} and (B) $PGA_{26}O_yPEG_z$.

3.5 Differential scanning calorimetry

3.5.1 PGA and saturated fatty acid grafted polyesters

The thermal properties of all polymers were investigated by differential scanning calorimetry. All samples have been heated from −65 °C up to 100 °C with a rate of 20 °C min^{-1} (see Fig. 8). PGA is an amorphous polymer with a

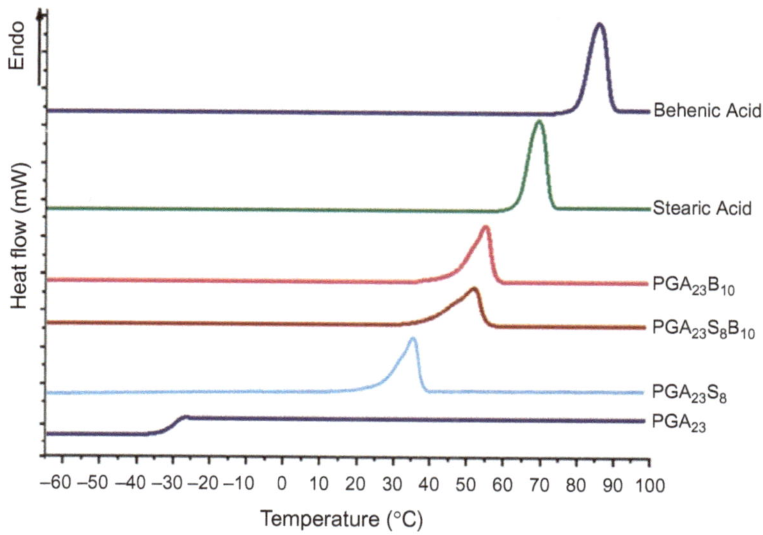

Fig. 8 DSC traces of PGA_{23}, $PGA_{23}S_8$, $PGA_{23}B_{10}$, $PGA_{23}S_8B_{10}$, stearic acid, and behenic acid recorded at a heating rate of 20 °C min^{-1}.

glass transition temperature T_g of −24.5 °C. All fatty acid grafted PGA show endothermal melting peaks which are related to the melting of the crystals formed by the grafted acyl chains. The DSC traces show an increase of T_m with increasing degree of grafting. The thermal behavior indicates a semi-crystalline structure. Higher substitution degrees lead to more perfect crystalline structures and, therefore, to higher melting temperatures. As a result, the grafted PGA_{23} with stearic acid side chains start to melt at 37 °C, which represents the body temperature. On the contrary, the grafted PGAs with behenic acyl side chains are solids at the body temperature (melting temperatures >40 °C). This leads to the assumption that the utilization of grafted PGA_{23} with behenic acid side chains might be more favorable for drug delivery systems (DDSs). The DSC traces of PGA_{23} and grafted PGA_{23} with behenic acid, and both stearic acid and behenic acid side chains are shown. For comparison, the DSC traces of both stearic acid and behenic acid are provided in Fig. 8, which shows that behenic acid has a melting temperature higher than that of stearic acid. It should be noted that a single but broadened melting endotherm occurs for $PGA_{23}S_8B_{10}$. The melting endotherm is located between the respective endotherms of $PGA_{23}S_8$ and $PGA_{23}B_{10}$. This might indicate that the two different side chains from the fatty acids crystallize into one crystal structure. Finally, this needs more studies, especially by X-ray crystallography.

3.5.2 Oleic acid and mPEG-Succinyl grafted PGA_{26}

The thermal properties of all polymers were determined by DSC. It is observed that the polymer backbone PGA_{26} and all oleic acid grafted polymers are amorphous since a melting endotherm of these polymers is not observed. The influence of the *cis*-bond at C_9 of oleic acid changes the properties of the graft copolymers fundamentally. The presence of this *cis*-bond makes it difficult for oleate grafted chains to pack into crystalline structures. Therefore, completely amorphous graft copolymers are formed. However, the glass transition temperature decreases with increasing degree of oleate grafting which is consistent with previous findings (Bilal et al., 2017). The DSC traces of PGA_{26} and $PGA_{26}O_y$ are shown in Fig. 9. Thus, the oleic acid side chains increase the flexibility of the respective polymers, which leads to a decrease of T_g with increasing amount of grafted oleic acyl chains. They act as internal plasticizers (Bilal et al., 2017).

The oleic acid grafted PGAs are then further grafted with mPEG17-Succinyl and mPEG23-Succinyl side chains. The DSC traces of mPEG-Succinyl grafted $PGA_{26}O_y$ are shown in Fig. 10. It is observed that after grafting with PEG side chains, all the polymers become semi-crystalline. The phenomenon of cold crystallization was also observed in some samples where the volume fraction of PEG is relatively small (Wunderlich, 1958). The cold crystallization temperatures of the products $PGA_{26}O_4PEG17_{16}$,

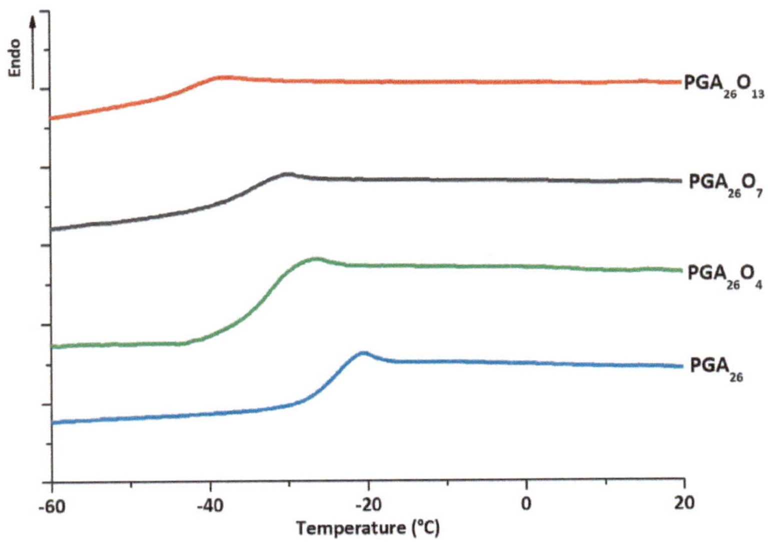

Fig. 9 DSC traces of PGA_{26} and oleic acid grafted $PGA_{26}O_y$ recorded at a heating rate of $10\,°C\,min^{-1}$.

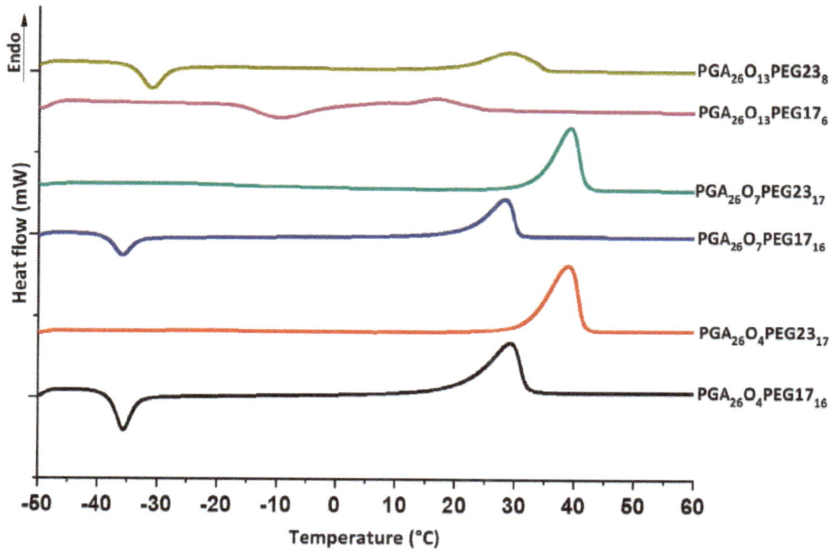

Fig. 10 DSC traces of graft copolymers ($PGA_{26}O_yPEG_z$) recorded at a heating rate of 10 °C min^{-1}.

$PGA_{26}O_7PEG17_{16}$, $PGA_{26}O_{13}PEG17_6$, and $PGA_{26}O_{13}PEG23_8$ are −35.6, −35.9, −9.4 and −30.8 °C, respectively. The reason of this cold crystallization is that in these samples during the cooling step with the selected cooling rate of 10 °C min^{-1}, the chains do not have enough time to organize themselves as the mobility is hindered by neighboring chains. Thus, during the heating scan, they organize themselves and achieve an ordered state (Wunderlich, 1958; Zhao, Pan, Ji, Cao, & Wang, 2016). The melting temperatures T_m, glass transition temperatures T_g, crystallinity, number average molar masses M_n and polydispersity index *PDI,* volume fraction of oleic acid side chains ϕ_{oleate}, and volume fraction of mPEG-Succinyl side chains ϕ_{mPEG} of all graft copolymers are given in Table 4. It is worth mentioning here that the glass transition temperature is only observed for oleic acid grafted polymers and not for PEG grafted polymers under given measurement conditions.

The crystallinity of the graft copolymers is calculated by the following relation.

$$X = \frac{\Delta H_m}{\Delta H_m^0 \cdot w_{PEG}} \tag{5}$$

Table 4 M_n, *PDI*, $\Phi_{oleic\ acyl}$, Φ_{mPEG}, T_g, and T_m of all grafted and ungrafted polymers.

Sample	M_n (g mol^{-1})	M_w/M_n (*PDI*)	Φ_{oleate}[a]	Φ_{PEG}[a]	T_g[b] (°C)	T_m[b] (°C)	Crystallinity[b] (%)
PGA_{26}	5300[c]	1.9[c]	–	–	−24	–	–
$PGA_{26}O_4$	6350[d]	1.7[c]	18	–	−33	–	–
$PGA_{26}O_7$	7250[d]	2.0[c]	0.28	–	−35	–	–
$PGA_{26}O_{13}$	8750[d]	1.9[c]	0.42	–	−42	–	–
$PGA_{26}O_4PEG17_{16}$	19,900[d]	2.7[e]	0.07	0.64	–	29.2	61
$PGA_{26}O_4PEG23_{17}$	25,000[d]	1.8[e]	0.05	0.73	–	38.9	67
$PGA_{26}O_7PEG17_{16}$	20,700[d]	1.8[e]	0.1	0.62	–	28.4	52
$PGA_{26}O_7PEG23_{17}$	25,800[d]	2.4[e]	0.08	0.71	–	39.2	66
$PGA_{26}O_{13}PEG17_6$	13,800[d]	1.9[e]	0.28	0.33	–	17	32
$PGA_{26}O_{13}PEG23_8$	17,500[d]	2.1[e]	0.22	0.47		29	53

[a]Volume fraction calculated from material studio software v 4.1.
[b]Obtained from DSC with a heating rate of 10 °C min^{-1}.
[c]Obtained from GPC using THF as eluent.
[d]Calculated on the basis of the degree of grafting obtained from ^{1}H NMR spectra.
[e]Obtained from GPC, using DMF (with 0.01 M LiBr) as eluent.

Here, w_{PEG} is the weight fraction of PEG chains in the graft copolymer, ΔH_m is the melting enthalpy of the corresponding PEG chains in the polymer, and ΔH_m^0 is the melting enthalpy of 100% crystalline PEG. The value of ΔH_m^0 is 197 J g^{-1} (Buckley & Kovacs, 1976).

3.6 Scanning electron microscopy

Due to their tunable properties and biodegradability, grafted PGAs are new alternatives to currently used lipids or polymer based excipients for drug delivery. The most promising applications are in the field of parenteral controlled release. The utilization as a new innovative matrix for drugs and therefore, the formulation of different drug delivery systems may be a possibility for the near future. The preparation of dimensionally stable microparticles by a solvent evaporation process using different fatty acids grafted PGAs has already been successful. One example is shown in Fig. 11. Here, the polymer has been diluted in methylene chloride and slowly injected into an aqueous solution of poly(vinyl alcohol) (0.5 wt%). After injection the remaining organic solvent has been removed with a rotary evaporator.

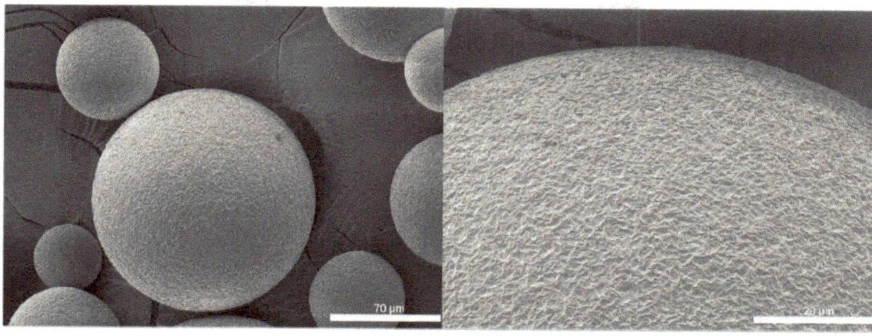

Fig. 11 Scanning electron micrographs of $PGA_{23}S_8B_{10}$ microparticles.

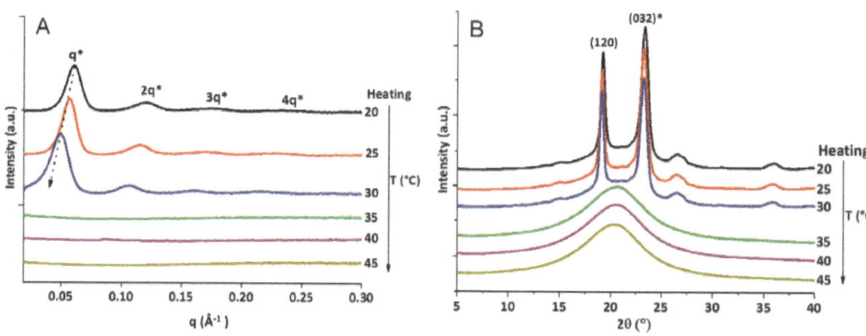

Fig. 12 (A) SAXS and (B) WAXS patterns of the bulk polymer $PGA_{26}O_4PEG23_{17}$ measured for different temperatures.

3.7 Small and wide angle X-ray scattering

Temperature dependent SAXS and WAXS traces of the bulk polymer, $PGA_{26}O_4PEG23_{17}$, are shown in Fig. 12. A typical pattern of a lamellar crystal phase in the SAXS region and the characteristic pattern of PEG in the WAXS region, respectively, confirm that the crystallinity is exclusively induced by PEG chains. Fig. 12A shows the SAXS pattern of the bulk polymer $PGA_{26}O_4PEG23_{17}$. It depicts the presence of the lamellar arrangement, as integer multiples of q^* are present. The first order lamellar peak appears at $q^* = 0.0513\,\text{Å}^{-1}$ with the corresponding d-spacing of 47.9 Å at 25 °C. The lamellar thickness value is calculated as 36 Å by considering the crystallinity of the polymer (X = 67%) obtained from DSC. By increasing the temperature, peaks slightly shift to lower q-values due to the increase of lamellar thickness (Zhang, Guo, & Xu, 2016), and further heating results in melting of the polymer. The melting temperature obtained from XRD is slightly lower than that measured by DSC. This could be because of different thermal histories of the two samples. In Fig. 12B, the temperature dependent WAXS pattern of the polymer is shown. The two characteristics peaks at $2\theta = 19.2°$ and $2\theta = 23.3°$ correspond to the (120) and (032)* Miller planes of monoclinic unit cell. It consists of four PEG chains with seven repeating units in a 7_2 helical structure (Takahashi, Sumita, & Tadokoro, 1973). The peak at $2\theta = 23.3°$ results not only from the (032) plane but also from $(\bar{1}32), (112), (\bar{2}12), (\bar{1}24), (\bar{2}04)$ and (004) planes as well (Lai, Hiltner, Baer, & Korley, 2012). The peak vanishes by increasing the temperature which confirms that the crystallinity in the polymer was induced by mPEG-Succinyl chains.

4. Application of multiple grafted polyesters

Out of all synthesized graft copolymers (see Table 4), only two polymers $PGA_{26}O_4PEG23_{17}$ and $PGA_{26}O_{13}PEG23_8$ were used further to investigate their properties as steric stabilizers for lyotropic liquid crystalline particles formed from glycerol monooleate (GMO) in water. The samples were selected on the basis of difference of volume fraction of PEG.

The basic requirement for steric stabilizers is that they must be water soluble and have a small critical aggregation concentration (CAC). The polymers mentioned fulfill both requirements. The CAC values for these polymers are obtained from tensiometry where an increase in polymer concentration causes a decrease in surface tension of the aqueous solutions (see Fig. 13). The CAC values are calculated to 1.45 and $2.2\,\mathrm{mg\,L^{-1}}$ for the polymers $PGA_{26}O_4PEG23_{17}$ and $PGA_{26}O_{13}PEG23_8$, respectively.

For cubosome preparation for SAXS measurement, 1, 3, 5, and 10 wt% of $PGA_{26}O_4PEG23_{17}$ with respect to GMO and 3, 5, and 10 wt% of $PGA_{26}O_{13}PEG23_8$ were used. Aqueous dispersions of GMO plus stabilizer at 10 wt% concentration were used for measurements. As reported before, GMO in excess water shows a *Pn3m* cubic phase (Briggs, Chung, Caffrey, & Phase, 1996). The GMO/water cubic phase could only incorporate as much as 1 wt% of $PGA_{26}O_4PEG23_{17}$. Increasing the concentration of $PGA_{26}O_4PEG23_{17}$ to >1 wt% leads to a change of the cubic symmetry from the cubic *Pn3m* to the cubic *Im3m* phase (see Fig. 14A). The lattice parameter of the *Im3m* cubic phase increases with increasing polymer concentration from 142 to 166 Å (with an increase in the diameter of the water channel

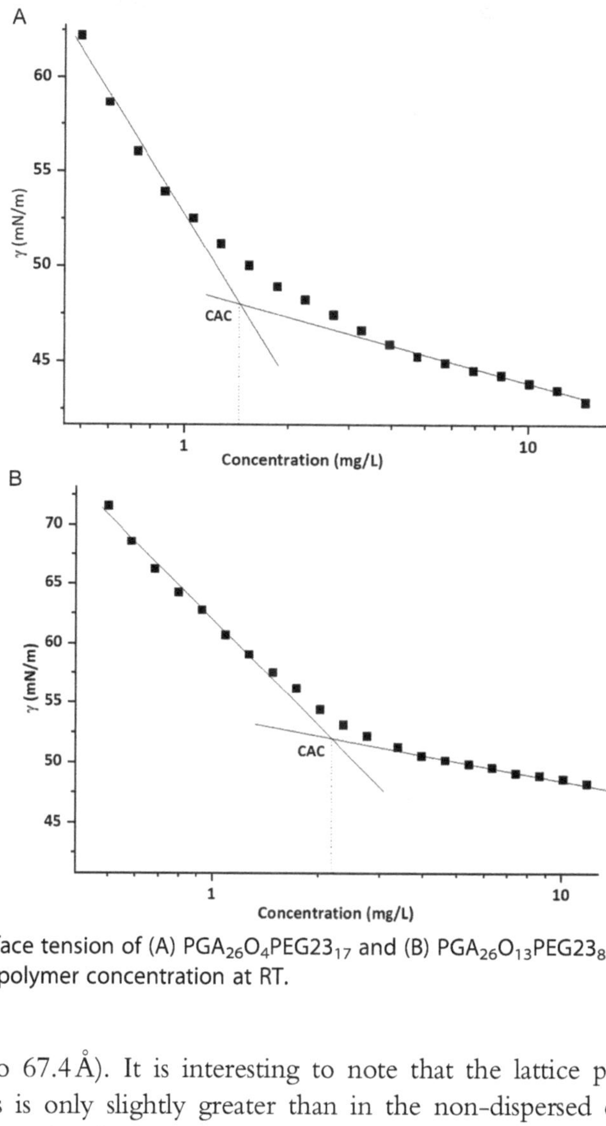

Fig. 13 Surface tension of (A) $PGA_{26}O_4PEG23_{17}$ and (B) $PGA_{26}O_{13}PEG23_8$ in water as a function of polymer concentration at RT.

from 48 to 67.4 Å). It is interesting to note that the lattice parameter of cubosomes is only slightly greater than in the non-dispersed cubic phase for the same ratio of polymer to GMO. In comparison to $PGA_{26}O_4PEG23_{17}$, when $PGA_{26}O_{13}PEG23_8$ is used as a steric stabilizer, surprisingly it retains the cubic *Pn3m* symmetry in all tested samples (see Fig. 14B). The lattice parameter of the *Pn3m* cubic phase increases from 115.3 to 121.5 Å (the diameter of the water channel increase from 48 to 60.6 Å) when the polymer $PGA_{26}O_{13}PEG23_8$ concentration increases from 3 to 10 wt% with respect to GMO The lattice parameter of the cubic *Pn3m* phase is 107 Å for the

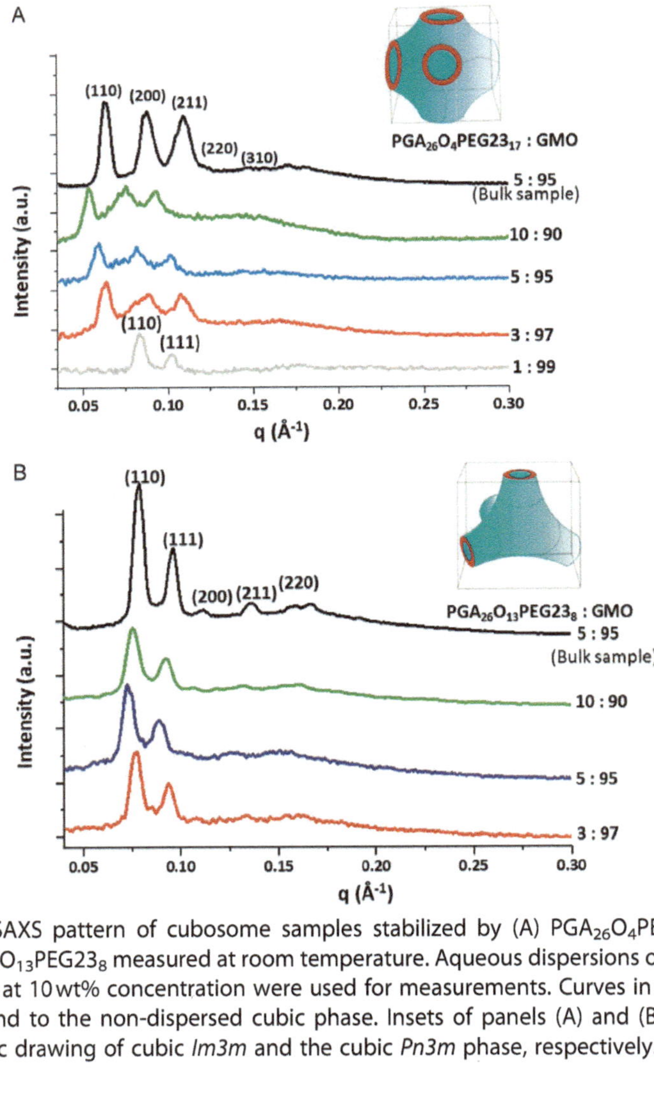

Fig. 14 SAXS pattern of cubosome samples stabilized by (A) $PGA_{26}O_4PEG23_{17}$ and (B) $PGA_{26}O_{13}PEG23_8$ measured at room temperature. Aqueous dispersions of GMO plus stabilizer at 10wt% concentration were used for measurements. Curves in black color correspond to the non-dispersed cubic phase. Insets of panels (A) and (B) show the schematic drawing of cubic *Im3m* and the cubic *Pn3m* phase, respectively.

GMO/water system ($d_w = 48$ Å). It is important to mention here that the sample with 1 wt% of $PGA_{26}O_{13}PEG23_8$ also shows a cubic *Pn3m* phase, but with this amount of stabilizer, only a small amount of GMO is possible to disperse which is not sufficient for SAXS measurement.

The two polymers behave differently when added to the GMO/water system. The polymer $PGA_{26}O_4PEG23_{17}$ induces a cubic phase transition from *Pn3m* to *Im3m*. In contrast, the $PGA_{26}O_{13}PEG23_8$ polymer retains the cubic *Pn3m* symmetry. The transition from the *Pn3m* cubic phase to the *Im3m* cubic phase indicates an excess of $PGA_{26}O_4PEG23_{17}$ molecules, which are saturated at the surface of the lipid rich phase and consequently form the *Im3m* cubic phase. This could be further explained by taking into account the critical packing parameter (Israelachvili, 2011). The mesophase transition could only be observed in the less curved phase (i.e., the *Im3m* cubic phase) when an amphiphile with a larger hydrophilic part compared to hydrophobic part is added (Fong, Le, & Drummond, 2012; van't Hag, Gras, Conn, & Drummond, 2017). On the contrary, when the polymer $PGA_{26}O_{13}PEG23_8$ which is less hydrophilic than $PGA_{26}O_4PEG23_{17}$ is used as stabilizer, the hydrophobic side chains penetrate into the lipid bilayer, while the hydrophilic part remain at the interface. This results in an increase of the lattice parameter from 107.1 to 121.6 Å (hence an increase in the diameter of the water channel from 48 to 60.6 Å) of the cubic *Pn3m* phase. A schematic overview of the cubic phase transition and the cubic phase swelling when polymers $PGA_{26}O_4PEG23_{17}$ or $PGA_{26}O_{13}PEG23_8$ are used as stabilizers is shown in Fig. 15. To find out the particle size of the cubosomes, the samples were diluted to make $2\,g\,L^{-1}$ concentrations for DLS

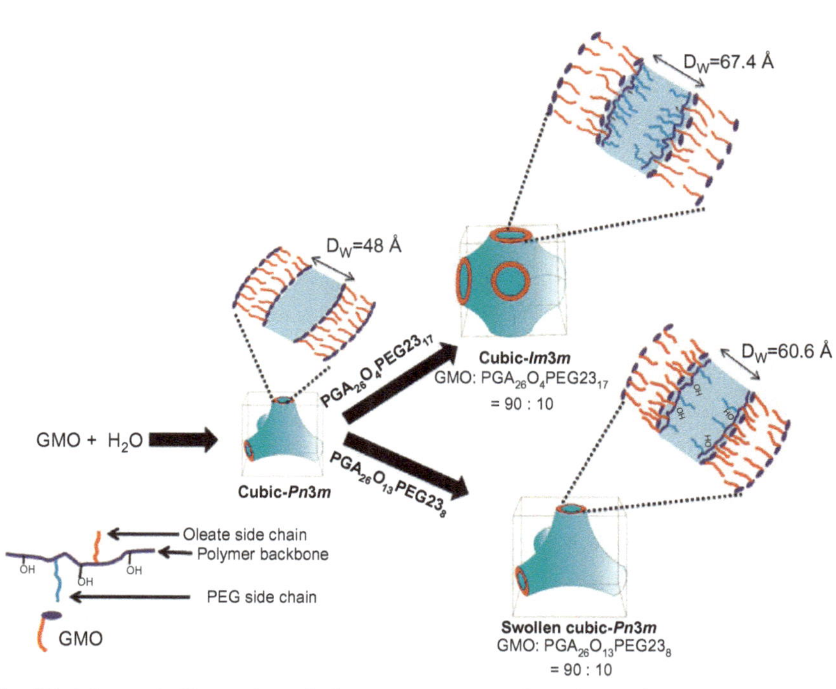

Fig. 15 Schematic illustration of phase transition and swelling of cubic phases when polymer $PGA_{26}O_4PEG23_{17}$ or $PGA_{26}O_{13}PEG23_8$ were used as steric stabilizers for cubosome dispersions of the system GMO/water.

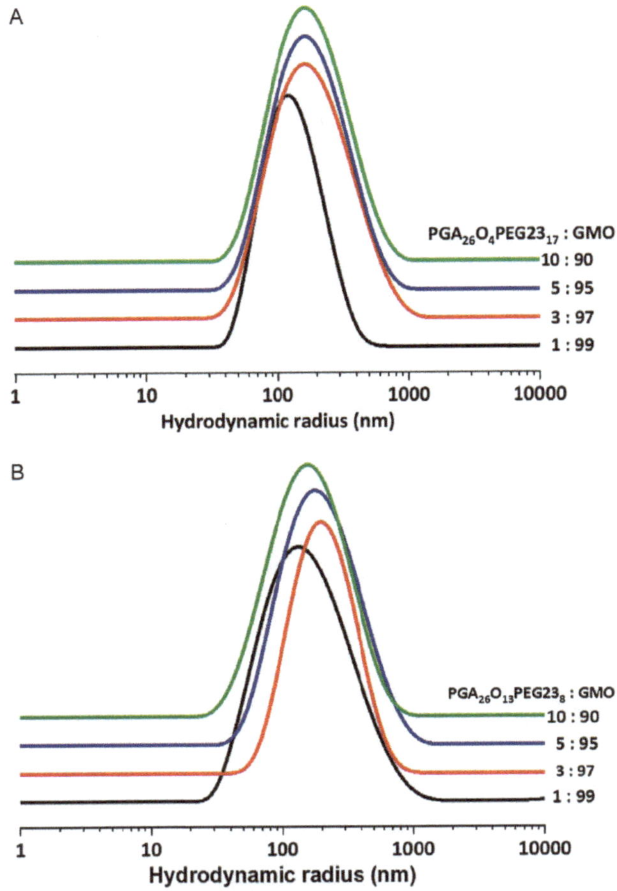

Fig. 16 The hydrodynamic radius distribution of all cubosome samples stabilized by polymer (A) $PGA_{26}O_4PEG23_{17}$ and (B) $PGA_{26}O_{13}PEG23_8$ in different ratios with respect to GMO measured at a scattering angle of 90°. The concentration of the samples was $2\,g\,L^{-1}$.

measurements. The dilution is necessary in order to minimize undesired multiple scattering effects and, furthermore, to minimize interparticle interactions (Schärtl, 2007). The particle size was determined for each sample for 7 different angles. The average hydrodynamic radius of all measured samples varies from 130 to 225 nm and the hydrodynamic radius distributions for each sample measured at an angle of 90° are shown in Fig. 16.

The cubosome dispersions stabilized with polymers $PGA_{26}O_4PEG23_{17}$ and $PGA_{26}O_{13}PEG23_8$ are stable for >2 weeks at room temperature. The stability of dispersions was identified by visual inspection, as no large aggregates were observed.

5. Conclusion

Poly(glycerol adipate) has been successfully synthesized using divinyl adipate or dimethyl adipate together with glycerol by enzymatic polycondensation with CAL-B as catalyst. PGA was further grafted with saturated (stearic, behenic, and lauric acid) and unsaturated (oleic acid) fatty acids. All PGAs grafted with stearic and behenic acids were semi-crystalline while those having oleic acid side chains were amorphous with glass transition temperatures decreasing with increasing degree of grafting. Oleic acid grafted PGAs were further modified with water soluble, low molar mass PEG chains. DSC measurements showed that the graft copolymers were semi-crystalline. The crystallinity was induced by the PEG chains. In some graft copolymers where the volume fraction of PEG is relatively small ($\phi_{mPEG} \leq 0.64$), the phenomenon of cold crystallization was observed. All the $PGA_xO_yPEG_z$ samples were water soluble. Two of the polymers, $PGA_{26}O_4PEG23_{17}$ and $PGA_{26}O_{13}PEG23_8$ having CAC values of 1.45 and 2.2 mg L^{-1}, respectively, were selected to investigate their ability as steric stabilizers of cubosomes. It was found that the incorporation of $PGA_{26}O_4PEG23_{17}$ to the GMO/water system caused a mesophase transition of the cubic phase from *Pn3m* to *Im3m*. On the contrary, when $PGA_{26}O_{13}PEG23_8$ was introduced into the GMO/water system, *Pn3m* cubic symmetry retained. It is assumed that the polymer $PGA_{26}O_4PEG23_{17}$ had a higher volume fraction of the hydrophilic part ($\phi_{mPEG} = 0.73$) in comparison to the polymer $PGA_{26}O_4PEG23_{17}$ ($\phi_{mPEG} = 0.47$). Hence, when it interacted with the lipid bilayer, it increased the volume of the head group leading to cubic phase transition from *Pn3m* to *Im3m*. The polymer $PGA_{26}O_4PEG23_{17}$ retained the cubic symmetry of *Pn3m* of the GMO/water system and caused only an increase of the lattice parameter and thus of the radius of the water channels. The cubosome dispersions were stable for more than 2 weeks at room temperature. From these findings, it was concluded that these multi-graft biocompatible copolymers could be used effectively to stabilize several lyotropic liquid crystal systems by controlling the hydrophobic to hydrophilic volume fractions.

Acknowledgments

This work was done in the frame of High Performance Center Chemical and Biosystems Technology Halle/Leipzig supported by the European Regional Development Fund (ERDF) and the Federal State Saxony-Anhalt. R.A. thanks the Deutsche Akademische Austauschdienst (DAAD) for a PhD scholarship (Research Grants—Doctoral Programs in Germany, 2017/18 (57299294)).

References

Besheer, A., Hertel, T. C., Kressler, J., Mäder, K., & Pietzsch, M. (2009). Enzymatically catalyzed HES conjugation using microbial transglutaminase: Proof of feasibility. *Journal of Pharmaceutical Sciences*, *98*, 4420–4428.

Bilal, M. H., Hussain, H., Prehm, M., Baumeister, U., Meister, A., Hause, G., et al. (2017). Synthesis of poly(glycerol adipate)-g-oleate and its ternary phase diagram with glycerol monooleate and water. *European Polymer Journal*, *91*, 162–175.

Bilal, M. H., Prehm, M., Njau, A. E., Samiullah, M. H., Meister, A., & Kressler, J. (2016). Enzymatic synthesis and characterization of hydrophilic sugar based polyesters and their modification with stearic acid. *Polymers*, *8*, 80.

Briggs, J., Chung, H., Caffrey, M., & Phase, T. T. (1996). The temperature-composition phase diagram and mesophase structure characterization of the monoolein/water system. *Journal de Physique II France*, *6*, 723–751.

Brosnan, S. M., Brown, A. H., & Ashby, V. S. (2013). It is the outside that counts: Chemical and physical control of dynamic surfaces. *Journal of the American Chemical Society*, *135*, 3067–3072.

Buckley, C. P., & Kovacs, A. J. (1976). Melting behaviour of low molecular weight poly(ethylene-oxide) fractions. *Colloid & Polymer Science*, *254*, 695–715.

Dong, Y.-D., Larson, I., Hanley, T., & Boyd, B. J. (2006). Bulk and dispersed aqueous phase behavior of phytantriol: Effect of vitamin E acetate and F127 polymer on liquid crystal nanostructure. *Langmuir*, *22*, 9512–9518.

Feng, C., Li, Y., Yang, D., Hu, J., Zhang, X., & Huang, X. (2011). Well-defined graft copolymers: From controlled synthesis to multipurpose applications. *Chemical Society Reviews*, *40*, 1282–1295.

Fong, C., Le, T., & Drummond, C. J. (2012). Lyotropic liquid crystal engineering–ordered nanostructured small molecule amphiphileself-assembly materials by design. *Chemical Society Reviews*, *41*, 1297–1322.

Hann, N. D. (1977). Effects of lithium bromide on the gel-permeation chromatography of polyester-based polyurethanes in dimethylformamide. *Journal of Polymer Science*, *15*, 1331–1339.

Hu, J., Gao, W., Kulshrestha, A., & Gross, R. A. (2006). "Sweet polyesters": Lipase-catalyzed condensation – polymerizations of alditols. *Macromolecules*, *39*, 6789–6792.

Israelachvili, J. (2011). Soft and biological structures. In J. N. Israelachvili (Ed.), *Intermolecular and surface forces* (pp. 535–576). San Diego: Academic Press.

Jbeily, M., Naolou, T., Bilal, M., Amado, E., & Kressler, J. (2014). Enzymatically synthesized polyesters with pendent OH groups as macroinitiators for the preparation of well-defined graft copolymers by atom transfer radical polymerization. *Polymer International*, *63*, 894–901.

Jérôme, C., & Lecomte, P. (2008). Recent advances in the synthesis of aliphatic polyesters by ring-opening polymerization. *Advanced Drug Delivery Reviews*, *60*, 1056–1076.

Jiang, Y., & Loos, K. (2016). Enzymatic synthesis of biobased polyesters and polyamides. *Polymers*, *8*, 243.

Kallinteri, P., Higgins, S., Hutcheon, G. A., Pourçain, C. B. S., & Garnett, M. C. (2005). Novel functionalized biodegradable polymers for nanoparticle drug delivery systems. *Biomacromolecules*, *6*, 1885–1894.

Kikuchi, H., Uyama, H., & Kobayashi, S. (2002). Lipase-catalyzed ring-opening polymerization of substituted lactones. *Polymer Journal*, *34*, 835–840.

Kim, D.-Y., & Dordick, J. S. (2001). Combinatorial array-based enzymatic polyester synthesis. *Biotechnology and Bioengineering*, *76*, 200–206.

Kline, B. J., Beckman, E. J., & Russell, A. J. (1998). One-step biocatalytic synthesis of linear polyesters with pendant hydroxyl groups. *Journal of the American Chemical Society*, *120*, 9475–9480.

Kulshrestha, A. S., Gao, W., & Gross, R. A. (2005). Glycerol copolyesters: Control of branching and molecular weight using a lipase catalyst. *Macromolecules*, *38*, 3193–3204.

Lai, C.-Y., Hiltner, A., Baer, E., & Korley, L. T. J. (2012). Deformation of confined poly(ethylene oxide) in multilayer films. *ACS Applied Materials & Interfaces*, *4*, 2218–2227.

Lu, C., & Zhong, W. (2010). Synthesis of propargyl-terminated heterobifunctional poly(ethylene glycol). *Polymers (Basel)*, *2*, 407–417.

Martin, B. D., Ampofo, S. A., Linhardt, R. J., & Dordick, J. S. (1992). Biocatalytic synthesis of sugar-containing poly(acrylate)-based hydrogels. *Macromolecules*, *25*, 7081–7085.

Mecerreyes, D., Humes, J., Miller, R. D., Hedrick, J. L., Detrembleur, C., Lecomte, P., et al. (2000). First example of an unsymmetrical difunctional monomer polymerizable by two living/controlled methods. *Macromolecular Rapid Communications*, *21*, 779–784.

Naolou, T., Busse, K., & Kressler, J. (2010). Synthesis of well-defined graft copolymers by combination of enzymatic polycondensation and "click" chemistry. *Biomacromolecules*, *11*, 3660–3667.

Naolou, T., Conrad, D., Busse, K., Mäder, K., & Kressler, J. (2013). Fatty acid modified poly(glycerol adipate)-polymeric analogues of glycerides. In C. Scholz & J. Kressler (Eds.), *Tailored polymer architectures for pharmaceutical and biomedical applications: Vol. 1135* (pp. 39–52): American Chemical Society.

Naolou, T., Jbeily, M., Scholtysek, P., & Kressler, J. (2013). Synthesis and characterization of stearoyl modified poly(glycerol adipate) containing ATRP initiator on its backbone. *Advanced Materials Research*, *812*, 1–11.

Naolou, T., Meister, A., Schöps, R., Pietzsch, M., & Kressler, J. (2013). Synthesis and characterization of graft copolymers able to form polymersomes and worm-like aggregates. *Soft Matter*, *9*, 10364–10372.

Olson, D. A., & Sheares, V. V. (2006). Preparation of unsaturated linear aliphatic polyesters using condensation polymerization. *Macromolecules*, *39*, 2808–2814.

Olsson, A., Lindström, M., & Iversen, T. (2007). Lipase-catalyzed synthesis of an epoxy-functionalized polyester from the suberin monomer cis-9,10-epoxy-18-hydroxyoctadecanoic acid. *Biomacromolecules*, *8*, 757–760.

Orafai, H., Kallinteri, P., Garnett, M., Huggins, S., Hutcheon, G., & Pourcain, C. (2008). Novel poly (glycerol-adipate) polymers used for nanoparticle making: A study of surface free energy. *Iranian Journal of Pharmaceutical Research*, 7, 11–19.

Pfefferkorn, D., Pulst, M., Naolou, T., Busse, K., Balko, J., & Kressler, J. (2013). Crystallization and melting of poly(glycerol adipate)-based graft copolymers with single and double crystallizable side chains. *Journal of Polymer Science Part B-Polymer Physics*, *51*, 1581–1591.

Puskas, J. E., Seo, K. S., & Sen, M. Y. (2011). Green polymer chemistry: Precision synthesis of novel multifunctional poly(ethylene glycol)s using enzymatic catalysis. *European Polymer Journal*, *47*, 524–534.

Rangelov, S., & Almgren, M. (2005). Particulate and bulk bicontinuous cubic phases obtained from mixtures of glyceryl monooleate and copolymers bearing blocks of lipid-mimetic anchors in water. *The Journal of Physical Chemistry B*, *109*, 3921–3929.

Rulev, A. Y. (2011). Aza-Michael reaction: Achievements and prospects. *Russian Chemical Reviews*, *80*, 197–218.

Schärtl, W. (2007). Sample preparation. In *Light scattering from polymer solutions and nanoparticle dispersions* (pp. 43–50). Berlin Heidelberg: Springer-Verlag.

Shoda, S., Uyama, H., Kadokawa, J., Kimura, S., & Kobayashi, S. (2016). Enzymes as green catalysts for precision macromolecular synthesis. *Chemical Reviews*, *116*, 2307–2413.

Sifontes Herrera, V. A., Oladele, O., Kordás, K., Eränen, K., Mikkola, J.-P., Murzin, D. Y., et al. (2011). Sugar hydrogenation over a Ru/C catalyst. *Journal of Chemical Technology & Biotechnology*, *86*, 658–668.

Spicer, P. T. (2004). Cubosomes: Bicontinuous liquid crystalline nanoparticles. In J. A. Schwarz, C. I. Contescu, & K. Putyera (Eds.), *Dekker encyclopedia of nanoscience and nanotechnology*: *Vol. 1*. New York: Marcel Dekker Inc.

Suksiriworapong, J., Taresco, V., Ivanov, D. P., Styliari, I. D., Sakchaisri, K., Junyaprasert, V. B., et al. (2018). Synthesis and properties of a biodegradable polymer-drug conjugate: Methotrexate-poly(glycerol adipate). *Colloids and Surfaces B: Biointerfaces*, *167*, 115–125.

Takahashi, Y., Sumita, I., & Tadokoro, H. (1973). Structural studies of polyethers. IX. Planar zigzag modification of poly(ethylene oxide). *Journal of Polymer Science Polymer Physics Edition*, *11*, 2113–2122.

Tanaka, A., Kohri, M., Takiguchi, T., Kato, M., & Matsumura, S. (2012). Enzymatic synthesis of reversibly crosslinkable polyesters with pendant mercapto groups. *Polymer Degradation and Stability*, *97*, 1415–1422.

Taresco, V., Creasey, R. G., Kennon, J., Mantovani, G., Alexander, C., Burley, J. C., et al. (2016). Variation in structure and properties of poly(glycerol adipate) via control of chain branching during enzymatic synthesis. *Polymer*, *89*, 41–49.

Therisod, M., & Klibanov, A. M. (1986). Facile enzymatic preparation of monoacylated sugars in pyridine. *Journal of the American Chemical Society*, *108*, 5638–5640.

Tsujimoto, T., Uyama, H., & Kobayashi, S. (2001). Enzymatic synthesis of cross-linkable polyesters from renewable resources. *Biomacromolecules*, *2*, 29–31.

Tsujimoto, T., Uyama, H., & Kobayashi, S. (2002). Enzymatic synthesis and curing of biodegradable crosslinkable polyesters. *Macromolecular Bioscience*, *2*, 329–335.

Uyama, H., Klegraf, E., Wada, S., & Kobayashi, S. (2000). Regioselective polymerization of sorbitol and divinyl sebacate using lipase catalyst. *Chemistry Letters*, *29*, 800–801.

Uyama, H., Kuwabara, M., Tsujimoto, T., & Kobayashi, S. (2003). Enzymatic synthesis and curing of biodegradable epoxide-containing polyesters from renewable resources. *Biomacromolecules*, *4*, 211–215.

Van't Hag, L., Gras, S. L., Conn, C. E., & Drummond, C. J. (2017). Lyotropic liquid crystal engineering moving beyond binary compositional space—Ordered nanostructured amphiphile self-assembly materials by design. *Chemical Society Reviews*, *46*, 2705–2731.

Varma, I. K., Albertsson, A.-C., Rajkhowa, R., & Srivastava, R. K. (2005). Enzyme catalyzed synthesis of polyesters. *Progress in Polymer Science*, *30*, 949–981.

Wang, Y.-F., Lalonde, J. J., Momongan, M., Bergbreiter, D. E., & Wong, C.-H. (1988). Lipase-catalyzed irreversible transesterifications using enol esters as acylating reagents: Preparative enantio- and regioselective syntheses of alcohols, glycerol derivatives, sugars and organometallics. *Journal of the American Chemical Society*, *110*, 7200–7205.

Warwel, S., Demes, C., & Steinke, G. (2001). Polyesters by lipase-catalyzed polycondensation of unsaturated and epoxidized long-chain α,ω-dicarboxylic acid methyl esters with diols. *Journal of Polymer Science: Part A: Polymer Chemistry*, *39*, 1601–1609.

Weiss, V. M., Naolou, T., Hause, G., Kuntsche, J., Kressler, J., & Mäder, K. (2012). Poly(-glycerol adipate)-fatty acid esters as versatile nanocarriers: From nanocubes over ellipsoids to nanospheres. *Journal of Controlled Release*, *158*, 156–164.

Williams, C. K. (2007). Synthesis of functionalized biodegradable polyesters. *Chemical Society Reviews*, *36*, 1573–1580.

Wu, R., Al-Azemi, T. F., & Bisht, K. S. (2008). Functionalized polycarbonate derived from tartaric acid: Enzymatic ring-opening polymerization of a seven-membered cyclic carbonate. *Biomacromolecules*, *9*, 2921–2928.

Wu, W.-X., Wang, N., Liu, B.-Y., Deng, Q.-F., & Yu, X.-Q. (2014). Lipase-catalyzed synthesis of azido-functionalized aliphatic polyesters towards acid-degradable amphiphilic graft copolymers. *Soft Matter*, *10*, 1199–1213.

Wunderlich, B. (1958). Theory of cold crystallization of high polymers. *The Journal of Chemical Physics*, *29*, 1395–1404.

Yu, Y., Wu, D., Liu, C., Zhao, Z., Yang, Y., & Li, Q. (2012). Lipase/esterase-catalyzed synthesis of aliphatic polyesters *via* polycondensation: A review. *Process Biochemistry*, *47*, 1027–1036.

Zhang, M., Guo, B.-H., & Xu, J. (2016). A review on polymer crystallization theories. *Crystals*, 7, article no. 4.

Zhang, Y.-R., Spinella, S., Xie, W., Cai, J., Yang, Y., Wang, Y.-Z., et al. (2013). Polymeric triglyceride analogs prepared by enzyme-catalyzed condensation polymerization. *European Polymer Journal*, *49*, 793–803.

Zhao, L., Pan, L., Ji, A., Cao, Z., & Wang, Q. (2016). Recrystallization of freezable bound water in aqueous solutions of medium concentration. *Chinese Physics B*, *25*, 075101.

CHAPTER FOUR

Tailored biosynthesis of polyhydroxyalkanoates in chemostat cultures

Véronique Amstutz, Nils Hanik, Julien Pott, Camila Utsunomia, Manfred Zinn*

HES-SO Valais-Wallis, Sion, Switzerland

*Corresponding author: e-mail address: manfred.zinn@hevs.ch

Contents

Abstract

Polyhydroxyalkanoates (PHAs) are accumulated intracellularly by many bacteria and serve as a carbon and energy storage compound. PHAs are polyesters of high molecular weight and can be isolated by solvent extraction and precipitation in antisolvents. The material properties of PHAs are of great interest due to the inherent biodegradability and excellent biocompatibility. To date, more than 150 different PHA monomers have been described in literature and it has been found that the monomeric unit composition significantly influences the physico-chemical properties of PHAs. The monomer

Methods in Enzymology, Volume 627
ISSN 0076-6879
https://doi.org/10.1016/bs.mie.2019.08.018

composition may be controlled to some extent by the choice of the PHA production strain but also by the cultivation conditions and the carbon substrate/PHA precursor supply. In previous studies, it has been shown that the most reproducible production method of PHA is the chemostat cultivation of suitable bacteria under multiple nutrient limited growth conditions. This chapter is dedicated to provide step-by step instructions to produce PHAs in a chemostat culture and specifically describes how the composition of PHA copolymers can be tailored during biosynthesis, as well as a set of analytical tools and methods to characterize PHAs.

1. Introduction

The success story of plastics started almost 100 years ago with the scientific description of the existence of polymeric materials by Staudinger (1920). As a consequence, polymer chemistry became a new field of research and helped to pave the way for the industrial production of plastics, particularly after the Second World War. A key motivation to use plastic materials was the fact that they were lighter than the original materials used and thus significantly helped in saving costs for transportation of goods. In addition, the polymer properties could be fine-tuned using plasticizers, colorants and fillers, and soon replaced more and more traditional materials like metal, glass, natural fibers, and wood during the past 50 years. Further growth of the plastic market was boosted by very cheap material and manufacturing costs, the wide diversification of polymerization substrates using cracking processes of naphtha, the isolation of the fractionation of fossil oil distillate, but also the ease of processing plastic materials in high numbers. In 2015, the world-wide production of plastic materials reached 322 million tons per year after an increase by 40% over the past 10 years (Europe, 2016). The largest growth was triggered by China and other Asian countries, whereas in Europe the plastic production was stable at around 58 million tons per year (Europe, 2016). The today's dominating plastic materials are polypropylene (PP), high and low density polyethylene (PE), polystyrene (PS), polyethylene terephthalate (PET), polyvinyl chloride (PVC), polyurethane (PUR), and acrylonitrile-butadiene-styrene (ABS). By far the largest field of application of plastics is packaging (39.9%), followed by building and construction materials (19.7%), automotive (8.9%), electronics (5.8%), agriculture (3.3%) and other applications (22.4%) including medical uses.

It has to be pointed out that the packaging materials comprise not only the largest volume of plastics, but also the application with the shortest lifetime. The official waste stream of plastics in Europe was 25.8 Mio t in 2014 (Europe, 2016). The largest amount (39.5%) of plastic waste was used for energy recycling, only 29.7% were recycled as material and still 30.8% ended up in landfills. The deposition in landfills is not favored since most plastics are recalcitrant to degradation in the environment, although recently, some enzyme activity has been demonstrated for the surface erosion of PET (Müller, Kleeberg, & Deckwer, 2001; Ribitsch et al., 2013; Wei & Zimmermann, 2017). The addition of oxo-active compounds to enhance catalytically the disintegration of plastic materials is not a valid solution (Portillo, Yashchuk, & Hermida, 2016). As a result, even more microplastic debris and microfibers would be produced that finally would become an additional burden to the ecosystem (Narancic & O'Connor, 2017).

Biodegradable and biocompatible polymers may therefore represent suitable alternatives when they meet the physico-chemical requirements of the application.

Valid candidates are the poly(hydroxyalkanoates) (PHAs, Fig. 1) that constitute a class of polyoxoesters of hydroxyalkanoates typically produced in the form of cytoplasmic inclusions (granules or carbonosomes) in specific Gram-positive and Gram-negative bacteria and extremophilic Archaea (Jendrossek, 2009; Jendrossek & Pfeiffer, 2014).

Typically, PHAs are classified into two groups depending on number of carbon atoms per monomer. The short-chain-length PHAs (scl-PHAs) have monomers of 3–5 carbons, whereas the medium-chain-length PHAs

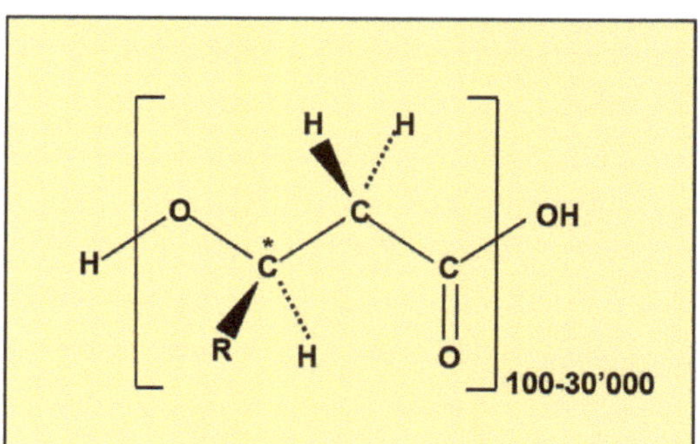

R = C_1-C_2 Short-chain-length PHA (sclPHA)

R = C_3-C_9 Medium-chain-length PHA (mclPHA)

Fig. 1 Polyhydroxyalkanoates (PHAs) are polyesters accumulated by many bacteria under specific growth conditions. The asterisks indicates a chiral center in *R* configuration, which is essential for the biodegradability of PHAs by intracellular and extracellular depolymerases.

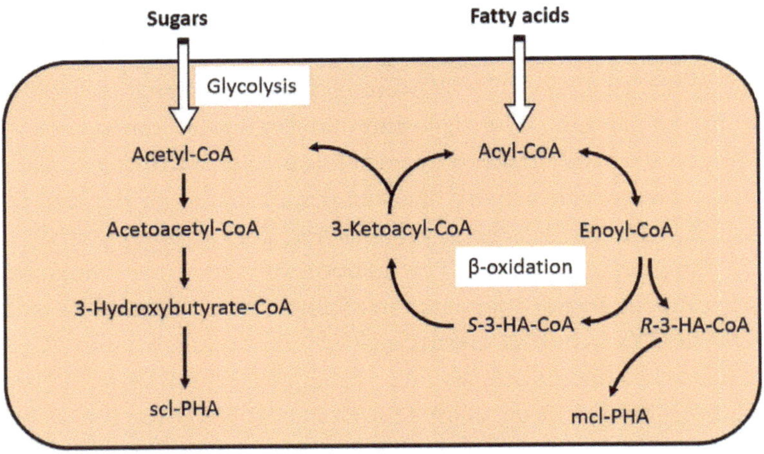

Fig. 2 Simplified scheme of the intracellular biosynthesis of scl- and mcl-PHAs from sugar or fatty acid substrates. *Figure redrawn from Numata, K., Morisaki, K., Tomizawa, S., Ohtani, M., Demura, T., Miyazaki, M., et al. (2013). Synthesis of poly- and oligo(hydroxyalkanoate)s by deep-sea bacteria, Colwellia spp., Moritella spp., and Shewanella spp.* Polymer Journal, 45*(10), 1094–1100. https://doi.org/10.1038/pj.2013.25.*

(mcl-PHAs) contain monomers with 6–14 carbons (Fig. 1). Fig. 2 gives a simplified schematic overview on the intracellular scl- and mcl-PHA biosynthesis from metabolized sugars or fatty acids.

In order to control the monomeric units during biosynthesis, clearly fatty acids are to be favored over sugars (Follonier et al., 2014; Follonier, Riesen, & Zinn, 2015; Hartmann et al., 2006; Zinn & Hany, 2005).

All chiral centers in the polymer backbone show (*R*)-configuration (Lageveen et al., 1988). The pure *R*-configuration of the monomers is guaranteed by the stereo-specificity of the synthetizing enzymes (Lee, 1996).

Probably the best studied PHA belongs to the scl-PHA class and is called poly(3-hydroxybutyrate) (PHB). PHB has the highest melting temperature with 180°C and a molecular weight up to 1.4×10^6 Da (Doi, Tamaki, Kunioka, & Soga, 1988). These properties seemed to be interesting enough for the British company Imperical Chemical Industry (ICI) (Holmes, 1985). However, PHB was found to be difficult to process and too brittle for many applications. These material properties could be improved with the production of a copolymer consisting of 3-hydroxybutyrate (3HB) and 3-hydroxyvalerate (3HV). Interestingly, the melting point could be lowered with an increase of the 3HV content and resulted in a softer plastic (Bloembergen, Holden, Hamer, Bluhm, & Marchessault, 1986). Under the tradename Biopol different copolymers with 3HV contents up to 12mol%

were successfully commercialized by Zeneca (ICI) and shampoo bottles could be produced from this polymer for Wella AG, a German shampoo producer. These bottles were fully degraded in the environment but also in sediments of lakes (Brandl & Püchner, 1991). In 1998, Zeneca was sold to Monsanto and already 1 year later the Biopol production was stopped for economic reasons. All rights were then sold to Metabolix, a company based in Boston, USA. Finally, in 2014 Metabolix gave up all PHA related activities due to unfavorable economics. To date, Biopol gained more interest due to several European research projects (e.g., http://www.phbottle.eu/ and www.synpol.org) and new collaborations have been established in industry (e.g., Danimer in collaboration with Nestlé) and further extension of the material properties by combining scl- with mcl-PHA monomers using genetic engineering could be achieved (Li, Yun, Gong, Zhao, & Zhang, 2005) and is currently commercialized by the Japanese company Kaneka.

As already indicated by the examples above, it is important to note that the material properties of the PHAs vary as a function of the type of side-chains, the molecular weight, and according to the distribution of different monomers leading to random copolymers, block-copolymers or homopolymers (Koller, Salerno, & Braunegg, 2013). Random copolymers can be biosynthesized when cells are growing on mixed or pure substrates. Fatty acids for instance are consumed through the β-oxidation cycle to generate energy, reducing the chain length by two carbon atoms per cycle and producing acetyl-CoA (Fig. 2) (Anderson, Haywood, & Dawes, 1990). Correspondingly, fatty acyl-CoA molecules with different chain lengths are available as substrates and end up in the polymerized form. In case of growth on octanoic acid poly(3-hydroxyoctanoate-*co*-3-hydroxyhexanoate) (PHO-HH) random copolymers are synthetized (Durner, Zinn, Witholt, & Egli, 2001).

The invention of the chemostat cultivation technique at the same time by Monod in France (Monod, 1950) and Novick & Szilard in USA (Novick & Szilard, 1950) opened the door to a new world in biotechnology. A chemostat is defined as a steady-state bioprocess, where a microbial culture is continuously supplied with nutrients at a fixed rate and concomitantly harvested to keep the culture volume constant. Thus, for the first time, cell physiology of microorganisms could be assessed in detail under reproducible growth conditions and consequently helped to extend the understanding of microbial growth. A great advantage of this cultivation method is the fact that particular growth conditions can be studied in a time independent way (the physiological parameters of cultures in steady-state do theoretically not change over time). Moreover, the specific growth rate and the growth-limiting nutrient can be set independently from each other by the selection of the dilution rate of the chemostat and by the medium composition, respectively.

To produce tailor-made copolymers, chemostat-based cultivation strategies have been developed (see Section 3). In a cultivation study comparing batch with chemostat cultivation, it could be shown that the polymer composition is constant over time for chemostat cultures which was not the case for batch cultivation (Hartmann, Hany, Witholt, & Zinn, 2010).

2. Biosynthesis of PHA in chemostat culture

2.1 Bioreactor equipment and set-up of a chemostat culture

A chemostat bioprocess can be established with most commercially available benchtop bioreactors that are equipped with pH and temperature control, a Rushton stirrer and an air sparging system. Usually, the number of ports to the bioreactor are a limiting factor and thus the following ports need to be available: pO_2 probe, pH probe, pH control feed (acid and/or base), antifoam probe and antifoam supply, sterile air supply through an immersed sparger, medium feed and harvest (see Fig. 3). In cases, where acidic substrates are added, the supply of acid as feed control is not necessary. The culture volume (V), also known as working volume, may be either kept constant by weight control (the whole bioreactor is placed on a balance thus controlling the effluent pump) or simply by an overflow tube connected to a continuously running harvest pump. Besides the bioreactor, at least two additional tanks with about a 10 × larger volume are required: One medium feed reservoir and a harvest tank that needs to be cooled to 4 °C during PHA collection in order to avoid PHA degradation. The medium feed as well as the solutions for pH and foam control may optionally be put on balances in order to monitor their consumption. In order to enable a constant supply of medium at a constant rate, the utilization of a peristaltic pump connected with the medium feed tank showed to be most reliable. With the insertion of a glass pipette (closed with a sterile filter) between medium feed tank and the pump, the feed rate can be checked (see Section 2.3).

2.2 Media composition and preparation

The medium plays a key role in the production of PHAs. It has been found that for precultures of *Cupriavidus necator* (typical strain for scl-PHA production)

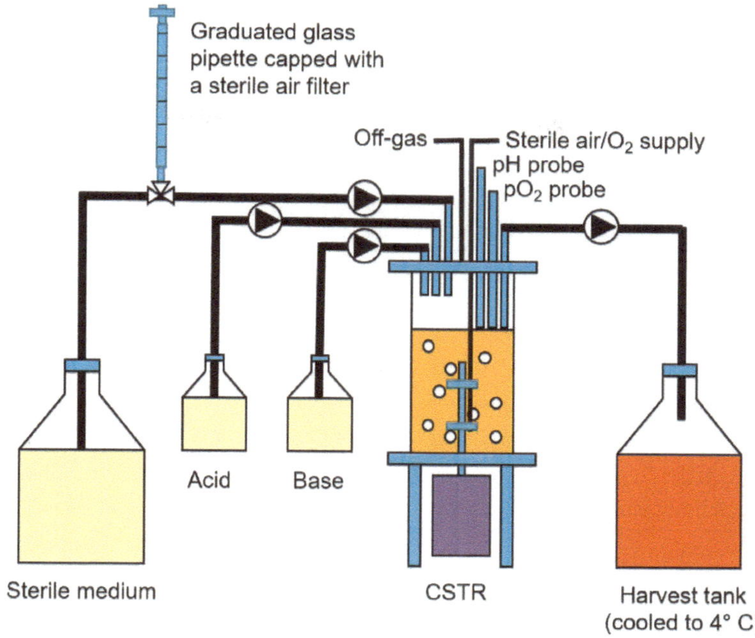

Fig. 3 Typical set-up of a chemostat with a continuous stirred tank reactor (CSTR) for the production of polyhydroxyalkanoates using an overflow tube for the control of the bioreactor volume. A graduated glass pipette closed with sterile air filter connected with the tubing between medium tank and feed pump allows the control of the medium flow.

and *Pseudomonas putida* GPo1 (typical strain for mcl-PHA production) the medium E was very useful (see Table 1). The medium XCC (Zinn et al., 2011) (Table 2) was designed for the production of PHA in a chemostat according to the medium design concept by Egli (2015), Egli, Mason, and Snozzi (1993), Egli and Zinn (2003), since it allowed the accurate control of the carbon to nitrogen ratio (C/N) in the medium feed while avoiding other, unwanted nutrient limitations.

It has to be noted that instead of using heat sterilization, medium filtration can be used. For such an approach Sartobran 300 filters with a pore size of 0.2 μm (Sartorius Stedim Biotech GmbH, Germany) are used to filter media to sterile medium tanks (e.g., 20 L Sartorius).

2.3 Start of chemostat cultivation

The start of a chemostat cultivation is a sequence of several steps:

Protocol Preculture

Table 1 Modified medium E used for batch cultivation (Lageveen et al., 1988).

Chemical	Amount [$g L^{-1}$]
$NaNH_4PO_4 * 4H_2O$	3.5
$K_2HPO_4 * 3H_2O$	7.5
KH_2PO_4	3.7
Tri-Na citrate dihydrate ($C_6H_8O_7 * 2H_2O$)	2.9
After heat sterilization the following nutrients were added to 1 L of medium	
$MgSO_4 * 7H_2O$	1 mL of 1 M sterile stock solution in H_2O
$FeSO_4 * 7H_2O$	1 mL of sterile 0.01 M stock solution in 1 M HCl
MT stock solution in 1 M HCl	1 mL of sterile MT stock solution in 1 M HCl
$MnCl_2 * 4H_2O$ (1.98 $g L^{-1}$)	
$CoSO_4 * 7H_2O$ (2.81 $g L^{-1}$)	
$CaCl_2 * 2H_2O$ (1.47 $g L^{-1}$)	
$CuCl_2 * 2H_2O$ (0.71 $g L^{-1}$)	
$ZnSO_4 * 7H_2O$ (0.29 $g L^{-1}$)	

Note: Medium E can also be prepared as Agar plates with addition of 15 $g L^{-1}$ of agar before heat sterilization.

1. Autoclave at 121 °C for 20 min a 0.5 L Erlenmeyer flask with baffles and a side arm connected to a tubing and needle for the connection to the bioreactor port.
2. Prepare 100 mL of medium E and adjust the pH to 7.1 with conc. NaOH and filter sterilize it into the sterile Erlenmeyer flask in a laminar flow bench.
3. Inoculate the medium with a suitable colony (or some colonies) from an agar plate (medium E with colonies of *C. necator* or *P. putida*).
4. Incubate in an orbital shaker (>100 rpm) at 30 °C for at least 12 h or an OD600 >1.5.

Protocol Bioreactor Preparation and Sterilization

1. Assemble the benchtop bioreactor with all probes and add a scale of working volume with a permanent pen (e.g., between 1.5 and 2.5 L each 0.1 L) on the glass cylinder.

Table 2 Medium XCC for the chemostat cultivation (Zinn et al., 2011).

Chemical	Amount [$g L^{-1}$]
$(NH_4)_2SO_4$	0.71 (corresponds to 150 $mg N L^{-1}$)
K_2HPO_4 * $3H_2O$	1.0
Carbon source	2.26 gC (corresponds to a C/N=15 $g g^{-1}$)
After heat sterilization the following nutrients were added to 1 L of medium	
$MgSO_4$ * $7H_2O$	1 mL of 1 M sterile stock solution in H_2O
$FeSO_4$ * $7H_2O$	1 mL of sterile 0.01 M stock solution in 1 M HCl
MT stock solution in 1 M HCl	1 mL of sterile MT stock solution in 1 M HCl
$MnCl_2$ * $4H_2O$ (1.98 $g L^{-1}$)	
$CoSO_4$ * $7H_2O$ (2.81 $g L^{-1}$)	
$CaCl_2$ * $2H_2O$ (1.47 $g L^{-1}$)	
$CuCl_2$ * $2H_2O$ (0.71 $g L^{-1}$)	
$ZnSO_4$ * $7H_2O$ (0.29 $g L^{-1}$)	

2. Fill the reactor with H_2O to the wanted volume of the bioreactor (e.g., 2.0 L) and switch on the stirrer to the wanted agitation speed (e.g., 1000 rpm) and aeration rate (ca. 1 volume air per volume of liquid per minute (vvm)).
3. Adjust the overflow tube in its position in order to barely touch the water surface.
4. Add 200 mL of H_2O and switch on the harvest pump (set to a high pumping rate). After 5 min switch off agitation and aeration and verify the working volume. Repeat this procedure until you are sure that the volume is controlled to the volume you wanted to have.
5. Empty the reactor, fill it again with medium E (without stock solutions), and mount the coverlid. Make sure that all sterile barriers (filters) are installed/sterilized according to the manufacturer's manual.
6. Remove all connected tubing that need to be sterilized separately (aeration and off-gas system, pH control set, etc.,) and replace the used silicone membranes of the connection ports with new ones.

7. Sterilize bioreactor depending on the type of reactor (in situ heat sterilization with a protection shield or ex situ in an autoclave) according to the manufacturer's protocol for at least 30 min. Sterilize all tubings to be required for the cultivation phase in an autoclave.
8. When cooled down to 30 °C, add the stock solutions in an aseptic way.
9. Connect the base and acid bottles with the appropriate tubing to the bioreactor (be careful to maintain sterility).
10. Switch on cooling water through the air outlet condenser and set temperature to 30 °C.
11. Supply air at a rate of approximately 1 vvm (volume of air per volume of medium per minute).
12. At 30 °C: take a sample from the bioreactor and measure the pH. Check that the off-line value corresponds to the pH indicated in the control system. If the difference between the off- and on-line measurements is <0.5 pH units, then simply adjust the slope of the pH controller such that the pH meter connected to the bioreactor reads the same value as the one measured off-line.
13. In order to calibrate the pO_2 probe, sparge the bioreactor with N_2 for the zero-point. Then, sparge the bioreactor with air until the pO_2 is stable. At this point set the value of the pO_2 to 100% (= air saturation).
14. The bioreactor may be left under these conditions overnight (possible to set the temperature at 15 °C to reduce evaporation of the medium).

Protocol Switch from Batch to Chemostat Cultivation

1. Samples of the preculture and of the bioreactor are taken in an aseptic way and analyzed for their purity under a microscope (phase contrast, 1000 × magnification with an oil objective). A contaminated preculture or bioreactor will not be used any further and needs to be decontaminated.
2. The growth conditions are verified and set to automatic control (pH to 7.0, aeration to 1 vvm, agitation to 1000 rpm, and temperature to 30 °C).
3. The tubing of the preculture in the Erlenmeyer flask is connected (in an as aseptic as possible way) to the bioreactor port and the complete preculture transferred to the bioreactor medium using a peristaltic pump.
4. A first sample is immediately taken and put on ice for further analysis (OD_{600nm}, flow cytometry). All process parameters are noted on a data sheet.
5. Typically, such batch culture may last for about 12 h.
6. When the optical density has reached an $OD_{600nm} = 2$, the medium feed tubing is connected to the bioreactor inlet port and the harvest

tubing connected to the port with the overflow tube for the control of the working volume. It has to be taken care that the outlet pump is running faster than the feed pump to make sure that the working volume in the bioreactor is quickly adjusted.

7. The dilution rate (D) is set in order to achieve a stable growth in the chemostat (ca. ½ of maximum specific growth rate of the microorganism). D is calculated by dividing the flow rate of the medium into the bioreactor by the working volume.
8. The flow rate of the medium into the reactor may be checked with the graduated glass pipette that is positioned between medium bag and peristaltic pump (see Fig. 2). After filling the pipette with medium from the medium tank, the connection to the medium tank is closed with a clamp, thus only leaving the connection between pipette and peristaltic pump. The time is measured that is needed to pump a certain volume (e.g., 10 mL) while emptying the pipette. This value can be used to adjust the pump speed in order to meet the desired flow rate. However, it may alter due to aging of the tubing that is in contact with the peristaltic pump and therefore needs to be adjusted once daily. *Note:* Make sure to put the clamp to the pipette tubing after each measurement in order to enable free medium flow to the culture.
9. Steady-state growth is usually achieved after five volume changes of the bioreactor and is typically confirmed by three consecutive measurements of the biomass (e.g., OD_{600nm} or flow cytometry) and other indirect parameters (e.g., pO_2).
10. With the achievement of steady-state growth conditions, the biomass collected in the harvest tank can be concentrated for the isolation of the PHA once a day (see protocol PHA purification).

3. Tailored composition of PHA in chemostat culture

As previously mentioned, the monomeric unit composition in the PHA has a significant influence on the material properties. Consequently, the fed carbon sources play a key role in tailoring the PHA composition. Best control of the monomeric unit composition can be achieved when all carbon sources are completely consumed (i.e., carbon (C) limitation). The highest PHA contents in chemostat cultures can be obtained under nitrogen (N) limitation. In previous studies for the scl- and mcl-PHA production, it has been shown that multiple nutrient limitations are most appropriate (Hartmann et al., 2006).

3.1 Multiple nutrient limited growth

Fig. 4 shows in a very simplified way for what dilution rates (under steady state growth conditions the dilution rate is equal to the specific growth rate) and for what carbon to nitrogen ratios in the medium feed a simultaneous carbon and nitrogen limited growth can be expected. Deviations to boundaries in Fig. 4 are possible and strain and medium dependent (lower capacity for PHA accumulation or higher maintenance energy consumption). The multiple nutrient limited growth regime can also be determined experimentally as described in Zinn et al. (2011), Zinn, Witholt, and Egli (2004).

3.2 Tailored biosynthesis

The tailoring of the PHA composition is of particular interest when the ratio between two monomeric units should be altered, e.g., the 3-hydroxyvalerate content in scl-PHA produced by *C. necator* (Zinn, Weilenmann, Hany, Schmid, & Egli, 2003) or the content of 3-hydroxyphenylvalerate in mcl-PHAs (Hartmann et al., 2004). Due to different affinities of the two substrates towards the PHA polymerase and different biochemical degradation pathways, the monomeric unit composition in PHA may differ significantly in comparison to the provided mixture of the substrate in the medium feed. Moreover, the growth conditions need to be exactly controlled in order

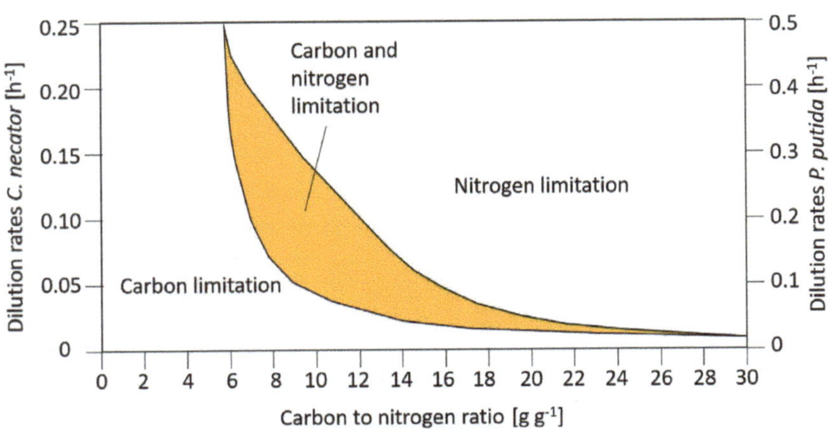

Fig. 4 Growth regimes for scl-PHA producers (*C. necator, left axis*) and mcl-producers (*P. putida, right axis*) as a function of the dilution rate in a chemostat and the carbon to nitrogen ratio in the feed medium. The boundary for simultaneous carbon and nitrogen limitation is only a rough estimate and may vary significantly with carbon substrate mixtures and other PHA producing microorganisms.

to guarantee a reproducible PHA production with respect to PHA quantity and monomeric unit composition. So far, best results could be obtained with fatty acids as carbon sources (Hartmann et al., 2004, 2006; Zinn, Witholt, & Egli, 2001).

4. Purification of PHAs

4.1 Isolation of cells and extraction of PHA

There are many different methods of PHA extraction described in literature (Furrer, Panke, & Zinn, 2007; Kosseva & Rusbandi, 2018). Below, we describe a method that is suitable for scl- and mcl-PHAs independent of their monomeric unit composition.

Equipment

- Fumehood
- Centrifuge (appropriate for volumes of culture broth)
- Stainless steel filter holder 142 mm, 2000 mL (Sartorius, Göttingen, Germany)
- Rotary evaporator (Büchi)
- Vacuum oven

Chemicals

- Dichloromethane (DCM) (Sigma-Aldrich)
- Activated charcoal powder (Sigma-Aldrich)
- Methanol (MeOH) (Sigma-Aldrich)

Protocol

1. Separate the biomass from the cooled culture broth of the harvest tank by centrifugation (ca. 5000 × *g* for 30 min, at 4 °C), the supernatant is discarded. The biomass is resuspended in 0.9% NaCl solution for washing and subjected to the same centrifugation conditions again. The supernatant is discarded. The pellet is collected and frozen at −20 °C for at least 24 h, then dried by lyophilization for at least 24 h and finally stored in a dry, airtight container until following steps are carried out.
2. Crush the freeze-dried biomass to make a powder.
3. Weigh the biomass in an enough large glass beaker.
4. Add the activated charcoal in volume with a beaker: 4 × the volume of biomass powder (in mL) giving the amount of activated charcoal powder in g (i.e., 4 mL of charcoal per 1 g of biomass).
5. Pour the DCM gently in the beaker with activated charcoal and biomass and stir gently at room temperature for 4 h. Amount of DCM

needed: 40 mL g^{-1} biomass (add some mL for rinsing). The beaker needs to be covered by a watch glass to reduce evaporation during this extraction phase.

6. Filter by pressure filtration using a stainless steel filter holder through a depth filter (optionally covered with an additional paper filter).
7. Remove DCM by rotary evaporation (40 °C, ambient pressure) until the polymer solution gets a viscous liquid. The distilled DCM can be collected for future use.
8. Prepare an ice bath and a beaker with MeOH preconditioned at 4 °C (10 × volume of the polymer solution).
9. Precipitation of PHA: Put the beaker with the cold MeOH into the ice bath and add drop by drop (or drizzle) the viscous polymer solution under gentle stirring with a magnet bar. When finished, let the precipitated PHA and MeOH mix stir for at least 3 h (remove the beaker from the ice bath to increase DCM evaporation).
10. After 3 h, filter the MeOH through a Buchner funnel fitted with a paper filter. Rinse the PHA with pure MeOH.
11. In case the precipitated PHA is still very wet, let it dry under the fume hood at room temperature for about 12 h.
12. Put the PHA inside a beaker (recover the most PHA that you can) and put the beaker in the vacuum oven (at room temperature) for at least 48 h.
13. The MeOH can be purified by distillation (Rotavap, ca. 40 °C, 340 mbar) and reutilized for future PHA precipitations.

Remark:

Activated charcoal is needed only when high purity is required (e.g., reduction of endotoxin levels in case of medical applications (Wampfler, Ramsauer, Kehl, Zinn, & Thöny-Meyer, 2010)). Without activated charcoal, the PHA recovery will be significantly better but with a reduced purity of ca. 90–95%.

5. Quantification and characterization of PHAs

5.1 Gas chromatography (GC)

In literature, there are different methods described in order to quantify PHA content in biomass (Baidurah et al., 2015; Betancourt, Yezza, Halasz, Tra, & Hawari, 2007; Braunegg, Sonnleitner, & Lafferty, 1978; Cui, Barford, & Renneberg, 2007; Furrer et al., 2007; Riis & Mai, 1988). In the following, we describe detailed methods that are used on a routine base in our laboratory to quantify scl-PHAs and mcl-PHAs.

5.1.1 Scl-PHA

Equipment

- Oven
- 10–15 mL Pyrex glass tubes with air-tight screwed caps containing PTFE covered rubber liner
- 0.45 μm PTFE filters
- Vacuum pump
- Syringes (2 mL capacity) with needles
- GC vials
- Pasteur glass pipette connected to a suction flask and a vacuum pump
- GC analyzer equipped with a BGD-Wax column (30 m length, 0.25 mm inner diameter and 0.25 μm film thickness, BGB Analytik)

Chemicals

- Solution of 15 vol% H_2SO_4 in methanol (MeOH)
- Solution of precisely 1 mg mL^{-1} of 2-ethyl-2-hydroxybutyric acid (internal standard) in dichloromethane (DCM)
- Solution of saturated NaCl (about 6 M in H_2O)
- Na_2SO_4 and Na_2CO_3 salts dehydrated
- Standards: PHA of high purity (or various PHA compounds if necessary)

Protocol

1. Prepare samples of biomass that contain about 4 mg polymer each (often 30–40 mg cell dry weight is suitable). They are obtained by centrifugation of a sample from the bioprocess (supernatant is discarded), followed by addition of 0.9 wt% NaCl solution to wash the pellet and a second centrifugation. Discard the supernatant and freeze the pellet at −20 °C for at least 24 h, then dry by lyophilization for at least 24 h and finally store the samples in a dry, airtight container.
2. Weigh precisely 30–40 mg of biomass samples in Pyrex glass tubes. The weight is used later on for calculation of the PHA content per g of biomass. Prepare as well a blank sample without biomass.
3. Add 2 mL of the DCM mixture and 2 mL of the MeOH in the Pyrex tubes.
4. The tubes need to be closed tightly and then vortexed. Place them in an oven and incubate at 100 °C for 2.5 h.
5. Remove the samples from the oven and let them cool down to room temperature.
6. Vortex and add 2 mL of saturated NaCl solution. Vortex again.

7. After phase separation, remove the upper phase (aqueous phase) with the aid of a Pasteur glass pipette connected to a suction flask and a vacuum pump.
8. Add a second time 2 mL of the saturated NaCl solution and remove the upper phase, after vortexing (as described above).
9. Open a syringe connected to a 0.45 μm PTFE syringe filter (removing the plunger) and add a scoop of dry Na_2SO_4 and of dry Na_2CO_3 salts.
10. Give about 1 mL of the sample into the syringe using a glass pipette (avoid to take the precipitated salt in the glass tubes, and importantly avoid the transfer of droplets from the water phase). Carefully place back the plunger into the syringe.
11. Filter the sample directly into a GC vial and seal it to be ready for gas chromatography.
12. Keep the samples at 4 °C until measurement.
13. The preparation of standards for the calibration curves is exactly the same, with the difference that known quantities of PHA are weighed in the Pyrex tubes (in the range of 4 mg of PHA).

The GC analysis is carried out as follows:

- A volume of 1 μL is injected at 250 °C with a split ratio of 1:10. The temperature is increased from 80 °C to 240 °C at a rate of 10 °C min^{-1}. Helium is used as carrier gas (3 mL min^{-1}). A flame ionization detector (FID) at 285 °C is used for detection.
- A calibration curve is built for each of the PHA standards. The ratio of the areas of the considered PHA compounds and the internal standard is used for recalculating the amount of PHA in the samples. Therefore, for each PHA compound expected, a calibration curve is necessary.
- The wt% of PHA is obtained by comparison of the mass of PHA present in the sample (calculated from the GC result) to the initial mass of freeze-dried biomass weighed in the Pyrex tube.

5.1.2 Mcl-PHA

Equipment

- Oven
- 10–15 mL Pyrex glass tubes with air-tight screwed caps containing PTFE covered rubber liner
- 0.45 μm PTFE filters
- Vacuum pump
- Syringes (2 mL capacity) with needles
- GC vials

- Pasteur glass pipette connected to a suction flask and a vacuum pump
- GC analyzer equipped with a BGD-Wax column (30 m length, 0.25 mm inner diameter and 0.25 μm film thickness, BGB Analytik)

Chemicals

- Solution of 1.3 M BF_3 (boron trifluoride) in methanol (MeOH)
- Solution of precisely about 2.5 $mg\,mL^{-1}$ of 2-ethyl-2-hydroxybutyric acid (internal standard) in dichloromethane (DCM)
- Solution of saturated NaCl (about 6 M in H_2O)
- Na_2SO_4 and Na_2CO_3 salts dehydrated
- Standards: PHA of high purity (or various PHA compounds if necessary)

Protocol

1. Prepare samples of biomass that contain about 4 mg polymer each (often 30–40 mg cell dry weight is suitable). They are obtained by centrifugation of a sample from the bioprocess (supernatant is discarded), followed by addition of 0.9 wt% NaCl solution to wash the pellet and a second centrifugation. Discard the supernatant and freeze the pellet at −20 °C for at least 24 h, then dry by lyophilization for at least 24 h and finally store the samples in a dry, airtight container.
2. Weigh precisely 30–40 mg of biomass samples in Pyrex glass tubes. The weight is used later on for calculation of the PHA content per g of biomass. Prepare as well a blank sample without biomass.
3. Add 2.5 mL of the DCM mixture. After vortexing, wait until dissolution of the polymer (standards) and good dispersion of the biomass (several minutes).
4. Add 2.5 mL of the MeOH mixture (containing BF_3) in the Pyrex tubes.
5. The tubes need to be closed tightly and then vortexed. Place them in an oven to incubate at 80 °C for 20 h.
6. Remove the samples from the oven and let them cool down to room temperature.
7. Vortex and add 2.5 mL of saturated NaCl solution. Vortexed again.
8. After phase separation, remove the upper phase (aqueous phase) with the aid of a Pasteur glass pipette connected to a suction flask and a vacuum pump.
9. Add again 2.5 mL of saturated NaCl solution, vortex and remove the upper phase as described above.
10. Open a syringe connected to a 0.45 μm PTFE syringe filter (removing the plunger) and add four scoops of dry Na_2SO_4 and two scoops of dry Na_2CO_3 salts.

11. Give about 1 mL of the sample into the syringe using a Pasteur glass pipette (avoid to take the precipitated salt in the glass tubes, and importantly avoid the transfer of droplets from the water phase). Carefully place back the plunger into the syringe.
12. Filter the sample directly into a GC vial and seal it to be ready for gas chromatography.
13. Keep the samples at 4 °C until measurement.
14. The preparation of standards for the calibration curves is exactly the same, with the difference that known quantities of PHA are weighed in the Pyrex tubes (in the range of 4 mg of PHA).

The GC analysis is carried out as follows:

- A volume of 1 μL is injected at 250 °C with a split ratio of 1:100 to reduce the harmful effect of the acid on the GC column. The temperature is increased from 80 °C to 240 °C at a rate of 10 °C min^{-1}. Helium is used as carrier gas (3 mL min^{-1}). A flame ionization detector (FID) at 285 °C is used for detection.
- A calibration curve is built for each of the PHA standards. The ratio of the areas of the considered PHA compounds and the internal standard is used for recalculating the amount of PHA in the samples. Therefore, for each PHA compound expected, a calibration curve is necessary.
- The wt% of PHA is obtained by comparison of the mass of PHA present in the sample (calculated from the GC result) to the initial mass of freeze-dried biomass weighed in the Pyrex tube.

5.2 Flow cytometry (FCM)

The intracellular PHA can be quantified by flow cytometry. The method described by Karmann, Follonier, Bassas-Galia, Panke, and Zinn (2016) allows the discrimination of lipophilic cell debris from intact cells containing PHA by the utilization of stained DNA as triggering signal. The fluorescence signal may be used for a quantification of the cellular PHA content when a calibration with GC is carried out.

Equipment

- Flow cytometer (e.g., BD Accuri C6 FCM, BD Bioscience, Erembodegern, Belgium) with excitation wavelengths at 488 nm and 640 nm and band filters 530/30 nm and 675/25 nm for detection emission signals.

Chemicals

- NaCl (Sigma-Aldrich)

- Tris(hydroxymethyl)-aminomethane (Tris) (Sigma-Aldrich)
- $MgCl_2$ (Sigma-Aldrich)
- Ethylenediaminetetraacetic acid (EDTA) (Sigma-Aldrich)
- Demineralized H_2O/MilliQ water
- BODIPY 493/503 (Sigma-Aldrich)
- SYTO 62 (Sigma-Aldrich)
- Dimethyl sulfoxide (DMSO, Sigma-Aldrich)

Protocol

1. Take around 1 mL of bioreactor sample and immediately put it on ice.
2. Dilute the sample to approximately 10^4 cells μL^{-1} with 50% PBS (Phosphate buffered saline PBS, per liter of demineralized H_2O: 9 g NaCl, 1.211 g Tris(hydroxymethyl)-aminomethane (Tris), and 0.952 g $MgCl_2$ with 1.461 g ethylenediaminetetraacetic acid (EDTA), adjust pH to 7.4).
3. Add 5 μL of staining mix (10 μg mL^{-1} BODIPY 493/503, 5 μM SYTO 62 in DMSO) to 500 μL diluted sample. BODIPY 493/503 is used as PHA stain and SYTO 62 as DNA stain.
4. Incubate the sample at 30 °C for 4–6 min while protecting the sample from light.
5. In a flow cytometer, a low flow rate of around 11 μL min^{-1} is selected and the trigger set on the fluorescence channel recording the DNA derived fluorescence for analyzing bacterial cells. The dyes are excited at 488 and 640 nm and the fluorescence signal is collected with a 530/30 nm and a 675/25 nm filter for BODIPY 493/503 and SYTO 62, respectively. Filtered Milli-Q water is used as sheath fluid.
6. Make a calibration curve using results from PHA derived fluorescence from FCM vs PHA content (wt%) from GC. A correlation (R^2) of 0.90–0.95 is expected. This is a rapid method for at line quantification of PHA production.

5.3 Nuclear magnetic resonance spectroscopy (^{1}H NMR)

Standard ^{1}H NMR spectroscopy is a straightforward method to analyze the polymer composition and purity within the limits of detection of the method.

Equipment

- ^{1}H NMR spectrometer (e.g., Bruker-400 (400 MHz) with Avance II console and Top-spin processing software)

Chemicals

- Deuterochloroform containing 1% (v/v) tetramethylsilan ($CDCl_3$, Sigma-Aldrich)

Sample preparation

1. Dissolve about 10–20 mg of purified PHA in 700 μL of $CDCl_3$ and transfer this solution into an NMR sample tube for the NMR analysis.

^1H NMR interpretation

Due to the large variety of polymercompositions it is difficult to give a general method to interpret the ^1H NMR results. However it should be noted that for scl-PHA the chemical shift of the methanetriyl group multiplet (chiral carbon in the monomer) between 5.0 and 5.5 ppm allows to calculate the fraction of different monomers with little or no signals interfering (one should make sure that samples are free of DCM with a chemical shift of 5.3 ppm). Different chemical shifts for the methyl groups between 0.5 and 1.5 ppm can also be interpreted for the calculation of monomer composition.

5.4 Gel permeation chromatography (GPC)

In practice, for the GPC measurements, a series of 10 polystyrene standards serve to establish a calibration curve which allows the determination of the molecular weight and the molecular weight distribution of the unknown sample relative to polystyrene. The detection with a refractive index detector only does not allow the determination of absolute molecular weights. For the detection of the separated PHA polymers, a refractive index detector is suitable. The use of an internal flow marker allows to compensate possible fluctuations between different runs.

Equipment

- Chromatographic system (e.g., System consisting of a Waters 717plus autosampling unit and a Waters 515 HPLC pump, equipped with an Agilent PLgel MiniMIX-C 250 × 4.6 mm column for separation and a PLgel 5 μm MiniMIX-C 50 × 4.6 mm as precolumn; all available from Agilent). A column oven set to 40 °C and an Agilent 1260 Infinity refractive index detector for the detection of the separated polymers.

Chemicals

- Chloroform ($CHCl_3$, Sigma-Aldrich)
- Butylated hydroxytoluene as internal flow marker (BHT, Sigma-Aldrich)
- Polystyrene (PS) standards within a Mw range of 500–3,000,000 Da (Polymer Standards Service)

Protocol

1. Dissolve 5 mg of the polymers (including polystyrene standards) in 1 mL of chloroform containing 0.5 mg mL^{-1} of butylated hydroxytoluene as internal flow marker 12–24 h prior to filtration to ensure complete dissolution.

2. Set the flow rate to $0.3\,\mathrm{mL\,min^{-1}}$ and temperature the column and the detector at 40 °C.
3. Inject 10 μL of sample or PS standard.
4. Using the BHT signal as the flow rate marker calculate average molecular weight ($\overline{M_w}$ and $\overline{M_n}$) and molecular weight distribution (PDI) based on the PS calibration values.

5.5 Differential scanning calorimetry (DSC)

DSC allows determining the changes in the specific heat capacity of different materials over a wide range of temperatures.

For the determination of the glass transition temperature (T_g) of polyhydroxyalkanoates, differential scanning calorimetry offers a suitable and convenient thermoanalytical method. The DSC instrument should be equipped with a sufficient cooling device to rapidly cool down the sample after initial melting. This way, crystallization is suppressed and the amount of amorphous material in the sample allows observing the glass transition during the second heating segment. The information on glass transition can also be withdrawn from the cooling segment of the thermoanalysis. However, to establish a sufficiently fast and well-controlled cooling rate is often very challenging, hence it is recommended to use the second heating segment instead. With respect to the determination of the melting temperature range, the below described temperature program needs to be adapted for each individual type of polymer.

Equipment

- Differential scanning calorimeter (e.g., DSC-3 +, Mettler Toledo with STARe acquisition and processing software)
- 40 μL Aluminum crucibles with lid (Mettler Toledo)
- Sealing press for crucibles (Mettler Toledo)

Protocol

1. Transfer about 10 mg of sample material into a 40 μL aluminum crucible for the DSC measurement.
2. Typically, the crucible is closed with a lid to avoid sample loss during handling and analysis.
3. Inert gases such as argon or nitrogen can be used as the method gas ($50\,\mathrm{mL\,min^{-1}}$ of nitrogen are recommended). An adequate sampling rate should be chosen (1 s recommended).
4. A suitable temperature program for a heating-cooling-heating cycle for the thermoanalysis may look as follows:

   First segment: 25–180 °C with a heating rate of $50\,\mathrm{K\,min^{-1}}$

Second segment: 180 °C for 3 min
Third segment: 180 to −80 °C with a cooling rate of 50 K min^{-1}
Fourth segment: −80 °C for 10 min
Fifth segment: −80 to 180 °C with a heating rate of 10 K min^{-1}

5. Transfer the sample into the DSC oven and start the temperature program.
6. Information about the T_g can be withdrawn from segment five of the above given example for a temperature program. An endothermal step in the thermogram indicates the glass transition of the sample and commonly step height and inflection point are estimated by standardized tangent methods (DIN 51007, ASTM E 1356, IEC 1006). If the information of step height is used, commonly expressed as J g^{-1} K^{-1} the amount of analyzed sample needs to be determined during the sample preparation with highest precision (0.001 mg recommended).

6. Conclusions

Polyhydroxyalkanoates are a very versatile set of polyesters that change their physico-chemical properties with their monomeric unit composition. These properties can be efficiently controlled during biosynthesis in many bacteria by the appropriate and continuous supply of precursor molecules during chemostat cultivation and concomitant multiple nutrient limited growth. Thus, no critical concentrations of substrates are to be detected in the supernatant, resulting in no toxic effects on cells. This means that many different substrates can be potentially used for only PHA biosynthesis without utilization for biomass formation. Appropriate analytical tools, in particular flow cytometry enable quick quantification of intracellular PHA contents. This method is also very useful to detect contaminations by other microorganisms and loss of productivity due to reduction of strain performance. The described PHA extraction and purification method, which is suitable for medical applications, should be replaced by more environmentally friendly methods in future. However, in our hands we found that—without exception—all biosynthesized PHAs can be extracted from freeze-dried biomass at a high yield using DCM, which is very important in this field of research.

Acknowledgments

We thank Monica Bassas-Galià, Simon Crelier, Rémy Dufresne, Roland Durner, Thomas Egli, Stéphanie Follonier, Patrick Furrer, Roland Hany, René Hartmann, Pascal Jacquemettaz, Franka Kalman, Stephanie Karmann, Luc Malbois, Fabrice Micaux, Umberto Piantini, Qun Ren, Stefan Willen, and Thierry Udrisard for their help in setting up basics of the herein described techniques.

References

Anderson, A. J., Haywood, G. W., & Dawes, E. A. (1990). Biosynthesis and composition of bacterial poly(hydroxyalkanoates). *International Journal of Biological Macromolecules, 12*(April), 102–105.

Baidurah, S., Kubo, Y., Kuno, M., Kodera, K., Ishida, Y., Yamane, T., et al. (2015). Rapid and direct compositional analysis of poly(3-hydroxybutyrate-*co*-3-hydroxyvalerate) in whole bacterial cells by thermally assisted hydrolysis and methylation-gas chromatography. *Analytical Sciences, 31*(2), 79–83. https://doi.org/10.2116/analsci.31.79.

Betancourt, A., Yezza, A., Halasz, A., Tra, H. V., & Hawari, J. (2007). Rapid microwave assisted esterification method for the analysis of poly-3-hydroxybutyrate in *Alcaligenes latus* by gas chromatography. *Journal of Chromatography A, 1154*(1–2), 473–476.

Bloembergen, S., Holden, D. A., Hamer, G. K., Bluhm, T. L., & Marchessault, R. H. (1986). Studies of composition and crystallinity of bacterial poly(ß-hydroxybutyrate-co-ß-hydroxyvalerate). *Macromolecules, 19*, 2865–2871.

Brandl, H., & Püchner, P. (1991). Biodegradation of plastic bottles made from 'Biopol' in an aquatic ecosystem under in situ conditions. *Biodegradation, 2*(4), 237–243. https://doi.org/10.1007/BF00114555.

Braunegg, G., Sonnleitner, B., & Lafferty, R. M. (1978). A rapid gas chromatographic method for the determination of poly-3-hydroxybutyric acid in microbial biomass. *European Journal of Applied Microbiology and Biotechnology, 6*, 29–37.

Cui, Y., Barford, J. P., & Renneberg, R. (2007). Gas chromatographic determination of poly(3-hydroxybutyrate) with alkaline hydrolysis and acid esterification. *Analytical Letters, 40*(15), 2915–2924.

Doi, Y., Tamaki, A., Kunioka, M., & Soga, K. (1988). Production of copolyesters of 3-hydroxybutyrate and 3-hydroxyvalerate by *Alcaligenes eutrophus* from butyric and pentanoic acids. *Applied Microbiology and Biotechnology, 128*, 330–334.

Durner, R., Zinn, M., Witholt, B., & Egli, T. (2001). Accumulation of poly[(*R*)-3-hydroxyalkanoates] in *Pseudomonas oleovorans* during growth in batch and chemostat culture with different carbon sources. *Biotechnololgy and Bioengineering, 72*(3), 278–288.

Egli, T. (2015). Microbial growth and physiology: A call for better craftsmanship. *Frontiers in Microbiology, 6*, 1–12. https://doi.org/10.3389/fmicb.2015.00287.

Egli, T., Mason, A., & Snozzi, M. (1993). Special issue—Kinetics, dynamics and physiology of microbial growth. *Antonie van Leeuwenhoek International Journal of General and Molecular Microbiology, 63*(3–4). 211–211.

Egli, T., & Zinn, M. (2003). The concept of multiple-nutrient-limited growth of microorganisms and its application in biotechnological processes. *Biotechnology Advances, 22*, 35–43.

Europe, P. (2016). *Platics—The facts 2016.* Retrieved from http://www.plasticseurope.org/Document/plastics--the-facts-2016-15787.aspx?Page=DOCUMENT&FolID=2.

Follonier, S., Goyder, M. S., Silvestri, A. C., Crelier, S., Kalman, F., Riesen, R., et al. (2014). Fruit pomace and waste frying oil as sustainable resources for the bioproduction of medium-chain-length polyhydroxyalkanoates. *International Journal of Biological Macromolecules, 71*, 42–52. https://doi.org/10.1016/j.ijbiomac.2014.05.061.

Follonier, S., Riesen, R., & Zinn, M. (2015). Pilot-scale production of functionalized mcl-PHA from grape pomace supplemented with fatty acids. *Chemical and Biochemical Engineering Quarterly, 29*(2), 113–121.

Furrer, P., Hany, R., Rentsch, D., Grubelnik, A., Ruth, K., Panke, S., et al. (2007). Quantitative analysis of bacterial medium-chain-length poly([*R*]-3-hydroxyalkanoates) by gas chromatography. *Journal of Chromatography A, 1143*(1–2), 199–206. https://doi.org/10.1016/j.chroma.2007.01.002.

Furrer, P., Panke, S., & Zinn, M. (2007). Efficient recovery of low endotoxin medium-chain-length poly([*R*]-3-hydroxyalkanoate) from bacterial biomass. *Journal of Microbiological Methods, 69*(1), 206–213. https://doi.org/10.1016/j.mimet.2007.01.002.

Hartmann, R., Hany, R., Geiger, T., Egli, T., Witholt, B., & Zinn, M. (2004). Tailored biosynthesis of olefinic medium-chain-length poly [(*R*)-3-hydroxyalkanoates] in *Pseudomonas putida* GPo1 with improved thermal properties. *Macromolecules*, *37*(18), 6780–6785.

Hartmann, R., Hany, R., Pletscher, E., Ritter, A., Witholt, B., & Zinn, M. (2006). Tailor-made olefinic medium-chain-length poly[(*R*)-3-hydroxyalkanoates] by *Pseudomonas putida* GPo1: Batch versus chemostat production. *Biotechnology and Bioengineering*, *93*(4), 737–746.

Hartmann, R., Hany, R., Witholt, B., & Zinn, M. (2010). Simultaneous biosynthesis of two copolymers in *Pseudomonas putida* GPo1 using a two-stage continuous culture system. *Biomacromolecules*, *11*(6), 1488–1493. https://doi.org/10.1021/bm100118t.

Holmes, P. A. (1985). Applications of PHB—A microbially produced biodegradable thermoplastic. *Physics in Technology*, *16*, 32. https://doi.org/10.1088/0305-4624/16/1/305.

Jendrossek, D. (2009). Polyhydroxyalkanoate granules are complex subcellular organelles (carbonosomes). *Journal of Bacteriology*, *191*(10), 3195–3202. https://doi.org/10.1128/jb.01723-08.

Jendrossek, D., & Pfeiffer, D. (2014). New insights in the formation of polyhydroxyalkanoate granules (carbonosomes) and novel functions of poly(3-hydroxybutyrate). *Environmental Microbiology*, *16*(8), 2357–2373. https://doi.org/10.1111/1462-2920.12356.

Karmann, S., Follonier, S., Bassas-Galia, M., Panke, S., & Zinn, M. (2016). Robust at-line quantification of poly(3-hydroxyalkanoate) biosynthesis by flow cytometry using a BODIPY 493/503-SYTO 62 double-staining. *Journal of Microbiological Methods*, *131*, 166–171. https://doi.org/10.1016/j.mimet.2016.10.003.

Koller, M., Salerno, A., & Braunegg, G. (2013). Polyhydroxyalkanoates: Basics, production and applications of microbial biopolyesters. In *Bio-based plastics* (pp. 137–170). John Wiley & Sons Ltd.

Kosseva, M. R., & Rusbandi, E. (2018). Trends in the biomanufacture of polyhydroxyalkanoates with focus on downstream processing. *International Journal of Biological Macromolecules*, *107*, 762–778. https://doi.org/10.1016/j.ijbiomac.2017.09.054.

Lageveen, R. G., Huisman, G. W., Preusting, H., Ketelaar, P., Eggink, G., & Witholt, B. (1988). Formation of polyesters by *Pseudomonas oleovorans*: Effect of substrates on formation and composition of poly-(*R*)-3-hydroxyalkanoates and poly-(*R*)-3-hydroxyalkenoates. *Applied and Environmental Microbiology*, *54*, 2924–2932.

Lee, S. Y. (1996). Bacterial polyhydroxyalkanoates. *Biotechnology and Bioengineering*, *49*(1), 1–14. https://doi.org/10.1002/(SICI)1097-0290(19960105)49:1<1::AID-BIT1>3.0.CO;2-P.

Li, J., Yun, H., Gong, Y. D., Zhao, N. M., & Zhang, X. F. (2005). Effects of surface modification of poly(3-hydroxybutyrate-co-3-hydroxyhexanoate) (PHBHHx) on physicochemical properties and on interactions with MC3T3-E1 cells. *Journal of Biomedical Materials Research Part A*, *75A*(4), 985–998.

Monod, J. (1950). La technique de culture continue; théorie et applications. *Annales de l'Institut Pasteur*, *79*(4), 390–410.

Müller, R.-J., Kleeberg, I., & Deckwer, W.-D. (2001). Biodegradation of polyesters containing aromatic constituents. *Journal of Biotechnology*, *86*(2), 87–95. https://doi.org/10.1016/S0168-1656(00)00407-7.

Narancic, T., & O'Connor, K. E. (2017). Microbial biotechnology addressing the plastic waste disaster. *Microbial Biotechnology*, *10*(5), 1232–1235. https://doi.org/10.1111/1751-7915.12775.

Novick, A., & Szilard, L. (1950). Description of the chemostat. *Science*, *112*, 715–716.

Portillo, F., Yashchuk, O., & Hermida, É. (2016). Evaluation of the rate of abiotic and biotic degradation of oxo-degradable polyethylene. *Polymer Testing*, *53*, 58–69. https://doi.org/10.1016/j.polymertesting.2016.04.022.

Ribitsch, D., Yebra, A. O., Zitzenbacher, S., Wu, J., Nowitsch, S., Steinkellner, G., et al. (2013). Fusion of binding domains to *Thermobifida cellulosilytica* cutinase to tune sorption characteristics and enhancing PET hydrolysis. *Biomacromolecules, 14*(6), 1769–1776. https://doi.org/10.1021/bm400140u.

Riis, V., & Mai, W. (1988). Gas chromatographic determination of poly-β-hydroxybutyric acid in microbial biomass after hydrochloric acid propanolysis. *Journal of Chromatography A, 445*, 285–289.

Staudinger, H. (1920). Über Polymerisation. *Berichte der deutschen chemischen Gesellschaft (A and B Series), 53*(6), 1073–1085. https://doi.org/10.1002/cber.19200530627.

Wampfler, B., Ramsauer, T., Kehl, K., Zinn, M., & Thöny-Meyer, L. (2010). Application of activated charcoal in the downstream processing of bacterial olefinic poly(3-hydroxyalkanoates). *Chimia, 64*(11), 784–788. https://doi.org/10.2533/chimia.2010.784.

Wei, R., & Zimmermann, W. (2017). Biocatalysis as a green route for recycling the recalcitrant plastic polyethylene terephthalate. *Microbial Biotechnology, 10*(6), 1302–1307. https://doi.org/10.1111/1751-7915.12714.

Zinn, M., Durner, R., Zinn, H., Ren, Q., Egli, T., & Witholt, B. (2011). Growth and accumulation dynamics of poly(3-hydroxyalkanoate) (PHA) in *Pseudomonas putida* GPo1 cultivated in continuous culture under transient feed conditions. *Biotechnology Journal, 6*(10), 1240–1252. https://doi.org/10.1002/biot.201100219.

Zinn, M., & Hany, R. (2005). Tailored material properties of polyhydroxyalkanoates through biosynthesis and chemical modification. *Advanced Engineering Materials*, 7(5), 408–411.

Zinn, M., Weilenmann, H. U., Hany, R., Schmid, M., & Egli, T. (2003). Tailored synthesis of poly([*R*]-3-hydroxybutyrate-*co*-3-hydroxyvalerate) (PHB/HV) in *Ralstonia eutropha* DSM 428. *Acta Biotechnologica, 23*(2–3), 309–316.

Zinn, M., Witholt, B., & Egli, T. (2001). Occurrence, synthesis and medical application of bacterial polyhydroxyalkanoate. *Advanced Drug Delivery Reviews, 53*(1), 5–21.

Zinn, M., Witholt, B., & Egli, T. (2004). Dual nutrient limited growth: Models, experimental observations, and applications. *Journal of Biotechnology, 113*, 263–279.

CHAPTER FIVE

Biocatalytic synthesis of polylactate and its copolymers by engineered microorganisms

So Young Choi[a,b,c,d,†], In Jin Cho[a,b,c,†], Youngjoon Lee[a,b,c], Seongjin Park[a,b,c], Sang Yup Lee[a,b,c,d,*]

[a]Department of Chemical and Biomolecular Engineering (BK21 Plus Program), Korea Advanced Institute of Science and Technology (KAIST), Daejeon, Republic of Korea
[b]Metabolic and Biomolecular Engineering National Research Laboratory and Institute for the BioCentury, KAIST, Daejeon, Republic of Korea
[c]Systems Metabolic Engineering and Systems Healthcare Cross-Generation Collaborative Laboratory, KAIST, Daejeon, Republic of Korea
[d]Applied Science Research Institute, KAIST, Daejeon, Republic of Korea
*Corresponding author: e-mail address: leesy@kaist.ac.kr

Contents

Abstract

Poly(lactate), also called poly(lactic acid) or poly(lactide) [PLA], has been one of the most attractive bio-based polymers since it possesses desirable material properties for its use in general performance plastics in addition to biodegradability and biocompatibility. PLA has been produced by biological and chemical hybrid process comprising microbial fermentation for lactate (LA) production followed by purification and chemical

† Authors with equal contribution.

Methods in Enzymology, Volume 627
ISSN 0076-6879
https://doi.org/10.1016/bs.mie.2019.04.032

polymerization process of LA. Recently, the direct one-step fermentative processes for production of PLA and several LA-containing polyesters have been developed by employing metabolically engineered microorganisms. Since natural microorganisms cannot produce the LA-containing polymers, several engineering strategies have been employed together based on the polyhydroxyalkanoate (PHA) biosynthesis system.

In this chapter, we summarize strategies and procedures on developing the engineered microorganisms producing PLA and its copolymers, cultivating the cells, and extracting the polymers from the cells. Focuses were given on construction of enzymatic polymerization process of LA: design of metabolic pathway for PLA by mimicking PHA biosynthetic pathway, examination of possible enzymes, and engineering of the enzymes for better performances. This synthetic pathway has been established in a microorganism producing LA that enabled one-step fermentative production of LA-containing polyesters from carbohydrates derived from renewable biomass. Polymer production has been further enhanced by implementing strain engineering to concentrate the metabolic fluxes toward PLA formation. In addition, various monomers such as glycolate, 2-hydroxybutyrate, and phenyllactate have been copolymerized with LA by the microbial system. These fermentative production systems developed by using the engineered microorganisms can be versatile and sustainable platforms for the production of LA-containing polyesters and other non-natural polymers.

Abbreviations

2H3MV	2-hydroxy-3-methylvalerate
2H4PhBA	2-hydroxy-4-phenylbutyrate
2HB	2-hydroxybutyrate
2HIC	2-hydroxyisocaproate
2HIV	2-hydroxyisovalerate
3H3PhPA	3-hydroxy-3-phenylpropionate
3HB	3-hydroxybutyrate
3HP	3-hydroxypropionate
4HB	4-hydroxybutyrate
4HBeA	4-hydroxybenzoate
4HMA	4-hydroxymandelate
4HPhLA	4-hydroxyphenyllactate
5HV	5-hydroxyvalerate
6HHx	6-hydroxyhexanoate
Bct	butyryl-CoA transferase
DTNB	5,5′-dithiobis-(2-nitrobenzoic acid)
GA	glycolate
GC	gas chromatography
HA	hydroxycarboxylic acid
HadA	isocaprenoyl-CoA:2-hydroxyisocaproate CoA-transferase
LA	lactate
LB	Luria-Bertani
LC	liquid chromatography
LC-MS	liquid chromatography-mass spectrometry
MA	mandelate

MCL	medium chain length
NMR	nuclear magnetic resonance
PA	polyamide
Pct	propionyl-CoA transferase
PE	polyethylene
PET	poly(ethylene terephthalate)
PHA	polyhydroxyalkanoate
PhLA	phenyllactate
PLA	poly(lactate)
PLGA	poly(lactate-*co*-glycolate)
PP	polypropylene
PTFE	poly(tetrafluoroethylene)
SCL	short chain length
TCA	trichloroacetic acid

1. Introduction

In recent years, our concerns on the fossil resource depletion and environmental pollution have been increasing. Also, pollution caused by the accumulation of plastic wastes has become one of serious problems threatening our environment, marine ecosystem and human health. Plastic is the one of most widely used materials in our lives as it has numerous applications such as food packaging, containers, automobiles, resins, additives, and medical devices. Most plastics are generally derived from petroleum and featured by their durable characteristic, which is attributed to their non-degradability. As huge amount of plastics are produced worldwide (about 335 million tons in 2016) (PlasticsEurope, https://www.plasticseurope.org) and disposed inappropriately, the plastic wastes have not been properly recycled (only about 10% of plastics have been recycled) and accumulated in landfills and ocean (Garcia & Robertson, 2017). In addition, the microplastics produced for certain applications and those generated through weathering process of plastic wastes have been reported to affect human health (do Sul & Costa, 2014). Thus, the urgent actions are required to address these problems. As one of promising alternatives, bio-based plastics have been extensively studied. Bio-based plastics generally mean the plastics made through the bio-based process using biomass instead of petroleum. However, those plastics made partially from biomass are also called bio-based plastics. It can include several well-known plastics: bio-based poly(ethylene terephthalate) [PET], bio-based polyethylene (PE), poly(trimethylene terephthalate) [PTT], bio-based polyamide (PA) (for examples, PA410, PA11, PA1010, and PA610), polycaprolactone (PCL), poly(butylene succinate) [PBS], poly(lactate) [PLA], and polyhydroxyalkanoate (PHA). These bioplastics are considered as sustainable and environmentally friendly materials that can reduce the emission of CO_2 as it is being captured by growing biomass and can be composted without emission of harmful chemicals.

For the production of bio-based polymers, establishing the efficient biorefinery which processes biomass to useful chemicals is the key. Microorganisms are generally used as cell factories that produce a wide range of chemicals used as monomers for polymer synthesis, or even polymers directly (Chung et al., 2015; Lee, Cho, Choi, & Lee, 2019; Lee, Kim, et al., 2019). Thus, development of efficient microbial cell factories determines the overall production cost of the bio-based polymers. Numerous approaches have been implemented to enhance the production performances of microbial strains. Especially, there have been significant advances in metabolic engineering technology that optimizes the metabolic fluxes while considering the overall bioprocess to produce chemicals of interest at industrial scale (Lee, Cho, et al., 2019, Lee, Kim, et al., 2019). Some of the key considerations include constructing efficient biosynthetic pathways, reducing the toxicity of intermediates, enhancing the tolerance toward the target chemicals, modulating the negative and positive regulations, balancing the cofactors required, reinforcing the metabolic fluxes toward a desired product, eliminating competing pathways, and optimizing the culture condition including carbon source, other nutritional components, nutrient feeding strategy, dissolved oxygen, and pH (Lee, Cho, et al., 2019; Lee & Kim, 2015). A variety of molecular biological and computational tools are employed to systematically analyze cell metabolism under different culture conditions and genetically manipulate the microbial strains to develop a cell factory possessing desired performance of high titer, yield and productivity (Chae, Choi, Kim, Ko, & Lee, 2017).

In this chapter, we describe detailed strategies and protocols for developing microorganisms by enzyme and metabolic engineering for the production of PLA and its copolymers. Together with the strategies for engineering microorganisms, we also present the procedures for cultivating cells and extracting the polymers from the cells. First, the key enzymes required for polymerizing lactate (LA) are discussed in detail. Then, the metabolic engineering strategies to improve fermentative production of LA-containing polyesters from renewable biomass are summarized for PLA and various copolymers.

1.1 Poly(lactate) [PLA]

PLA is a bio-based and biodegradable thermoplastic polyester. It was first synthesized via condensation of lactic acid by distillation of water in the early 1800s. Until 1990s, several methods for synthesizing PLA were developed by the companies, Dupont and Cargill. Cargill developed the process for producing high molecular weight PLA. Based on this technology, PLA further was commercially produced by a joint venture of Cargill and Dow Chemical, named as Cargill Dow Polymer LLC, which is currently known as NatureWorks LLC in the United States (Sin, Rahmat, & Rahman, 2012; Vink, Rabago, Glassner, & Gruber, 2003). PLA has been the most attractive biopolymer since it has biodegradability, biocompatibility, clarity, and superior barrier properties along with suitable material properties for general performance plastics (Drumright, Gruber, & Henton, 2000; Vink et al., 2003). The mechanical properties of PLA are similar with several petroleum-based polymers such as polystyrene (PS), polypropylene (PP), and PET (Auras, Harte, & Selke, 2004). Currently, PLA can be processed for various application types such as 3D printing filaments, fiber, film, cup, food container, packaging, textiles, medical uses such as surgical sutures, and tissue scaffolds. In 2018, the global production capacities of PLA reached about 217,000 tons and is the fourth most produced bioplastic followed by bio-based PET, bio-based PA, and starch blends. The production is expected to increase by 60% in 2023 compared to 2018 (https://www.european-bioplastics.org/market). Most of the commercial PLA are produced by NatureWorks. The company is producing about 150,000 tons of PLA annually. China's Hisun Biomaterials is the second largest company on producing PLA with an annual capacity of 15,000 tons. The Netherlands Synbra Technology also annually produces 5000 tons of PLA by using LA provided by Purac, which produces 75,000 tons of LA per year. In 2010, Galactic, a Belgian LA manufacturer, inaugurated PLA production plant having capacity of 1500 tons per year in Belgium by establishing the joint venture with Total Petrochemicals (Futerro) and is planning to expand its capacity in the future.

The monomer LA exists as L- and D-forms having different optical activity. LA can be synthesized through two processes: chemical synthesis from the raw materials based on fossil fuels such as coal, oil, natural gas, and fermentative synthesis from the carbohydrates such as corn and sugar cane (Benninga, 1990; Drumright et al., 2000). The chemically synthesized LA is a racemic mixture of D- and L-form of LA. On the other hand, the biologically synthesized LA by fermenting microorganism can be almost pure

L-form (99.9%). Thus, the most of LA have been produced through the bio-processes. Based on the LA produced by fermentation, several processes for polymerization have been developed. PLA can be synthesized by the polycondensation of LA or ring-opening polymerization (ROP) of a cyclic dimer of LA, lactide. Since the direct condensation of LA has several disadvantages such as the tricky reaction conditions (vacuum condition with high temperature), and difficulties in removing water (by-product of the reaction) that leads to low molecular weight of PLA, ROP is mainly used for PLA production. Based on the ROP, NatureWorks has developed large-scale process that enables low cost and continuous production of high molecular weight PLA (Fig. 1). In this process, aqueous LA is condensed to yield low molecular weight PLA (about 1000–5000 Da) which is subsequently converted to a mixture of lactide stereoisomers using tin catalyst. The lactide mixture is purified by vacuum distillation, and finally lactide is polymerized to form high molecular weight PLA (above 100 kDa) through ROP reaction (Castro-Aguirre, Iñiguez-Franco, Samsudin, Fang, & Auras, 2016) (Fig. 1).

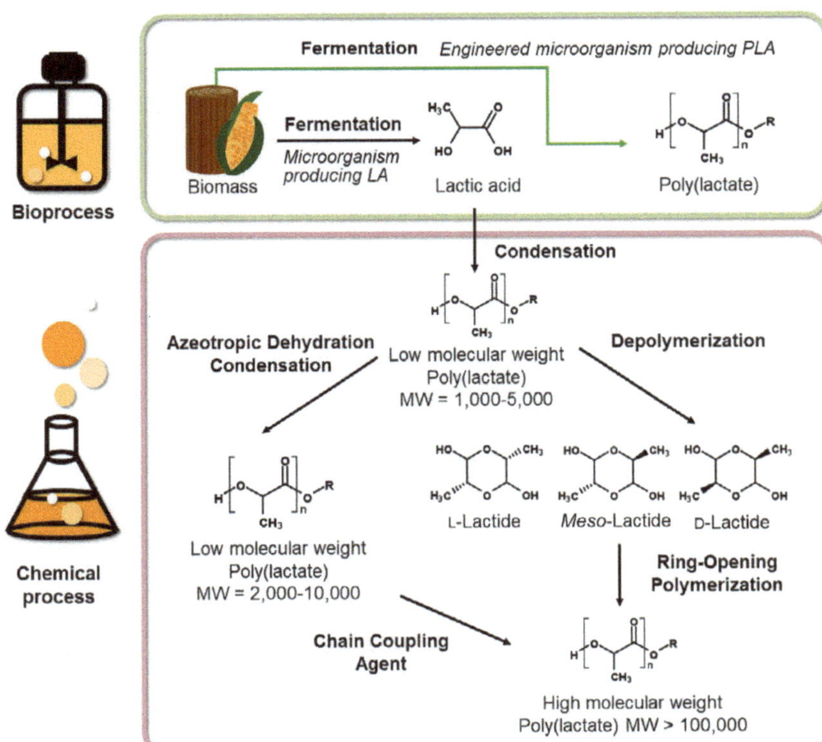

Fig. 1 Production process of PLA.

1.2 Polyhydroxyalkanoate (PHA)

PHAs are a group of polyesters composed of various hydroxycarboxylic acids (HAs) and accumulated as intracellular granules by numerous microorganisms when they are deficient in important elements such as nitrogen (Lee, 1996; Madison & Huisman, 1999). The native PHA producers can accumulate and/or decompose PHA *in vivo* according to their circumstances. PHAs play important roles in metabolism and fitness of the cells as the carbon, and energy storage materials. Regarding the materials properties, PHAs possess comparable mechanical and thermal properties with the petroleum-based commodity plastics such as PE and PP. Therefore, PHAs can be processed for various types of products including packaging, adhesives, fabrics, and films (Bugnicourt, Cinelli, Lazzeri, & Alvarez, 2014). In addition, their biodegradability and biocompatibility make PHAs versatile that can be also used for biomedical applications such as absorbable sutures and implants (Valappil, Misra, Boccaccini, & Roy, 2006).

To date, more than 300 different bacteria have been reported to produce PHAs (Muhammadi, Afzal, & Hameed, 2015). Owing to such a wide range of microbial platforms, PHAs can be produced in a sustainable way using various feedstocks such as lignocellulosic hydrolysates derived from agricultural and industrial wastes, glycerol generated from biodiesel production as a waste, waste stream, and C1 carbon sources (methanol, methane, and carbon dioxide) (Favaro, Basaglia, & Casella, 2019). Therefore, PHAs have attracted much attention as sustainable and environmentally friendly alternatives to the petroleum-based synthetic plastics. Up to now, more than 160 kinds of HAs have been identified as monomer constituents of PHAs in the natural and recombinant microorganisms (Agnew & Pfleger, 2013; Steinbüchel & Valentin, 1995). The monomers encompass 3-, 4-, 5-, and 6-hydroxyacyl-CoAs of saturated and unsaturated, and aliphatic and aromatic HAs. Such diversity allows PHAs to have diverse material properties according to the monomer type and composition. For examples, the short chain length (SCL) PHAs such as poly(3-hydroxybutyrate) [poly(3HB)] show rigid and stiff properties while medium chain length (MCL) PHAs consisting of C6-C14 3-hydroxyacyl-CoAs show elastomeric properties with relatively low tensile strength and high elongation to break (Lee, 1996; Muhammadi et al., 2015).

The biosynthetic pathways of PHAs in natural and engineered microorganisms are composed of two steps: (1) hydroxyacyl-CoAs are generated through inherent or engineered metabolic pathways, and (2) the hydroxyacyl-CoAs are polymerized by the polymerase, PHA synthase. Microorganism can synthesize various hydroxyacyl-CoAs through their metabolic pathways (inherent or engineered) that can be related to the central metabolism and others such as glycolysis, TCA cycle, fatty acid degradation and synthesis pathways, and so on (Tan et al., 2014). In addition, PHA synthases have been known for their promiscuous activities that can accept a wide range of substrates. They prefer using 3-hydroxyacyl-CoAs as substrates, but also show activities for 4-, 5-, and 6-hydroxyacyl-CoAs. Thus, many researchers have tried to manipulate microbial strains to synthesize various types of PHAs with desired material properties for various applications (Park, Kim, Kim, Lee, & Lim, 2012). As one of successful examples, the non-natural polyesters like PLA and various LA-containing polymers have been synthesized by engineering the PHA biosynthetic system. The details of the engineering strategies and procedures are described below.

2. Key enzymes for polymerization of lactate

As described above, LA has been commercially produced by microbial fermentation of sugars obtained from renewable biomass (Castro-Aguirre et al., 2016). LA is one of the most abundant HAs in organisms and is mainly synthesized from pyruvate by LA dehydrogenase (LdhA) (Datta & Henry, 2006). However, there has been no report on microorganism producing PLA due to the absence of metabolic pathways for polymerizing LA in nature. Thus, engineering was focused on developing the enzymatic pathways polymerizing LA based on the PHA biosynthetic pathway: LA is converted into lactyl-CoA (as PHA synthase utilizes the CoA thioester form as a substrate), and then lactyl-CoA is polymerized by PHA synthase. The two enzymatic steps are discussed below. As a different approach, several lipases have been reported to mediate the polycondensation of LA in organic solvents (Lassalle & Ferreira, 2008). The reaction condition of lipase (use of organic solvents) makes it difficult to be applied for *in vivo* PLA synthesis. Thus, the details of lipase-based polymerization are not discussed here.

2.1 Synthesis of lactyl-CoA

To convert LA into lactyl-CoA, propionyl-CoA transferase (Pct) was selected based on the previous reports showing Pcts from several microorganisms can transfer CoA from propionyl-CoA or acetyl-CoA to LA (Selmer, Willanzheimer, & Hetzel, 2002). The activities of CoA transferase on LA could be demonstrated by *in vitro* assay using a colorimetric reagent and liquid chromatography-mass spectrometry (LC-MS) analysis (the detailed procedures are summarized below). The two Pcts from *Clostridium propionicum* and *Megasphaera elsdenii* could convert LA to lactyl-CoA using acetyl-CoA as a CoA donor (Taguchi et al., 2008; Yang et al., 2010). However, when they were expressed in *E. coli,* the cell growth was inhibited that can result in inefficient *in vivo* lactyl-CoA synthesis. Thus, *C. propionicum* Pct was evolved though random mutagenesis by error-prone PCR and screening processes (Yang et al., 2010). The two Pct variants, Pct532 (Ala243Thr, and one silent nucleotide mutation of A1200G) and Pct540 (Val193Ala, and four silent nucleotide mutations of T78C, T669C, A1125G, and T1158C), were constructed. Although the variants showed similar or slightly lower specific activities for LA compared to the wild type Pct, they were proven to be beneficial for the *in vivo* PLA production system with better cell growth (Yang et al., 2010).

Recently, other CoA transferases have been employed to generate lactyl-CoA *in vivo.* Butyryl-CoA transferases (Bcts) from *Roseburia* sp., *Eubacterium hallii, Faecalibacterium prausnitzii*, and *Anaerostipes caccae* were reported to show activities on LA in addition to 2-hydroxybutyrate (2HB), and 3HB (David et al., 2017) (Table 1). Furthermore, isocaprenoyl-CoA:2-hydroxyisocaproate (2HIC) CoA-transferase (HadA) from *Clostridium difficile* was also found to be capable of activating LA into lactyl-CoA in addition to its original substrate, 2HIC (Mizuno et al., 2018; Yang et al., 2018). These CoA transferases can activate not only LA, but also other HAs. Thus, if PHA synthase having activities for both LA and second HA monomer is employed, the copolymer can be synthesized. *C. propionicum* Pct can activate 3HB, 2HB, 3-hydroxypropionate (3HP), glycolate (GA), 4-hydroxybutyrate (4HB), 5-hydroxyvalerate (5HV), 6-hydroxyhexanoate (6HHx), and 2-hydroxyisovalerate (2HIV) as well as LA (Choi et al., 2016; Park, Kim, et al., 2012; Park, Lee, et al., 2012; Tran & Charles, 2015; Yang et al., 2011, 2010). *C. difficile* HadA showed activities for a broad range of substrates; aliphatic HAs such as 2HIV, 2-hydroxy-3-methylvalerate (2H3MV), 2HIC, GA, 2HB, 3HB, 4HB, 5HV, and 6HHx and aromatic HAs such as mandelate (MA), 4-hydroxymandelate (4HMA), phenyllactate (PhLA), 4-hydroxyphenyllactate (4HPhLA), 2-hydroxy-4-phenylbutyrate (2H4PhBA), 3-hydroxy-3-phenylpropionate (3H3PhPA), and 4-hydroxybenzoate (4HBeA) (Mizuno et al., 2018; Yang et al., 2018). Other CoA transferases that are not mentioned above can be also applied for the microbial PLA production. For example, Pct from *Cupriavidus necator* H16 might be employed since it showed *in vitro* activities toward LA with 3HB, 3HP, and 4HB and has not yet been employed for the polymer synthesis (Table 1).

Table 1 Enzymes for CoA activation of LA.

Enzyme type	Source	Expression host and comments[a]	Reference
Propionyl-CoA transferase (Pct)	*Clostridium propionicum*	*Escherichia coli* *Cupriavidus necator* *Sinorhizobium meliloti* Mutants Pct532, Pct540[b] Activities on 3HB, 3HP, 2HB, GA, 4HB, 5HV, 6HHx and 2HIV	Yang et al. (2010) Yang et al. (2011) Tran and Charles (2015) Park et al. (2013) Choi et al. (2016)
Propionyl-CoA transferase (Pct)	*Megasphaera elsdenii*	*Escherichia coli* *Corynebacterium glutamicum* Activities on 3HB, 2HB and GA	Taguchi et al. (2008) Song et al. (2012)
Propionyl-CoA transferase (Pct)	*Clostridium perfringens*	*Escherichia coli* Activities on 3HB	Kim et al. (2016)
Propionyl-CoA transferase (Pct)	*Cupriavidus necator* H16	*Escherichia coli* Activities on 3HB, 4HB and 3HP	Lindenkamp, Schürmann, and Steinbüchel (2013) Volodina, Schürmann, Lindenkamp, and Steinbüchel (2014)
Butyryl-CoA transferase (Bct)	*Roseburia* sp.	*Escherichia coli* Activities on 3HB and 2HB	David et al. (2017)
Butyryl-CoA transferase (Bct)	*Eubacterium hallii*	*Escherichia coli* Activities on 3HB and 2HB	David et al. (2017)
Butyryl-CoA transferase (Bct)	*Faecalibacterium prausnitzii*	*Escherichia coli* Activities on 3HB and 2HB	David et al. (2017)
Butyryl-CoA transferase (Bct)	*Anaerostipes caccae*	*Escherichia coli* Activities on 3HB and 2HB	David et al. (2017)
Isocaprenoyl-CoA:2HIC CoA-transferase (HadA)	*Clostridium difficile*	*Escherichia coli* Activities on 2HIV, 2H3MV, 2HIC, GA, 2HB, 3HB, 4HB, 5HV, 6HHx, MA, 4HMA, PhLA, 4HPhLA, 2H4PhBA, 3H3PhPA, 4HBeA	Yang et al. (2018) Mizuno, Enda, Saika, Hiroe, and Tsuge (2018)

[a]The HAs other than LA used as substrates of each enzyme are provided.
[b]Pct532, Pct mutant containing Ala243Thr and a silent mutation of A1200G; Pct540, Pct mutant containing Val193Ala and four silent nucleotide mutations of T78C, T669C, A1125G, and T1158C.

2.1.1 *In vitro enzyme assay*

The protocol is based on the previous studies (Choi et al., 2016; Jung, Kim, Park, & Lee, 2010; Jung & Lee, 2011; Yang et al., 2010, 2018).

2.1.1.1 Buffers and reagents

- Potassium phosphate buffer (pH 7.5)
- LA
- Acetyl-CoA
- 5,5′-Dithiobis-(2-nitrobenzoic acid) (DTNB)
- Oxaloacetate
- Citrate synthase
- Methanol
- Trichloroacetic acid (TCA)
- Ammonium hydroxide

2.1.1.2 DTNB assay (Fig. 2A)

1. 880 μL of 11.11 mM LA in 50 mM potassium phosphate buffer (pH 7.5) and 10 μL of 100 mM acetyl-CoA in 50 mM potassium phosphate buffer (pH 7.5) are mixed with 15 μg of purified CoA transferase.
2. The solution is incubated at 30 °C for 30 min, and then shifted at 95 °C for 10 min to inactivate the CoA transferase.
3. 10 μL of 100 mM oxaloacetate in 50 mM potassium phosphate buffer (pH 7.5) and 100 μL of 5 mM DTNB in 50 mM potassium phosphate buffer (pH 7.5) are mixed to the reaction mixture.
4. 5 μg of citrate synthase is added, and the reaction mixture is incubated at room temperature for 10 min.
5. Absorbance at 412 nm wavelength of the mixture is measured to quantify the remaining acetyl-CoA. The activity of CoA transferase can be calculated by the difference between the remaining acetyl-CoA concentration and initial acetyl-CoA concentration, reaction time and the amount of the CoA transferase (in this case, 15 μg).

2.1.1.3 LC-MS analysis

To directly detect the lactyl-CoA generated from the reaction, lactyl-CoA is purified from the assay mixture by using OASIS HLB SPE cartridge, and then subjected to LC-MS analysis as below.

1. OASIS HLB SPE cartridge is washed by loading 1 mL of methanol and subsequently 1 mL of 0.15% (w/v) TCA methanol solution.
2. 1 mL of reaction mixture of step 2 of Section 2.1.1.2 is loaded on the cartridge.

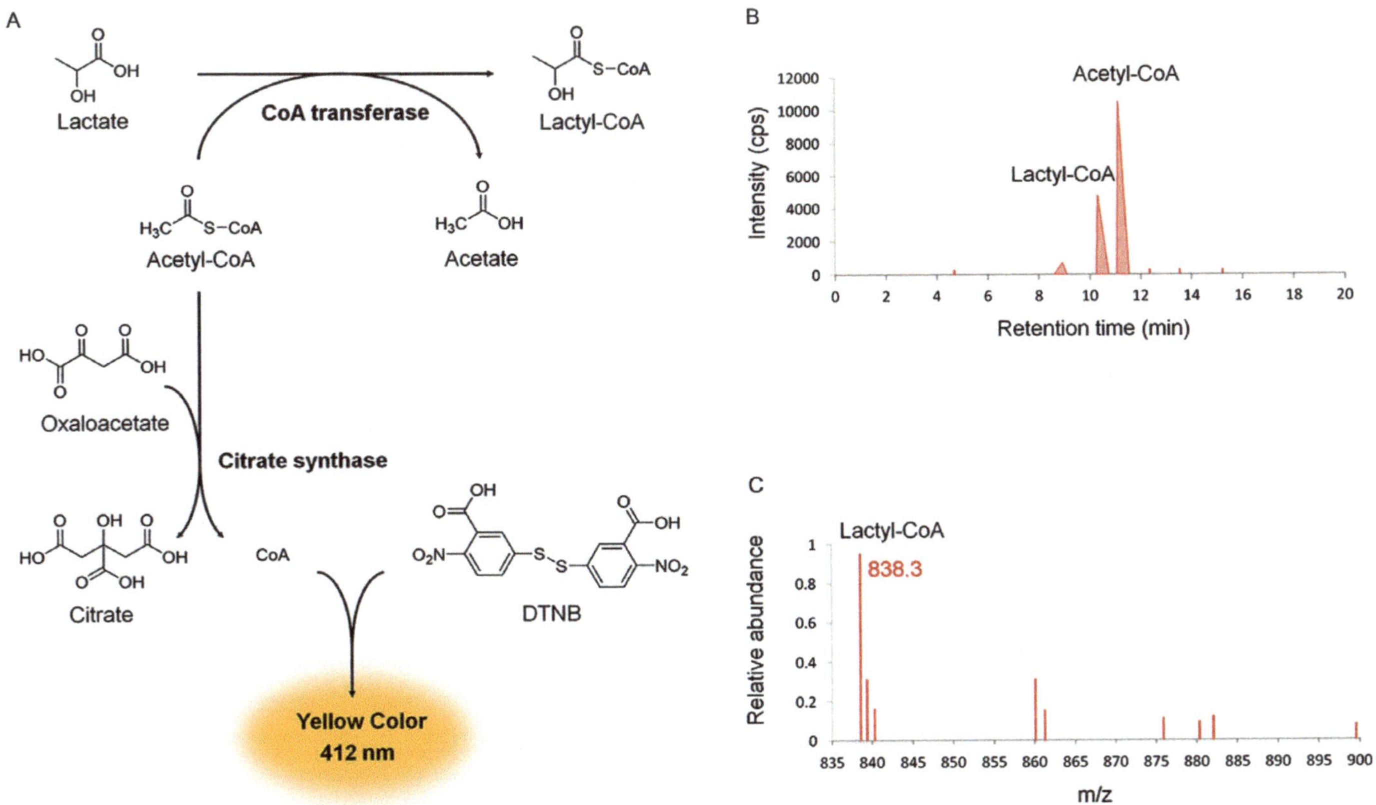

Fig. 2 Schematic diagram of *in vitro* CoA transferase assay. (A) Residual acetyl-CoA after the reaction of CoA transferase subsequently reacts with oxaloacetate to yield citrate and free CoA by citrate synthase. Absorbance of 412 nm wavelength light upon the addition of DTNB allows the quantification of free CoA. (B) LC analysis of the reaction mixture. Lactyl-CoA and acetyl-CoA can be detected at the different retention time. (C) LC-MS analysis of the reaction mixture. The mass peak of 838.3 corresponds to lactyl-CoA. *Panels (B) and (C) are schematically redrawn based on Yang, J. E., Park, S. J., Kim, W. J., Kim, H. J., Kim, B. J., Lee, H., et al. (2018). One-step fermentative production of aromatic polyesters from glucose by metabolically engineered* Escherichia coli *strains.* Nature Communications, *9, 79.*

3. The cartridge is washed by loading 1 mL of 0.15% (w/v) TCA methanol solution.
4. The lactyl-CoA bound to the cartridge is eluted by using 1% (v/v) ammonium hydroxide methanol solution.
5. The eluted solution is completely dried by using a vacuum centrifuge.
6. The dried pellet is dissolved with distilled water and is directly injected to LC-MS (Fig. 2B and C).

2.1.1.4 Notes

- As oxaloacetate is unstable in aqueous solution, it should be stored in a powder form. The oxaloacetate stock solution is prepared just before the essay and is placed in ice bath.
- The procedures for preparing enzyme are encouraged to be kept cold to prevent the loss of enzyme activity.
- All buffers are encouraged to be filtered by 0.45 nm PTFE filter to avoid impurities.

2.2 Polymerization of lactyl-CoA into PLA

To fulfill *in vivo* polymerization of LA, various PHA synthases from different microbial strains have been examined whether they can incorporate lactyl-CoA into PHAs, usually poly(3HB) since it is a representative PHA and can be synthesized relatively easily. The PHA synthase assay can be performed by either *in vivo* or *in vitro* ways in which Pct (or other suitable CoA transferase) is used together for supplying lactyl-CoA to PHA synthase (see Section 2.1 and protocols below).

As natural PHA synthases showed no or only marginal activities on lactyl-CoA (Cho, Park, Lee, & Jung, 2006), enzyme engineering was performed to develop PHA synthase variants with enhanced activity and specificity for lactyl-CoA. The PHA synthases from *Pseudomonas* MBEL sp. 6-19 and *Pseudomonas* sp. 61-3 were selected and mutated through site-directed mutagenesis of the amino acid residues which were reported to affect enzyme substrate specificities and activities (Taguchi et al., 2008; Yang et al., 2010). The four amino acid residues (Glu130, Ser325, Ser477, and Gln481) were found to have significant influence on the enzyme activity for lactyl-CoA (Taguchi et al., 2008; Yang et al., 2011, 2010). Especially, the substitutions of Glu130Asp, Ser325Thr, Ser477Arg/His/Phe/Gly, and Gln481Lys/Met were effective for *in vivo* production of PLA and LA-containing copolymers. All the PHA synthase variants used for polymerizing lactyl-CoA are summarized in Table 2.

Table 2 Enzymes for polymerization of lactyl-CoA.

Enzyme	Source	Mutation	Expression host and comments[a]	Reference
Class II PHA synthase	*Pseudomonas* sp. MBEL 6-19	E130D/Q481K	*Escherichia coli*	Yang et al. (2010)
		E130D/Q481M	*Cupriavidus necator*	Yang et al. (2011)
		E130D/S325T/Q481K	*Sinorhizobium meliloti*	Tran and Charles (2015)
		E130D/S325T/Q481M	Activities on CoA thioesters of 3HB, 2HB, GA, 4HB, 5HIV, 6HHA and 2HIV	Park, Jang, et al. (2013)
		E130D/S477R/Q481K		Park et al. (2015)
		E130D/S477R/Q481M		
		E130D/S477H/Q481K		
		E130D/S477H/Q481M		
		E130D/S477F/Q481K		
		E130D/S477F/Q481M		
		E130D/S477G/Q481K		
		E130D/S477G/Q481M		
		E130D/S325T/S477R/Q481K		
		E130D/S325T/S477R/Q481M		
		E130D/S325T/S477H/Q481K		
		E130D/S325T/S477H/Q481M		
		E130D/S325T/S477F/Q481K		
		E130D/S325T/S477F/Q481M		
		E130D/S325T/S477G/Q481K		
		E130D/S325T/S477G/Q481M		

Class II PHA synthase	*Pseudomonas chlororaphis*	E130D/Q481K	*Escherichia coli* Activities on CoA thioester of 3HB	Yang et al. (2011)
		E130D/S477F/Q481K		
		E130D/S325T/Q481K		
		E130D/S325T/S477F/Q481K		
		E130D/S325T/S477G/Q481K		
Class II PHA synthase	*Pseudomonas* sp. 61-3	E130D/Q481K	*Escherichia coli* *Corynebacterium glutamicum* Activities on CoA thioesters of 3HB, GA and 2HB	Yang et al. (2011) Taguchi et al. (2008) Yamada et al. (2010)
		E130D/S477F/Q481K		
		E130D/S325T/Q481K		
		E130D/S325T/S477F/Q481K		
		E130D/S325T/S477G/Q481K		
		S325T/Q481K		
		S325T/F392S/Q481K		
Class II PHA synthase	*Pseudomonas putida* KT2440	E130D/Q481K	*Escherichia coli* Activities on CoA thioester of 3HB	Yang et al. (2011)
		E130D/S477F/Q481K		
		E130D/S325T/Q481K		
		E130D/S325T/S477F/Q481K		
		E130D/S325T/S477G/Q481K		

Continued

Table 2 Enzymes for polymerization of lactyl-CoA.—cont'd

Enzyme	Source	Mutation	Expression host and comments[a]	Reference
Class II PHA synthase	*Pseudomonas resinovorans*	E130D/Q481K	*Escherichia coli* Activities on CoA thioester of 3HB	Yang et al. (2011)
		E130D/S477F/Q481K		
		E130D/S325T/Q481K		
		E130D/S325T/S477F/Q481K		
		E130D/S325T/S477G/Q481K		
Class II PHA synthase	*Pseudomonas aeruginosa* PAO1	E130D/Q481K	*Escherichia coli* Activities on CoA thioester of 3HB	Yang et al. (2011)
		E130D/S477F/Q481K		
		E130D/S325T/Q481K		
		E130D/S325T/S477F/Q481K		
		E130D/S325T/S477G/Q481K		
Class II PHA synthase	*Pseudomonas* sp. SG4502	S324T/Q480K	*Escherichia coli* Activities on CoA thioester of 3HB	Tajima et al. (2012)
Class II PHA synthase	*Pseudomonas stutzeri* 1317	S325T/Q481K	*Escherichia coli* Activities on CoA thioesters of 3HB and 3HP	Ren et al. (2017)
		E130D/S325T/F392S/S477G/Q481K		
		E130D/S325T/S477G/Q481K		

Class I PHA synthase	*Cupriavidus necator* H16	A510C	*Escherichia coli* *Cupriavidus necator* Activities on CoA thioester of 3HB	Ochi et al. (2013) Park, Jang, et al. (2013)
		A510F		
		A510G		
		A510H		
		A510I		
		A510K		
		A510L		
		A510M		
		A510N		
		A510Q		
		A510R		
		A510S		
		A510T		
		A510V		
		A510W		
		A510K/S506G		

[a]The hydroxyacyl-CoAs other than lactyl-CoA used as substrates of each enzyme are provided.

Similarly, other PHA synthases from different *Pseudomonas* strains also were engineered through site-directed mutagenesis on the four amino acids residues and the resulting variants showed increased substrate specificity toward lactyl-CoA (Ren et al., 2017; Tajima et al., 2012; Yang et al., 2011). Besides, *C. necator* PHA synthase variants harboring amino acid substitutions at Ser506 and Ala510, corresponding to Ser477 and Gln481 of *Pseudomonas* PHA synthase could synthesize LA-containing polymers (Park, Jang, et al., 2013). The multiple sequence alignment of PHA synthases is shown in Fig. 3.

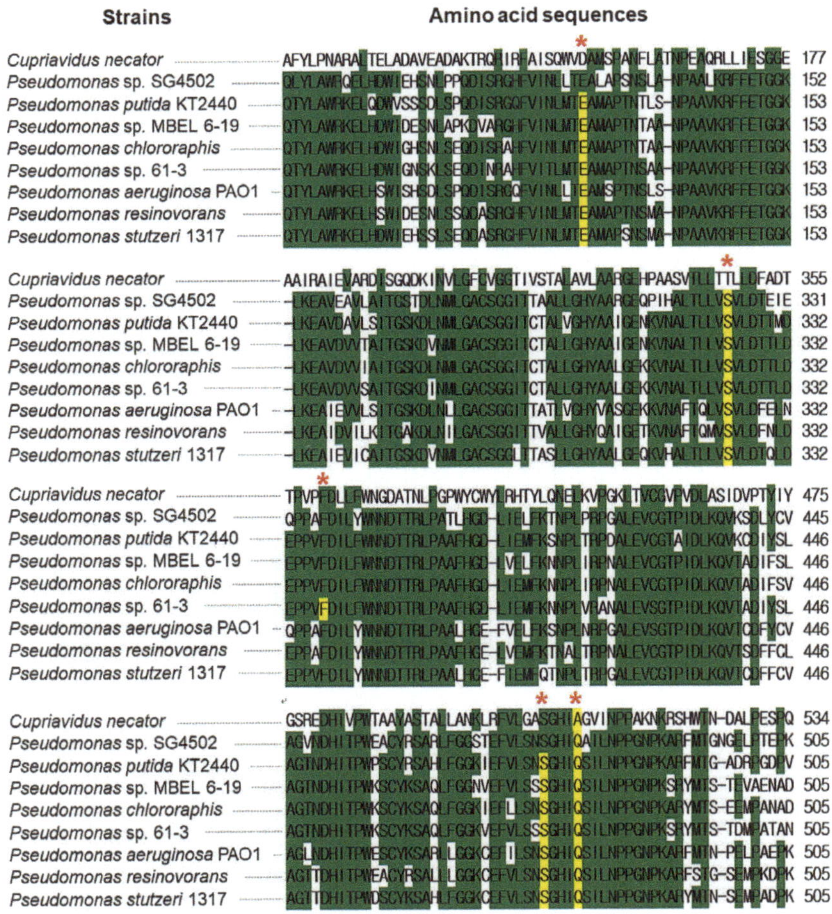

Fig. 3 Multiple sequence alignment of PHA synthases used for polymerizing LA. The important amino acid residues for LA polymerizing activity are marked with an asterisk (*). For details, see Table 2.

2.2.1 *In vivo polymerization*

The protocol is based on the previous studies (Jung et al., 2010; Jung & Lee, 2011; Yang et al., 2010).

2.2.1.1 Buffers and reagents

- Luria-Bertani (LB) medium
- Trace metal solution: 10 g $FeSO_4 \cdot 7H_2O$, 2 g $CaCl_2$, 2.2 g $ZnSO_4 \cdot 7H_2O$, 0.5 g $MnSO_4 \cdot 4H_2O$, 1 g $CuSO_4 \cdot 5H_2O$, 0.1 g $(NH_4)_6Mo_7O_{24} \cdot 4H_2O$, 0.02 g $Na_2B_4O_7 \cdot 10H_2O$ in 1 L of 0.5 M HCl solution
- MR medium: 6.67 g KH_2PO_4, 4 g $(NH_4)_2HPO_4$, 0.8 g citric acid, 5 mL trace metal solution in 930 mL of water
- Glucose
- 3HB
- $MgSO_4 \cdot 7H_2O$
- Antibiotics according to the plasmid employed (e.g., ampicillin)
- Methanol
- PHA solution: 15% (v/v) sulfuric acid, 0.9% (w/v) benzoic acid in methanol
- Chloroform

2.2.1.2 Cell culture and harvest

1. A host strain is transformed by the plasmid containing genes of CoA transferase and PHA synthase. The information on possible enzymes is provided in Tables 1 and 2.
2. The transformed strain is pre-cultured in a test tube containing 5 mL of LB medium added 5 μL of 100 gL^{-1} ampicillin stock solution. The cells are cultivated overnight at 30 °C with 200 rpm rotary shaking.
3. 93 mL of MR medium, 4 mL of 500 gL^{-1} glucose stock solution, 1 mL of 100 gL^{-1} 3HB stock solution, 1 mL of 80 gL^{-1} $MgSO_4 \cdot 7H_2O$ stock solution, and 100 μL of 100 gL^{-1} ampicillin stock solution are mixed in an autoclaved 250 mL Erlenmeyer flask.
4. 1 mL of the pre-culture (step 2) is inoculated to the flask of step 3. The cells are cultivated at 30 °C with 200 rpm rotary shaking for 72–96 h.
5. The cell cultures are centrifuged at 4000 × *g* and 4 °C for 10 min. After centrifugation, the supernatant is discarded.
6. The cell pellet is resuspended by distilled water and centrifuged like as step 5, and the supernatant is discarded. This step is repeated once more to eliminate the residual medium completely (cell washing).

2.2.1.3 Notes

- The host strain capable of producing LA should be used.
- All the materials used for cell culture (medium, stock solution, and flask) should be sterilized by autoclaving. Glucose and $MgSO_4·7H_2O$ stock solutions are encouraged to be separately autoclaved. For 3HB and antibiotics stock solutions, filtering sterilization is encouraged than autoclaving to avoid thermal degradation.
- A microbial strain may require certain supplements to be grown in a defined medium such as MR medium. For example, *E. coli* XL1-Blue strain, a thiamine auxotroph, should be cultivated in the thiamine included MR medium.

2.2.1.4 Gas chromatography (GC) analysis (Fig. 4)

1. The washed cell pellet from step 6 of Section 2.2.1.2 is dried at 70° C oven.
2. About 30–100 mg of the dried cell pellet is mixed with 1 mL of PHA solution and 2 mL of chloroform.
3. The mixture is incubated at 95 °C overnight for methanolysis reaction.
4. 1 mL of distilled water is added to the mixture and is vortexed thoroughly.
5. The mixture is left over 3 h until the phase separation is completed.
6. The organic phase at the bottom is collected, and then injected to GC. By GC analysis, the polymer content in cells and monomer composition can be calculated.

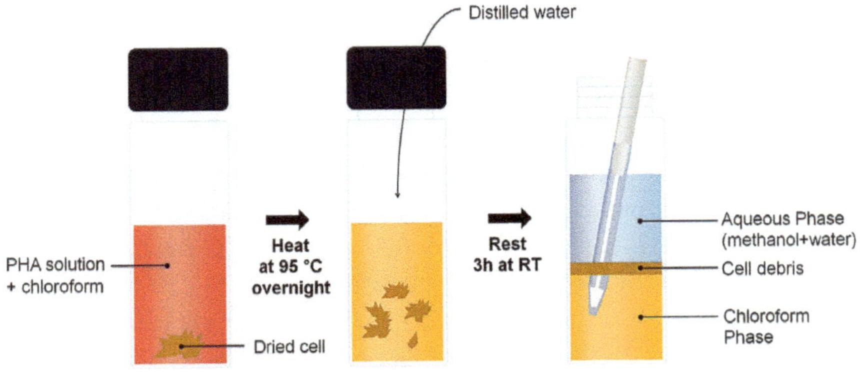

Fig. 4 Procedures for GC sample preparation.

2.2.1.5 Polymer extraction and analysis (Fig. 5)

1. The washed cell pellet from step 6 of Section 2.2.1.2 is freeze-dried for 24h.
2. The freeze-dried cells are subjected to the soxhlet apparatus using chloroform as an extraction solvent (Fig. 5A). The chloroform in a round flask is heated at 55°C for 14h.
3. The chloroform solution in the round flask is filtered through a cellulose filter by using a filter holder and a vacuum pump.
4. The filtered solution is concentrated by using an evaporator.
5. Ice-cold methanol is added to the solution of step 4 and incubated at 4°C overnight. When methanol is added, the polymer is shown as a white precipitate (or slightly yellow).
6. The precipitate is collected by filtering the mixture of step 5.
7. The polymer is dried in room temperature or is freeze-dried for 24h.
8. The dried polymer is characterized by nuclear magnetic resonance (NMR) analysis. Fig. 5B and C show the NMR spectra of PLA.

2.2.1.6 Notes

- In step 2 of Section 2.2.1.4, the freeze-dried cell (step 1 of Section 2.2.1.5) can be also used instead of the dried cell in hot oven.
- The GC analysis can be performed for the extracted polymer of step 7 in Section 2.2.1.5 with same procedures in which the dried polymer is used instead of the dried cell.

2.2.2 *In vitro* polymerization

The protocol is based on Tajima et al. (2012).

2.2.2.1 Buffers and reagents

- Sodium phosphate buffer (pH7.5)
- LA
- 3HB
- Acetate
- ATP
- CoA
- $MgCl_2$
- Bovine serum albumin
- Pyrophosphatase
- Acetyl-CoA synthase

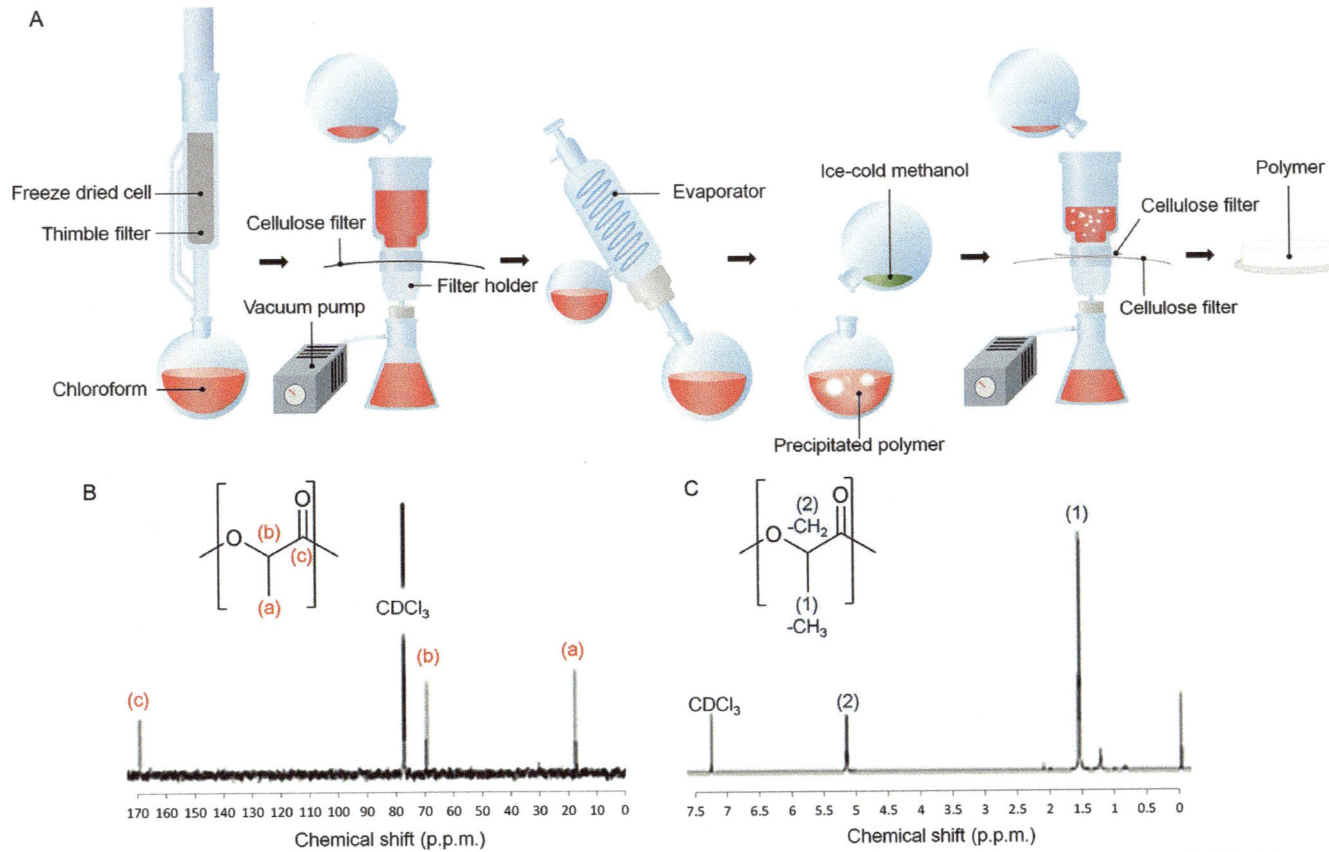

Fig. 5 Procedures for polymer extraction and NMR analysis. (A) Overall scheme of polymer extraction using chloroform solvent. (B) ^{13}C NMR spectrum of PLA in chloroform-d ($CDCl_3$). Each NMR peak corresponds to the carbon of PLA labeling with same alphabet. (C) ^{1}H NMR spectrum of PLA in chloroform-d ($CDCl_3$). Each NMR peak corresponds to the hydrogen of PLA labeling with same number. *Panels (B) and (C) are reproduced from Jung, Y. K., Kim, T. Y., Park, S. J., & Lee, S. Y. (2010). Metabolic engineering of* Escherichia coli *for the production of polylactic acid and its copolymers.* Biotechnology and Bioengineering, 105, *161–171 with permission.*

- Purified CoA transferase and PHA synthase in 100mM sodium phosphate buffer (pH 7.5)
- Methanol
- Chloroform

2.2.2.2 Reaction

1. Reaction medium is prepared with 200mM of LA and 3HB, 10mM acetate, 30mM ATP, 2mM CoA, 10mM $MgCl_2$, 0.2mgmL^{-1} of bovine serum albumin, 0.4U pyrophosphatase, 0.2mgmL^{-1} of acetyl-CoA synthase, and 0.2mgmL^{-1} of CoA transferase.
2. The reaction starts upon the addition of purified PHA synthase at a concentration of 0.5mgmL^{-1}. The reaction mixture is made up to 5mL and is incubated at 30°C for 72h.
3. 5mL of chloroform is thoroughly mixed with the reaction mixture of step 2, and then is incubated at 70°C for 3h.
4. The solution is cooled to room temperature. Then, the chloroform phase is filtered by 0.2μm PTFE membrane to remove impurities.
5. The filtered solution is concentrated to about 1mL by using an evaporator.
6. 10mL of methanol is added to the solution to precipitate polymer. The mixture is stored at 4°C for 16h.
7. The precipitate is collected by filtering using 0.2μm PTFE membrane.
8. The polymer is dried and used for further analysis.

3. Microbial strain development for the one-step production of LA-containing polyester

The enzymatic polymerization of LA has been enabled by employing Pct (or Bct, HadA) and engineered PHA synthase (see Section 2). It can simplify the current bio-chemical hybrid process for PLA production (described in Section 1) to one-step fermentative production. The engineered microorganisms equipped with both LA biosynthetic and LA polymerizing synthetic pathways can be employed as cell factories that converts inexpensive biomass-derived carbon sources such as glucose and xylose into PLA through their metabolism.

Compared to the *in vitro* synthesis of PLA by using CoA transferase and engineered PHA synthase (or lipase-based synthesis which is not discussed here), the *in vivo* synthesis of PLA using the microbial strain has several advantages. The *in vivo* synthesis system is generally more robust compared to the *in vitro* system since the cells can retain the stability of the enzymes and intermediates (metabolites) against a certain range of condition variations such as pH and temperature. In addition, the microbial system does not require enzyme purification and cofactor regeneration, which are ones of the most costly processes (Dudley, Karim, & Jewett, 2015; Hodgman & Jewett, 2012). Especially, as large-scale fermentation process of LA has been well established, the lab-scale *in vivo* PLA synthesis system can be easily transferred to large-scale production.

E. coli has mainly been used for production of non-natural polyesters including PLA since it is relatively easy to engineer with various synthetic and systems biology tools. Also it has been already demonstrated as an efficient host for production of natural PHAs including poly(3HB). Recently, other bacterial strains have been engineered for production of LA-containing polyesters. The selection criteria are based on their unique characteristics. For examples, *C. necator*, a native producer of PHAs, has been selected since it possesses good metabolic and cellular characteristics suitable for polymer production (Park, Jang, et al., 2013). In addition, *Corynebacterium glutamicum* has been employed as it is a "generally recognized as safe (GRAS)" strain which is advantageous since polymer produced by *C. glutamicum* does not require additional endotoxin purification steps (Song et al., 2012).

In this section, we summarize the metabolic engineering strategies implemented for efficient production of PLA and various copolymers. For PLA, each of the genetic manipulations reported is described in detail. In addition, we discuss how the microbial strains have been further engineered to produce various LA-copolymers.

3.1 PLA

To optimize the metabolic fluxes toward PLA, engineering targets can be rationally selected based on the available information on the metabolic pathways and involved enzymes. For examples, *ldhA* gene encoding LA dehydrogenase converting pyruvate to LA can be selected as an amplification target to promote LA biosynthesis. The amplification has been implemented by either ways: replacing the native *ldhA* promoter with *trc* promoter that can more strongly express *ldhA* gene compared to the native promoter with removal of native regulations on the native promoter or overexpressing *ldhA* gene based on a plasmid system (Jung et al., 2010; Jung & Lee, 2011; Park, Jang, et al., 2013; Song et al., 2012). It can be also considered as a common metabolic engineering strategy to inactivate the metabolic pathways that can decrease the metabolic fluxes toward LA. In this context, *pflB* gene encoding pyruvate formate-lyase was deleted from the chromosome to increase the level of pyruvate by blocking the conversion of pyruvate into formate in the expectation that the increased pool of pyruvate can be driven to LA (Choi et al., 2016; Jung & Lee, 2011). The conversion of acetyl-CoA to acetate was also inactivated by knocking out the corresponding gene (*ackA*) to increase the intracellular level of acetyl-CoA as it is CoA donor for lactyl-CoA synthesis (Jung et al., 2010).

In another approach, the engineering targets can be suggested by *in silico* simulation using genome-scale metabolic model. It can identify chromosomal genes to be deleted or overexpressed for increasing PLA production rates considering the cell growth rate. In one study, two target genes, *adhE* gene encoding acetaldehyde/alcohol dehydrogenase and *acs* gene encoding acetyl-CoA synthetase, were selected from the simulation results. Indeed, the knock-out of *adhE* and overexpression of *acs* by replacing the native promoter to stronger promoter were effective for production of PLA and poly(LA-*co*-3HB) (Jung et al., 2010). The manipulations conducted are summarized in Table 3 and Fig. 6. The production performance of the microbial strain can be effectively enhanced by implementing the multiple manipulations simultaneously. For example, the engineered *E. coli* equipped with PLA biosynthetic pathway can only produce poly(3HB-*co*-30 mol% LA) when 2 $g L^{-1}$ of 3HB is supplemented in the medium, however, poly(3-HB-*co*-LA) containing 60 mol% of LA can be produced when the strain was further engineered by deleting *ackA* and *ppc* genes and replacing the native promoter of *ldhA*. In addition, when manipulations of all five genes (*ackA*, *ppc*, *ldhA*, *adhE*, and *acs*) were performed in *E. coli* equipped with PLA biosynthetic pathway, the homo PLA could be produced to polymer content of 4–11 wt% solely from glucose (Jung et al., 2010).

3.2 Other copolymers

One of the key advantages of the microbial production system is relative easiness of incorporating several other co-monomers, which results in production of various LA-containing copolymers possessing diverse material properties. As above mentioned, PHA synthase is well known for its broad substrate availability. Therefore, it is possible to produce various LA-containing copolymers by employing the engineered PHA synthase which accepts both substrates, lactyl-CoA and other hydroxyacyl-CoAs that

Table 3 Metabolic engineering for enhanced production of LA-containing polyesters.

Gene	Enzyme	Engineering type	Engineering rationale	Reference
ldhA	D-LA dehydrogenase	Replacement of the native promoter with *trc* promoter	To promote LA synthesis	Jung et al. (2010) Jung and Lee (2011) Choi et al. (2016)
		Overexpression by plasmid	To promote LA synthesis	Song et al. (2012) Park, Jang, et al. (2013)
ppc	Phosphoenolpyruvate carboxylase	Deletion of chromosomal gene	To enhance level of pyruvate, a precursor of LA	Jung et al. (2010)
ackA	Acetate kinase	Deletion of chromosomal gene	To enhance level of acetyl-CoA	Jung et al. (2010)
adhE	Acetaldehyde/alcohol dehydrogenase	Deletion of chromosomal gene	To increase PLA production rate with proper cell growth rate (predicted by *in silico* simulation)	Jung et al. (2010) Jung and Lee (2011) Choi et al. (2016)
acs	Acetyl-CoA synthetase	Replacement of the native promoter with *trc* promoter	To increase PLA production rate (predicted by *in silico* simulation)	Jung et al. (2010) Jung and Lee (2011)
pflB	Pyruvate formate-lyase	Deletion of chromosomal gene	To prevent by-product formations	Jung and Lee (2011) Choi et al. (2016)
pflA	Pyruvate formate-lyase activating enzyme	Deletion of chromosomal gene	To enhance level of pyruvate, a precursor of LA	Nduko, Matsumoto, Ooi, and Taguchi (2013)
frdABCD	Fumarate reductase	Deletion of chromosomal gene	To prevent by-product formations	Jung and Lee (2011)
frdB	Fumarate reductase (subunit)	Deletion of chromosomal gene	To prevent by-product formations	Choi et al. (2016)
lacI	Transcriptional repressor of Lac operon	Deletion of chromosomal gene	To constitutively express genes under *trc* promoter without inducer	Jung and Lee (2011)
poxB	Pyruvate oxidase	Deletion of chromosomal gene	To enhance level of pyruvate, a precursor of LA	Choi et al. (2016)
dld	D-LA dehydrogenase	Deletion of chromosomal gene	To prevent conversion of LA to pyruvate	Choi et al. (2016)

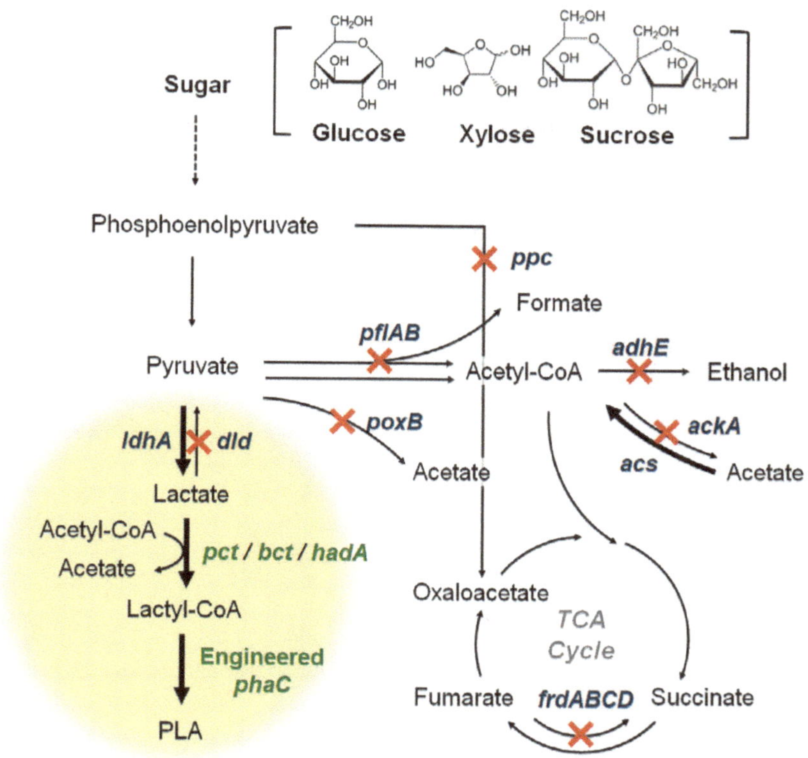

Fig. 6 Metabolic engineering for the production of PLA from sugars. The cross mark indicates chromosomal gene deletion and the bold arrow indicates overexpression of the gene. The details of each manipulation are summarized in Table 3.

can be provided *in vivo* through corresponding metabolic pathways of metabolically engineered host strain. Thus, there is no need to separately produce and prepare co-monomers and optimize the reaction condition, which can increase of the overall production costs. Various natural PHA constituents such as 3HB (Jung et al., 2010; Taguchi et al., 2008), MCL-HAs (Shozui, Matsumoto, Motohashi, Yamada, & Taguchi, 2010), 3HP (Ren et al., 2017) and 4HB (Choi et al., 2016) have been successfully polymerized as copolymers with LA. In addition, the non-natural monomers of 2HA as well as LA, GA (Choi et al., 2016; Li et al., 2016), 2HB (Choi et al., 2017; Park, Jang, et al., 2013; Park, Kang, et al., 2013; Park, Kim, et al., 2012, Park, Lee, et al., 2012), 2HIV (Choi et al., 2016; Yang et al., 2016), 2HIC (Mizuno et al., 2018), 2H3MV (Mizuno et al., 2018), PhLA (Mizuno et al., 2018; Yang et al., 2018), MA (Yang et al., 2018), and 4HPhLA (Yang et al., 2018) can be incorporated into copolymers with LA (Fig. 7).

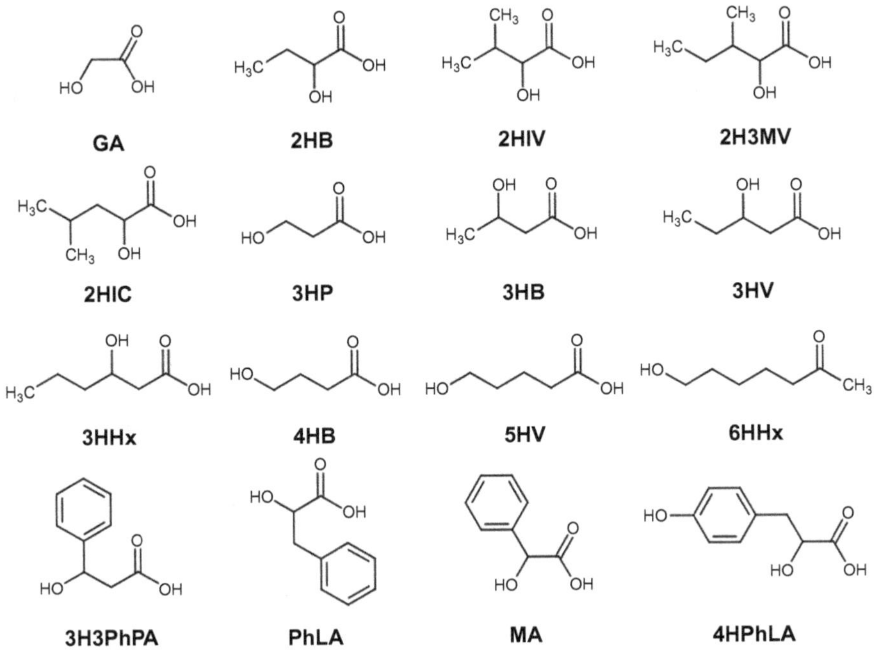

Fig. 7 Molecules identified as co-monomers of LA-copolymers produced by recombinant microorganism.

If the activities of PHA synthase are confirmed for the co-monomers, the copolymers can be produced *in vivo* by two ways: (1) by externally supplying the precursors of co-monomers which can be converted into CoA thioester forms through the inherent or engineered metabolisms, (2) by constructing the metabolic pathways for generating the co-monomers from unrelated carbon sources (Fig. 8). The metabolic pathways and key engineering strategies for generating HA monomers are summarized in Table 4. The CoA transferase and PHA synthase used for the copolymer synthesis are described in Tables 1 and 2. The microorganism can be engineered to implement different monomer synthesis modules simultaneously in various combinations to produce various copolymers such as poly(LA-*co*-GA) [PLGA], poly(LA-*co*-GA-*co*-3HB), and poly(LA-*co*-GA-*co*-3HB-*co*-4HB) (Choi et al., 2016; Li et al., 2017, 2016). In the case of copolymers, the metabolic engineering can become more complicated as relative metabolic fluxes between monomers have to be considered. The study of developing *E. coli* strain producing PLGA is a representative case (Choi et al., 2016). To modulate the monomer fractions of PLGA produced by recombinant *E. coli*, the genetic

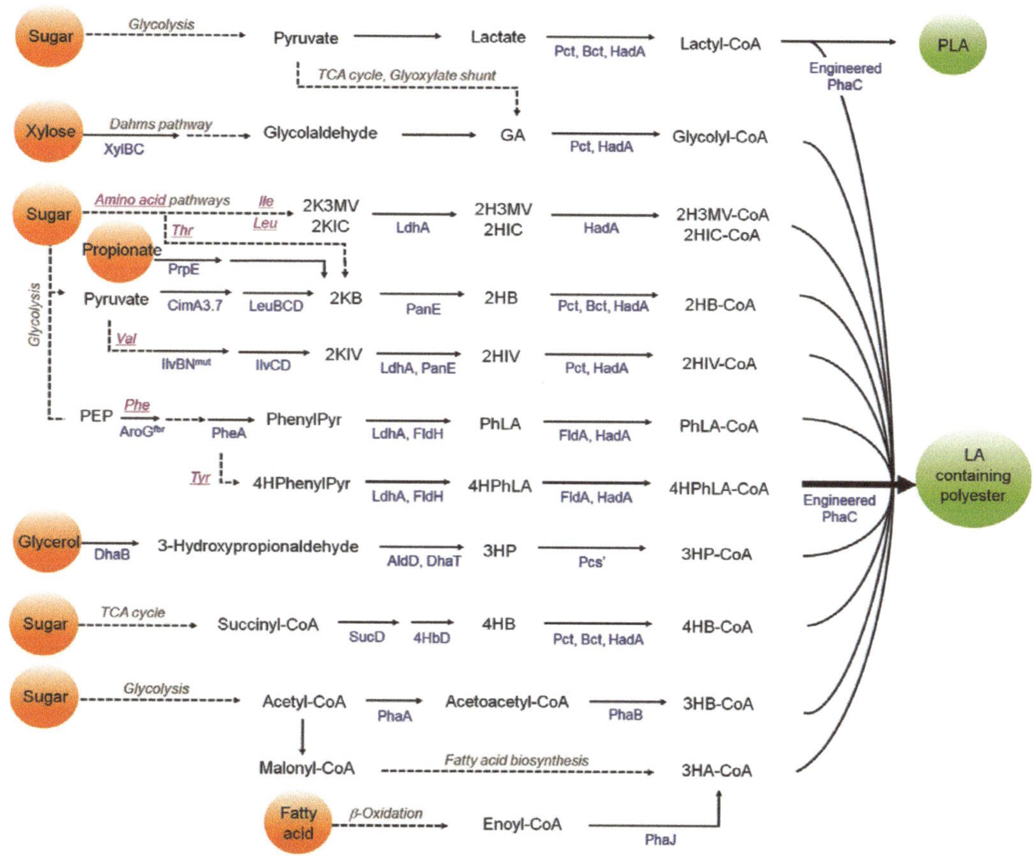

Fig. 8 See figure legend on next page.

manipulations have been conducted for concentrating the metabolic fluxes toward both LA and GA in a step-wise manner. It resulted in production of PLGA having the ratios of LA:GA, approximately 9:1, 8:2, 7:3, and 5:5 (Choi et al., 2016)

4. Summary and conclusion

To synthesize PLA by one-step fermentative production from renewable biomass, the non-natural metabolic pathway for the polymerization of LA was designed and constructed based on the well-known microbial polyester, PHA biosynthetic pathway. In this new biosynthetic pathway of PLA, LA can be converted into lactyl-CoA by Pct, Bct, or HadA and then, the lactyl-CoA can be polymerized into PLA by the engineered PHA synthase. The key breakthrough in constructing this pathway was that the native PHA synthases which possess only ignorable activity on lactyl-CoA were successfully engineered to accept lactyl-CoA with enhanced substrate specificity and activities. By employing the two key enzymes of CoA transferase and engineered PHA synthase, the microorganisms have been further developed for the platform strains which efficiently convert the cheap carbohydrates such as glucose into PLA *in vivo*. Several chromosomal gene targets were logically selected and deleted to inactivate the by-product formations. In addition, by the aid of the *in silico* simulation analyzing the metabolic fluxes at genome-scale, PLA could be more efficiently produced. The resulting

Fig. 8 Metabolic pathways for the production of LA-containing polyesters. The enzymes shown are: 4HbD, 4HB dehydrogenase; AldD, aldehyde dehydrogenase; AroG^{fbr}, feed-back inhibition resistant mutant of 3-deoxy-D-arabino-heptulosonate-7-phosphate synthase; CimA3.7, citramalate synthase mutant; DhaB, glycerol dehydratase; DhaT, 1,3-propanediol dehydrogenase; FldA, cinnamoyl-CoA:PhLA CoA-transferase; FldH, PhLA dehydrogenase; IlvBN^{mut}, acetolactate synthase isoenzyme I mutant; IlvC, ketol-acid reductoisomerase; IlvD, dihydroxy-acid dehydratase; LdhA, 2HA dehydrogenase; LeuBCD, 3-isopropylmalate dehydrogenase and 3-isopropylmalate dehydratase; PanE, 2-dehydropantoate 2-reductase; Pcs′, propionyl-CoA synthetase; PhaA, acetyl-CoA acetyltransferase; PhaB, acetoacetyl-CoA reductase; PhaC, PHA synthase; PhaJ, enoyl-CoA hydratase; PheA, chorismate mutase/prephenate dehydratase; SucD, succinate semialdehyde dehydrogenase; XylBC, xylose dehydrogenase and xylonolactonase. The metabolites shown are: 2K3MV, 2-keto-3-methylvalerate; 2KB, 2-ketobutyrate; 2KIC, 2-ketoisocaproate; 2KIV, 2-ketoisovalerate; 4HPhenylPyr, 4-hydroxyphenylpyruvate; PEP, phosphoenolpyruvate; PhenylPyr, phenyllpyruvate. For the others, see Abbreviations section.

Table 4 Metabolic engineering for production of various LA-containing copolymers.

Comonomer	Pathway	Strategies[a]	Reference
3HB	Poly(3HB) biosynthesis pathway	Constructing the poly(3HB) pathway by expressing *C. necator phaAB*	Taguchi et al. (2008) Jung et al. (2010) Jung and Lee (2011)
3HP	Glycerol-based 3HP biosynthesis pathway	Constructing the synthetic pathway by expressing *K. pneumoniae dhaB*, *P. putida* KT2442 *aldD, dhaT, C. aurantiacus pcs'*	Ren et al. (2017)
3HV	Propionate dependent 3HV biosynthesis pathway	Feeding propionate and expressing *C. necator phaAB*	Shozui, Matsumoto, Nakai, Yamada, and Taguchi (2010)
3HV	β-Oxidation pathway and fatty acid *de novo* biosynthesis pathway	Feeding valerate and expressing *P. aeruginosa phaJ4*	Shozui et al. (2011)
3HHx	β-Oxidation pathway and fatty acid *de novo* biosynthesis pathway	Feeding butyrate and expressing *P. aeruginosa phaJ4*	Shozui, Matsumoto, Motohashi, et al. (2010)
2HB	Citramalate pathway	Constructing the synthetic pathway by expressing *M. jannaschii cimA3.7, E. coli leuBCD, L. lactis* subsp. *lactis* Il1403 *panE*	Park, Kim, et al. (2012), Park, Lee, et al., (2012)
	Propionyl-CoA pathway	Feeding propionate and expressing *C. necator panE, prpE*	Park, Kang, et al. (2013)
	Amino acid (L-threonine) biosynthesis pathway	2HB was produced through the inherent metabolic pathways	Choi et al. (2016)
GA	Dahms pathway	Constructing the Dahms pathway by expressing *C. crescentus xylBC*	Choi et al. (2016)
	Glyoxylate shunt pathway	Overexpressing the *E. coli ycdW*, *aceAK* and knocking out *glcD* to prevent the GA conversion	Li et al. (2016)

Continued

Table 4 Metabolic engineering for production of various LA-containing copolymers.—cont'd

Comonomer	Pathway	Strategies[a]	Reference
2HIV	Amino acid (L-valine) biosynthesis pathway	Constructing the 2HIV pathway by expressing *L. lactis subsp. lactis* Il1403 *panE* and strengthening the metabolic fluxes by overexpressing *ilvCD* and feedback inhibition resistant mutant *E. coli* $ilvBN^{mut}$	Choi et al. (2016) Yang et al. (2016)
2HIC 2H3MV PhLA	Amino acid (L-leucine/L-isoleucine/L-phenylalanine) biosynthesis pathways	Constructing the pathway by expressing *C. difficile ldhA*	Mizuno et al. (2018)
PhLA 4HPhLA	Amino acid (L-phenylalanine/L-tyrosine) biosynthesis pathways	Constructing the pathway by expressing *C. botulinum fldH* and strengthening the metabolic fluxes by expressing the feedback inhibition resistant mutants *E. coli* $AroG^{fbr}$, $PheA^{fbr}$	Yang et al. (2018)
4HB	Succinyl-CoA dependent 4HB biosynthesis pathway	Constructing the synthetic pathway by expressing *C. kluyveri sucD, 4hbD*	Choi et al. (2016) Li, Qiao, Che, and Stephanopoulos (2017)

[a]The strain names are: *C. necator*, *Cupriavidus necator*; *K. pneumoniae*, *Klebsiella pneumoniae*; *P. putida*, *Pseudomonas putida*; *C. aurantiacus*, *Chloroflexus aurantiacus*; *P. aeruginosa*, *Pseudomonas aeruginosa*; *M.* jannaschii, *Methanococcus jannaschii*; *E.* coli, *Escherichia coli*; *L. lactis* subsp. *lactis*, *Lactococcus lactis* subsp. *lactis*; *C. crescentus*, *Caulobacter crescentus*; *C.* difficile, *Clostridium difficile*; *C. botulinum*, *Clostridium botulinum*; *C. kluyveri*, *Clostridium kluyveri*. The metabolic pathways are shown in Fig. 8.

recombinant *E. coli* strain was applied for the production of various LA-containing copolymers by additionally equipping the metabolic pathways for co-monomers.

This microbial system has several advantages over the current processes consisted of fermentation followed by chemical process for the polymerization. First, it simplifies the current multi-step PLA production process into one-step direct fermentative production using renewable biomass. There is no need for preparing/producing extremely pure lactide for the ROP. In addition, there is no need to worry about the presence of unreacted residual lactide after polymerization and metal catalyst employed, which all complicate the PLA processing. This is particularly important for the medical applications of polymers that require complete removal of the catalysts.

Second, as discussed in Section 3.2, the microbial system opens up the possibility of producing various copolymers having more diverse material properties. All the monomers can be produced and used for polymer synthesis in this microbial system without further chemical processing of other monomers. For example, when considering PLGA, the current process become more expensive than PLA because the additional processes for GA (such as in the form of glycolide) are needed. On the other hand, in microbial system, PLGA can be produced without additional process step from PLA by developing microbial strain producing PLGA.

Third, purification of polymers is simple and cost-effective when one-step fermentative process is employed. Simple solvent extraction (described in Section 2.2.1.5) or even mild alkaline digestion can lead to purification of the polymer produced by metabolically engineered microorganisms.

Despite the above advantages, the current one-step fermentative production system of PLA has a disadvantage: the yield and productivity is still low to have price competitiveness. For PLA homopolymer, the polymer contents were below 11 wt%. The titer and productivity were about $0.6\,gL^{-1}$ and $0.02\,gL^{-1}\,h^{-1}$, respectively (Yang, Choi, Shin, Park, & Lee, 2013; Jung & Lee, 2011). On the other hand, the polymer content, titer, and productivity of poly(3HB) have been reported to reach about 90 wt%, $157\,gL^{-1}$, and $5.13\,gL^{-1}\,h^{-1}$, respectively (Wang & Lee, 1997a, 1997b). Thus, the production efficiencies of this microbial PLA synthesis need to be much improved for future commercialization. In light of the many successful examples of microbial production systems that have greatly increased compound production through applying systems metabolic engineering strategies, a variety of strains manipulation strategies can greatly enhance the production performances of the microbial strains. In addition, since the 3-dimensional protein structures of PHA synthases, which play a key role in this microbial system, have recently been discovered (Kim, Choi, et al., 2017; Kim, Kim, Choi, Lee, & Kim, 2017; Wittenborn, Jost, Wei, Stubbe, & Drennan, 2016), the rational engineering of PHA synthase is now possible. Therefore, it is expected that the microorganism can be a versatile and powerful platform for production of PLA, LA-copolymers, and other non-natural polymers.

Acknowledgments

This work was supported by the Technology Development Program to Solve Climate Changes (Systems Metabolic Engineering for Biorefineries) from the Ministry of Science and ICT through the National Research Foundation (NRF) of Korea (NRF-2012M1A2A2026556 and NRF-2012M1A2A2026557) to S.Y.L.

References

Agnew, D. E., & Pfleger, B. F. (2013). Synthetic biology strategies for synthesizing polyhydroxyalkanoates from unrelated carbon sources. *Chemical Engineering Science*, *103*, 58–67.

Auras, R., Harte, B., & Selke, S. (2004). An overview of polylactides as packaging materials. *Macromolecular Bioscience*, *4*, 835–864.

Benninga, H. (1990). A history of lactic acid making: A chapter in the history of biotechnology. *Vol. 11*. Springer Science & Business Media.

Bugnicourt, E., Cinelli, P., Lazzeri, A., & Alvarez, V. A. (2014). Polyhydroxyalkanoate (PHA): Review of synthesis, characteristics, processing and potential applications in packaging. *Express Polymer Letters*, *8*, 791–808.

Castro-Aguirre, E., Iñiguez-Franco, F., Samsudin, H., Fang, X., & Auras, R. (2016). Poly (lactic acid)—Mass production, processing, industrial applications, and end of life. *Advanced Drug Delivery Reviews*, *107*, 333–366.

Chae, T. U., Choi, S. Y., Kim, J. W., Ko, Y.-S., & Lee, S. Y. (2017). Recent advances in systems metabolic engineering tools and strategies. *Current Opinion in Biotechnology*, *47*, 67–82.

Cho, J. H., Park, S. J., Lee, S. Y., & Jung, Y. K., (2006). Cells or plants that can produce polylactate or its copolymers and to a method for preparing polylactate or its copolymers using the same. WO/2006/126796.

Choi, S. Y., Kim, W. J., Yu, S. J., Park, S. J., Im, S. G., & Lee, S. Y. (2017). Engineering the xylose-catabolizing Dahms pathway for production of poly (D-lactate-co-glycolate) and poly(D-lactate-co-glycolate-co-D-2-hydroxybutyrate) in *Escherichia coli*. *Microbial Biotechnology*, *10*, 1353–1364.

Choi, S. Y., Park, S. J., Kim, W. J., Yang, J. E., Lee, H., Shin, J., et al. (2016). One-step fermentative production of poly(lactate-co-glycolate) from carbohydrates in *Escherichia coli*. *Nature Biotechnology*, *34*, 435–440.

Chung, H., Yang, J. E., Ha, J. Y., Chae, T. U., Shin, J. H., Gustavsson, M., et al. (2015). Bio-based production of monomers and polymers by metabolically engineered microorganisms. *Current Opinion in Biotechnology*, *36*, 73–84.

Datta, R., & Henry, M. (2006). Lactic acid: Recent advances in products, processes and technologies? - review. *Journal of Chemical Technology & Biotechnology: International Research in Process, Environmental & Clean Technology*, *81*, 1119–1129.

David, Y., Joo, J. C., Yang, J. E., Oh, Y. H., Lee, S. Y., & Park, S. J. (2017). Biosynthesis of 2-hydroxyacid-containing polyhydroxyalkanoates by employing butyryl-coa transferases in metabolically engineered *Escherichia coli*. *Biotechnology Journal*, *12*, 1700116.

do Sul, J. A. I., & Costa, M. F. (2014). The present and future of microplastic pollution in the marine environment. *Environmental Pollution*, *185*, 352–364.

Drumright, R. E., Gruber, P. R., & Henton, D. E. (2000). Polylactic acid technology. *Advanced Materials*, *12*, 1841–1846.

Dudley, Q. M., Karim, A. S., & Jewett, M. C. (2015). Cell-free metabolic engineering: Biomanufacturing beyond the cell. *Biotechnology Journal*, *10*, 69–82.

Favaro, L., Basaglia, M., & Casella, S. (2019). Improving polyhydroxyalkanoate production from inexpensive carbon sources by genetic approaches: A review. *Biofuels, Bioproducts and Biorefining*, *13*, 208–227.

Garcia, J. M., & Robertson, M. L. (2017). The future of plastics recycling. *Science*, *358*, 870–872.

Hodgman, C. E., & Jewett, M. C. (2012). Cell-free synthetic biology: Thinking outside the cell. *Metabolic Engineering*, *14*, 261–269.

Jung, Y. K., Kim, T. Y., Park, S. J., & Lee, S. Y. (2010). Metabolic engineering of *Escherichia coli* for the production of polylactic acid and its copolymers. *Biotechnology and Bioengineering*, *105*, 161–171.

Jung, Y. K., & Lee, S. Y. (2011). Efficient production of polylactic acid and its copolymers by metabolically engineered *Escherichia coli*. *Journal of Biotechnology*, *151*, 94–101.

Kim, Y. J., Chae, C. G., Kang, K. H., Oh, Y. H., Joo, J. C., Song, B. K., et al. (2016). Biosynthesis of lactate-containing polyhydroxyalkanoates in recombinant *Escherichia coli* by employing new CoA transferases. *Korean Society for Biotechnology and Bioengineering Journal*, *31*, 27–32.

Kim, Y.-J., Choi, S. Y., Kim, J., Jin, K. S., Lee, S. Y., & Kim, K.-J. (2017). Structure and function of the N-terminal domain of *Ralstonia eutropha* polyhydroxyalkanoate synthase, and the proposed structure and mechanisms of the whole enzyme. *Biotechnology Journal*, *12*, 1600649.

Kim, J., Kim, Y. J., Choi, S. Y., Lee, S. Y., & Kim, K.-J. (2017). Crystal structure of *Ralstonia eutropha* polyhydroxyalkanoate synthase C-terminal domain and reaction mechanisms. *Biotechnology Journal*, *12*, 1600648.

Lassalle, V. L., & Ferreira, M. L. (2008). Lipase-catalyzed synthesis of polylactic acid: An overview of the experimental aspects. *Journal of Chemical Technology & Biotechnology: International Research in Process, Environmental & Clean Technology*, *83*, 1493–1502.

Lee, S. Y. (1996). Bacterial polyhydroxyalkanoates. *Biotechnology and Bioengineering*, *49*, 1–14.

Lee, Y., Cho, I. J., Choi, S. Y., & Lee, S. Y. (2019). Systems metabolic engineering strategies for non-natural microbial polyester production. *Biotechnology Journal*, 1800426.

Lee, S. Y., & Kim, H. U. (2015). Systems strategies for developing industrial microbial strains. *Nature Biotechnology*, *33*, 1061–1072.

Lee, S. Y., Kim, H. U., Chae, T. U., Cho, J. S., Kim, J. W., Shin, J. H., et al. (2019). A comprehensive metabolic map for production of bio-based chemicals. *Nature Catalysis*, *2*, 18–33.

Li, Z. J., Qiao, K., Che, X. M., & Stephanopoulos, G. (2017). Metabolic engineering of *Escherichia coli* for the synthesis of the quadripolymer poly(glycolate-co-lactate-co-3-hydroxybutyrate-co-4-hydroxybutyrate) from glucose. *Metabolic Engineering*, *44*, 38–44.

Li, Z. J., Qiao, K., Shi, W., Pereira, B., Zhang, H., Olsen, B. D., et al. (2016). Biosynthesis of poly(glycolate-co-lactate-co-3-hydroxybutyrate) from glucose by metabolically engineered *Escherichia coli*. *Metabolic Engineering*, *35*, 1–8.

Lindenkamp, N., Schürmann, M., & Steinbüchel, A. (2013). A propionate CoA-transferase of *Ralstonia eutropha* H16 with broad substrate specificity catalyzing the CoA thioester formation of various carboxylic acids. *Applied Microbiology and Biotechnology*, *97*, 7699–7709.

Madison, L. L., & Huisman, G. W. (1999). Metabolic engineering of poly (3-hydroxyalkanoates): From DNA to plastic. *Microbiology and Molecular Biology Reviews*, *63*, 21–53.

Mizuno, S., Enda, Y., Saika, A., Hiroe, A., & Tsuge, T. (2018). Biosynthesis of polyhydroxyalkanoates containing 2-hydroxy-4-methylvalerate and 2-hydroxy-3-phenylpropionate units from a related or unrelated carbon source. *Journal of Bioscience and Bioengineering*, *125*, 295–300.

Muhammadi, S., Afzal, M., & Hameed, S. (2015). Bacterial polyhydroxyalkanoates-eco-friendly next generation plastic: Production, biocompatibility, biodegradation, physical properties and applications. *Green Chemistry Letters and Reviews*, *8*, 56–77.

Nduko, J. M., Matsumoto, K. I., Ooi, T., & Taguchi, S. (2013). Effectiveness of xylose utilization for high yield production of lactate-enriched P(lactate-co-3-hydroxybutyrate) using a lactate-overproducing strain of *Escherichia coli* and an evolved lactate-polymerizing enzyme. *Metabolic Engineering*, *15*, 159–166.

Ochi, A., Matsumoto, K. I., Ooba, T., Sakai, K., Tsuge, T., & Taguchi, S. (2013). Engineering of class I lactate-polymerizing polyhydroxyalkanoate synthases from *Ralstonia eutropha* that synthesize lactate-based polyester with a block nature. *Applied Microbiology and Biotechnology*, *97*, 3441–3447.

Park, S. J., Jang, Y.-A., Lee, H., Park, A.-R., Yang, J. E., Shin, J., et al. (2013). Metabolic engineering of *Ralstonia eutropha* for the biosynthesis of 2-hydroxyacid-containing polyhydroxyalkanoates. *Metabolic Engineering*, *20*, 20–28.

Park, S. J., Jang, Y. A., Noh, W., Oh, Y. H., Lee, H., David, Y., et al. (2015). Metabolic engineering of *Ralstonia eutropha* for the production of polyhydroxyalkanoates from sucrose. *Biotechnology and Bioengineering*, *112*, 638–643.

Park, S. J., Kang, K.-H., Lee, H., Park, A.-R., Yang, J. E., Oh, Y. H., et al. (2013). Propionyl-CoA dependent biosynthesis of 2-hydroxybutyrate containing polyhydroxyalkanoates in metabolically engineered *Escherichia coli*. *Journal of Biotechnology*, *165*, 93–98.

Park, S. J., Kim, T. W., Kim, M. K., Lee, S. Y., & Lim, S. C. (2012). Advanced bacterial polyhydroxyalkanoates: Towards a versatile and sustainable platform for unnatural tailor-made polyesters. *Biotechnology Advances*, *30*, 1196–1206.

Park, S. J., Lee, T. W., Lim, S. C., Kim, T. W., Lee, H., Kim, M. K., et al. (2012). Biosynthesis of polyhydroxyalkanoates containing 2-hydroxybutyrate from unrelated carbon source by metabolically engineered *Escherichia coli*. *Applied Microbiology and Biotechnology*, *93*, 273–283.

Ren, Y., Meng, D., Wu, L., Chen, J., Wu, Q., & Chen, G. Q. (2017). Microbial synthesis of a novel terpolyester P(LA-co-3HB-co-3HP) from low-cost substrates. *Microbial Biotechnology*, *10*, 371–380.

Selmer, T., Willanzheimer, A., & Hetzel, M. (2002). Propionate CoA-transferase from *Clostridium propionicum*: Cloning of the gene and identification of glutamate 324 at the active site. *European Journal of Biochemistry*, *269*, 372–380.

Shozui, F., Matsumoto, K. I., Motohashi, R., Sun, J., Satoh, T., Kakuchi, T., et al. (2011). Biosynthesis of a lactate (LA)-based polyester with a 96 mol% LA fraction and its application to stereocomplex formation. *Polymer Degradation and Stability*, *96*, 499–504.

Shozui, F., Matsumoto, K. I., Motohashi, R., Yamada, M., & Taguchi, S. (2010). Establishment of a metabolic pathway to introduce the 3-hydroxyhexanoate unit into LA-based polyesters via a reverse reaction of β-oxidation in *Escherichia coli* LS5218. *Polymer Degradation and Stability*, *95*, 1340–1344.

Shozui, F., Matsumoto, K. I., Nakai, T., Yamada, M., & Taguchi, S. (2010). Biosynthesis of novel terpolymers poly (lactate-co-3-hydroxybutyrate-co-3-hydroxyvalerate)s in lactate-overproducing mutant *Escherichia coli* JW0885 by feeding propionate as a precursor of 3-hydroxyvalerate. *Applied Microbiology and Biotechnology, 85,* 949–954.

Sin, L. T., Rahmat, A. R., & Rahman, W. A. W. A. (2012). *Polylactic acid: PLA biopolymer technology and applications.* William Andrew.

Song, Y., Matsumoto, K. I., Yamada, M., Gohda, A., Brigham, C. J., Sinskey, A. J., et al. (2012). Engineered *Corynebacterium glutamicum* as an endotoxin-free platform strain for lactate-based polyester production. *Applied Microbiology and Biotechnology, 93,* 1917–1925.

Steinbüchel, A., & Valentin, H. E. (1995). Diversity of bacterial polyhydroxyalkanoic acids. *FEMS Microbiology Letters, 128,* 219–228.

Taguchi, S., Yamada, M., Matsumoto, K. I., Tajima, K., Satoh, Y., Munekata, M., et al. (2008). A microbial factory for lactate-based polyesters using a lactate-polymerizing enzyme. *Proceedings of the National Academy of Sciences, 105,* 17323–17327.

Tajima, K., Han, X., Satoh, Y., Ishii, A., Araki, Y., Munekata, M., et al. (2012). In vitro synthesis of polyhydroxyalkanoate (PHA) incorporating lactate (LA) with a block sequence by using a newly engineered thermostable PHA synthase from *Pseudomonas* sp. SG4502 with acquired LA-polymerizing activity. *Applied Microbiology and Biotechnology, 94,* 365–376.

Tan, G. Y., Chen, C. L., Li, L., Ge, L., Wang, L., Razaad, I., et al. (2014). Start a research on biopolymer polyhydroxyalkanoate (PHA): A review. *Polymers, 6,* 706–754.

Tran, T. T., & Charles, T. C. (2015). Genome-engineered *Sinorhizobium meliloti* for the production of poly (lactic-co-3-hydroxybutyric) acid copolymer. *Canadian Journal of Microbiology, 62,* 130–138.

Valappil, S. P., Misra, S. K., Boccaccini, A. R., & Roy, I. (2006). Biomedical applications of polyhydroxyalkanoates, an overview of animal testing and in vivo responses. *Expert Review of Medical Devices, 3,* 853–868.

Vink, E. T., Rabago, K. R., Glassner, D. A., & Gruber, P. R. (2003). Applications of life cycle assessment to NatureWorks™ polylactide (PLA) production. *Polymer Degradation and Stability, 80,* 403–419.

Volodina, E., Schürmann, M., Lindenkamp, N., & Steinbüchel, A. (2014). Characterization of propionate CoA-transferase from *Ralstonia eutropha* H16. *Applied Microbiology and Biotechnology, 98,* 3579–3589.

Wang, F., & Lee, S. Y. (1997a). Production of poly (3-hydroxybutyrate) by fed-batch culture of filamentation-suppressed recombinant *Escherichia coli. Applied and Environmental Microbiology, 63,* 4765–4769.

Wang, F., & Lee, S. Y. (1997b). Poly (3-Hydroxybutyrate) production with high productivity and high polymer content by a fed-batch culture of *Alcaligenes latus* under nitrogen limitation. *Applied and Environmental Microbiology, 63,* 3703–3706.

Wittenborn, E. C., Jost, M., Wei, Y., Stubbe, J., & Drennan, C. L. (2016). Structure of the catalytic domain of the class I polyhydroxybutyrate synthase from *Cupriavidus necator. Journal of Biological Chemistry, 291,* 25264–25277.

Yamada, M., Matsumoto, K. I., Shimizu, K., Uramoto, S., Nakai, T., Shozui, F., et al. (2010). Adjustable mutations in lactate (LA)-polymerizing enzyme for the microbial production of LA-based polyesters with tailor-made monomer composition. *Biomacromolecules, 11,* 815–819.

Yang, J. E., Choi, S. Y., Shin, J. H., Park, S. J., & Lee, S. Y. (2013). Microbial production of lactate-containing polyesters. *Microbial Biotechnology, 6,* 621–636.

Yang, T. H., Jung, Y. K., Kang, H. O., Kim, T. W., Park, S. J., & Lee, S. Y. (2011). Tailor-made type II *Pseudomonas* PHA synthases and their use for the biosynthesis of polylactic acid and its copolymer in recombinant *Escherichia coli. Applied Microbiology and Biotechnology, 90,* 603–614.

Yang, T. H., Kim, T. W., Kang, H. O., Lee, S.-H., Lee, E. J., Lim, S.-C., et al. (2010). Biosynthesis of polylactic acid and its copolymers using evolved propionate CoA transferase and PHA synthase. *Biotechnology and Bioengineering*, *105*, 150–160.

Yang, J. E., Kim, J. W., Oh, Y. H., Choi, S. Y., Lee, H., Park, A. R., et al. (2016). Biosynthesis of poly (2-hydroxyisovalerate-co-lactate) by metabolically engineered *Escherichia coli*. *Biotechnology Journal*, *11*, 1572–1585.

Yang, J. E., Park, S. J., Kim, W. J., Kim, H. J., Kim, B. J., Lee, H., et al. (2018). One-step fermentative production of aromatic polyesters from glucose by metabolically engineered *Escherichia coli* strains. *Nature Communications*, *9*, 79.

Further reading

Niaounakis, M. (2014). *Biopolymers: Processing and products*. William Andrew.

CHAPTER SIX

Enzymatic synthesis and modification of high molecular weight DNA using terminal deoxynucleotidyl transferase

Sonal Deshpande[a,†], Yunqi Yang[b,†], Ashutosh Chilkoti[a,b,*], Stefan Zauscher[a,b,*]

[a]Department of Biomedical Engineering, Duke University, Durham, NC, United States
[b]Department of Mechanical Engineering and Materials Science, Duke University, Durham, NC, United States
*Corresponding authors: e-mail address: chilkoti@duke.edu; zauscher@duke.edu

Contents

† Contributed equally to this work.

Methods in Enzymology, Volume 627
ISSN 0076-6879
https://doi.org/10.1016/bs.mie.2019.07.044

Abstract

The recognition that nucleic acids can be used as polymeric materials led to the blossoming of the field of DNA nanotechnology, with a broad range of applications in biotechnology, biosensors, diagnostics, and drug delivery. These applications require efficient methods to synthesize and chemically modify high molecular weight DNA. Here, we discuss terminal deoxynucleotidyl transferase (TdT)-catalyzed enzymatic polymerization (TcEP) as an alternative to conventional enzymatic and solid-phase DNA synthesis. We describe biochemical requirements for TcEP and provide step-by-step protocols to carry out TcEP in solution and from surfaces.

1. Introduction

Beyond their role as the genetic material of living things, nucleic acids (DNA and RNA) are now being used to create synthetic polymeric materials for use in nanotechnology (Bae, Kocabey, & Liedl, 2019; Liu, Jiang, Wang, & Ding, 2019; Suo et al., 2019; Zhao et al., 2019), biosensing (Abolhasan, Mehdizadeh, Rashidi, Aghebati-Maleki, & Yousefi, 2019; Mason, Tang, Li, Xie, & Li, 2018; Tjong, Yu, Hucknall, Rangarajan, & Chilkoti, 2011) and drug delivery (Liu et al., 2019; Mathur & Medintz, 2019). Nucleic acids are versatile due to their customizable sequences and specific molecular recognition. Applications of DNA and RNA nanotechnology often require chemical and enzymatic polynucleotide modifications such as the incorporation of fluorescent groups, reactive groups including azides and amines, biofunctional groups such as biotin, and sugar modifications such as 2′-O-methyl (2′-OMe) or 2′-deoxy-2′-fluoro-ribonucleotide (2′-F) to bestow RNA with nuclease stability for RNA interference (RNAi) and CRISPR-based genome editing (Yin et al., 2017).

Modifications of DNA can be introduced chemically by using solid-phase synthesis or enzymatically by using a DNA polymerase together with unnatural deoxynucleotidyl triphosphates (dNTPs). Solid-phase synthesis is widely used when large quantities are desired, but is limited by the oligonucleotide length that can be produced (~200 nt) and by the compatibility of functional groups with the chemical reactions used in oligonucleotide synthesis (Rothlisberger & Hollenstein, 2018). Enzymatic DNA synthesis methods such as PCR (Rittié & Perbal, 2008; Terpe, 2013; Vosberg, 1989), primer extension (Carey, Peterson, & Smale, 2013) and rolling circle amplification (RCA) (Dean, Nelson, Giesler, & Lasken, 2001) typically use polymerases that have limited tolerance for chemically modified dNTPs

(Houlihan, Arangundy-Franklin, & Holliger, 2017) and that require a template to catalyze primer extension. In contrast, terminal deoxynucleotidyl transferase (TdT) is a template-independent polymerase which tolerates chemically modified dNTPs and thus is suitable for the synthesis of chemically modified, high molecular weight DNA. This mechanism is harnessed in TdT catalyzed enzymatic polymerization (TcEP) for the synthesis of single-stranded polynucleotides (Tang, Navarro Jr., Chilkoti, & Zauscher, 2017) (Fig. 1).

TdT is a member of the X family of DNA polymerases, whose members add dNTPs randomly to the 3′-hydroxyl group (3′-OH) of single-stranded DNA (Bollum, 1960). Biologically, TdT contributes to the diversity of immunoglobulins and T cell receptors (TCR) by participating in the rearrangement of variable (V), diversity (D), and joining (J) segments within immunoglobulin and TCR genes, as briefly described next.

The vertebrate adaptive immune system consists of B cells and T cells that express a vast repertoire of antigen receptors (Alberts et al., 2002; Owen, Punt, Stranford, Jones, & Kuby, 2013). Binding of these receptors to antigens on the surfaces of pathogens or foreign bodies triggers a cascade of responses that activates the immune system. Rather than encoding the huge number of antigen-binding receptors with individual genes, cells

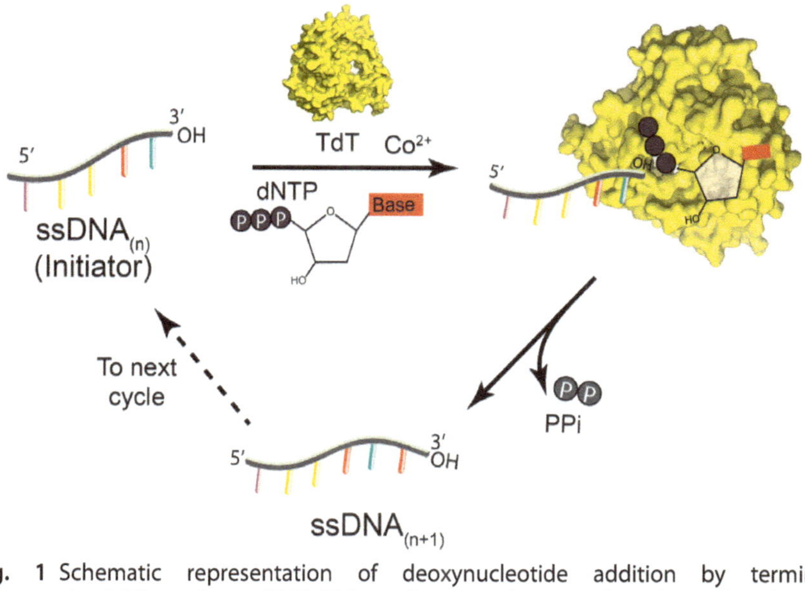

Fig. 1 Schematic representation of deoxynucleotide addition by terminal deoxynucleotidyl transferase (TdT). This mechanism is harnessed in TdT catalyzed enzymatic polymerization (TcEP). n = number of nucleotides (degree of polymerization).

use a gene rearrangement process called V(D)J recombination. This process is orchestrated by polymerases, nucleases, and ligases. TdT participates in the second stage of the process, following a double-stranded break at a highly conserved recombination signal sequence (RSS) that is induced by the recombination-activating gene (RAG) enzyme and the endonuclease, Artemis (Jung & Alt, 2004; Nick-McElhinny et al., 2005). Cleavage at the RSS by RAG results in the formation of a hairpin structure following transesterification of the 3′-OH of a nicked strand with a phosphorous on the opposite strand. Artemis opens the hairpin and TdT adds ~1–10 random nucleotides (Schatz, Oettinger, & Schlissel, 1992). TdT preferentially adds dGTP and dCTP, resulting in a GC-rich region in these genes. This GC bias promotes efficient annealing of single-stranded DNA in the subsequent ligation.

As we show and discuss in more detail in Section 3, TdT catalyzed enzymatic polymerization can be harnessed for the synthesis of high molecular weight single-stranded homo- and co-polynucleotides in solution and from surfaces (Chow, Lee, Zauscher, & Chilkoti, 2005; Gu, Tang, Aritome, & Zauscher, 2017; Tang et al., 2014; Tang et al., 2017; Tjong et al., 2011). These polyelectrolytes hold great promis for applications ranging from drug delivery to biosensors. Very recently it was shown that TdT can be employed in the synthesis of sequenced polynucleotides, overcoming some of the limitations of current solid phase olionucleotide synthesis (Lee, Kalhor, Goela, Bolot, & Church, 2019).

2. Biochemical requirements for TdT mediated polymerization

A TdT-catalyzed enzymatic polymerization reaction consists of an oligodeoxynucleotide as the initiator, natural or unnatural dNTPs, TdT, and buffer supplemented with a metal ion cofactor (Fig. 1) (Bollum, 1960, 2006; Kato, Goncalves, Houts, & Bollum, 1967). This section will describe each of these components in TcEP reactions in more detail.

2.1 Initiators

An oligodeoxynucleotide with a free 3′-hydroxyl terminus is essential for initiation of polymerization by TdT. Generally, TdT prefers either a single-stranded DNA (ssDNA) initiator or double-stranded DNA (dsDNA) with a four-nucleotide 3′ overhang (Delarue et al., 2002; Fowler & Suo, 2006). The presence of a secondary structural motif at the 3′ terminus can hinder access of the 3′-OH to the catalytic site of the enzyme and limit nucleotide addition (Delarue et al., 2002). Even single-stranded homo-oligonucleotides such as poly(dC) and poly(dG) can form secondary structures (i-motifs and G-quadruplexes, respectively) that impede polymerization. Even though poly(dA) can form secondary structures, it is an efficient TcEP initiator, as is poly(dT), which does not form secondary structure. The minimum length for efficient initiation by TdT is typically a trideoxynucleotide (Bollum, 1974; Tjong et al., 2011), though Hayes et al. successfully used a dinucleotide as initiator (Hayes, Mitchell, Ratliff, Schwartz, & Williams, 1966). However, for such short initiators (<4 nt), 5′ phosphorylation is required. Initiation from blunt-ended dsDNA is also possible. However, in that case, the reaction buffer must be supplemented with Co^{2+} rather than Mg^{2+} (Roychoudhury, Jay, & Wu, 1976). Interestingly, TdT can extend an initiator that has a single ribonucleotide 5′-monophosphate (rNMP) at the 3′ terminus, but in this case the reaction rate is reduced and the number of nucleotides added is limited (Roychoudhury, 1972).

2.2 Deoxynucleoside triphosphates (dNTPs)

TdT can incorporate a wide range of natural and unnatural dNTPs (Table 1), where the preferred order of incorporation by TdT is dGTP > dCTP > dTTP > dATP (Berdis & McCutcheon, 2007; Chang et al., 1988; Wan et al., 2013). However, as mentioned above, initiator extension in the presence of only dGTP or dCTP is limited by the formation of secondary structures of oligo(dG) and oligo(dC) (Bollum, 1974). Poly(dA) and poly(dT) ssDNA homopolymers as long as 8 kb have been synthesized by our group using dA_{10} and dT_{10} initiators and dATP and dTTP monomers, respectively (Tang et al., 2014; Tang et al., 2017; Tjong et al., 2011). We observed that the ratio of monomer (dNTP) to initiator concentration (M/I ratio) as well as the incubation time provide control over the polynucleotide length (molecular weight) (Tang et al., 2017).

Unnatural nucleotides can be used to endow synthesized DNA with desirable functionalities. A variety of unnatural dNTPs have been used successfully in TdT-mediated polymerization, including dNTPs modified with biotin (Flickinger et al., 1992; Gavrieli et al., 1992), alkyne (Gu et al., 2018; Winz et al., 2015), digoxigenin (Schmitz et al., 1991), or fluorescent dyes (Guerra, 2006; Igloi, 1998; Tjong et al., 2011) as well as amine allyl (Kumar et al., 1988) and dideoxy (Guerra, 2006) dNTPs (Table 1).

Table 1 Examples of unnatural dNTP substrates for TdT and their applications.

Modification	**dNTP**	**Application**
Biotin (Flickinger, Gebeyehu, Buchman, Haces, & Rashtchian, 1992)	Biotin-11-dUTP	Labeling DNA probes *In situ* labeling of cells (Gavrieli, Sherman, & Ben-Sasson, 1992)
Amine (Kumar, Tchen, Roullet, & Cohen, 1988)	Aminoallyl-dUTP	Tuning nuclease resistance (Gu, Oweida, Yingling, Chilkoti, & Zauscher, 2018)
Azide (Winz, Linder, Andre, Becker, & Jaschke, 2015)	2′-N_3–2-dA/C/G/UTP 5-*PEG*N_3-dUTP	
Alkyne (Winz et al., 2015)	5-ethynyl-dUTP	
Dibenzocyclooctyne (DBCO) (Gu et al., 2018)	DBCO-dUTP	
Digoxigenin (Schmitz, Walter, Seibl, & Kessler, 1991)	DIG-dUTP	
Dideoxy dNTPs (Guerra, 2006)	Cy™3-ddUTP	
Fluorescent dyes (Guerra, 2006; Igloi, 1998; Tjong et al., 2011)	Fluorescein-12-ddUTP Cy5-UTP	
Photocrosslinkers	4-thio-2′-dUTP	Studying DNA structure and interactions with proteins (Tauraitė, Jakubovska, Dabužinskaitė, Bratchikov, & Meškys, 2017)

The ability of TdT to incorporate these unnatural nucleotides depends on the type and position of the dNTP modification and must be evaluated on a case-by-case basis. DNA polymerases generally tolerate modifications at the C5 position of pyrimidines (T and C) and at the C7 position of purines (A and G) (Fig. 2) that do not hinder interactions with the enzyme or affect Watson-Crick base pairing (Bergen et al., 2012; Obeid, Baccaro, Welte, Diederichs, & Marx, 2010). However, dNTPs modified with bulky hydrophobic groups such as dibenzylcyclooctyne (DBCO) or large fluorophores such as Cy3 or fluorescein often require a hydrophilic linker such as poly(ethylene glycol) (PEG) to position these groups away from the enzyme's active site. A study of how TdT tolerates modifications at

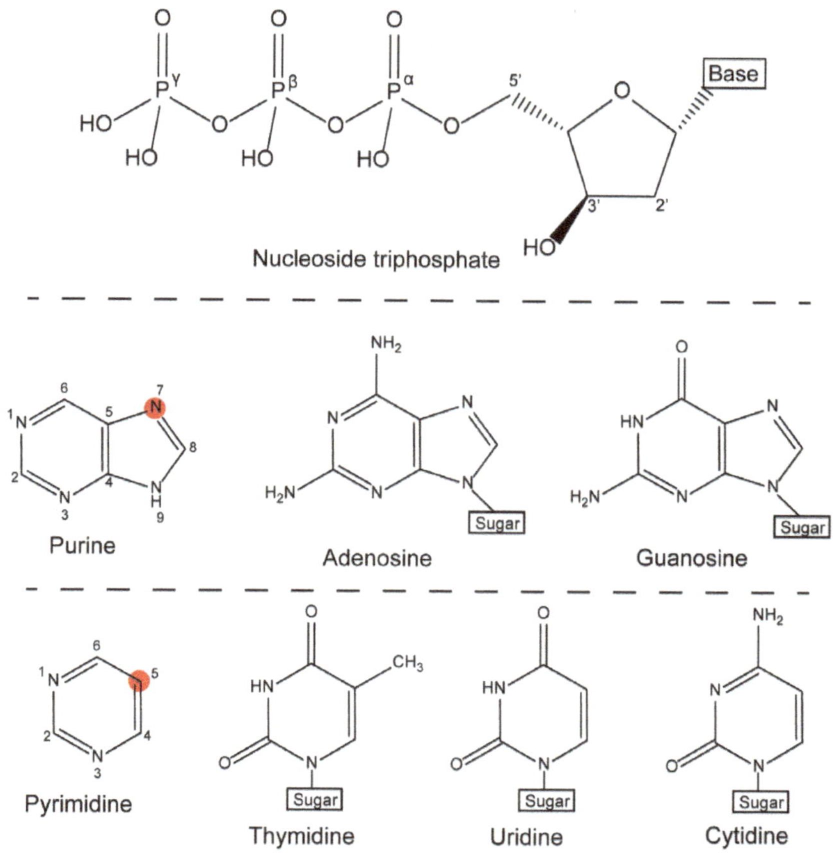

Fig. 2 Generic structure of deoxynucleotide triphosphates and nitrogen bases of nucleic acids. Positions marked in red indicate the location for chemical modifications that are most easily tolerated by DNA polymerases.

position 2 of dATP found that bulky modifications such as phenyl groups have an inhibitory effect on TdT activity, while smaller modifications such as Cl, NH_2, CH_3, vinyl, and ethynyl groups are readily incorporated (Matyašovský, Perlíková, Malnuit, Pohl, & Hocek, 2016). Interestingly, dNTPs with bulky, hydrophobic modifications impart nuclease stability to the modified ssDNA (Gu et al., 2018). Like most of the DNA polymerases, TdT poorly facilitates the addition of ribonucleotide triphosphates (NTPs). However, substituting ribose with arabinose altogether inhibits polymerization (Hansbury et al., 1970; Roychoudhury et al., 1976).

Palluk et al. recently demonstrated the use of covalent TdT-dNTP conjugates for *de novo* (template-free) DNA synthesis (Fig. 3) (Palluk et al., 2018). In template-free DNA synthesis, TdT incorporates the dNTP into

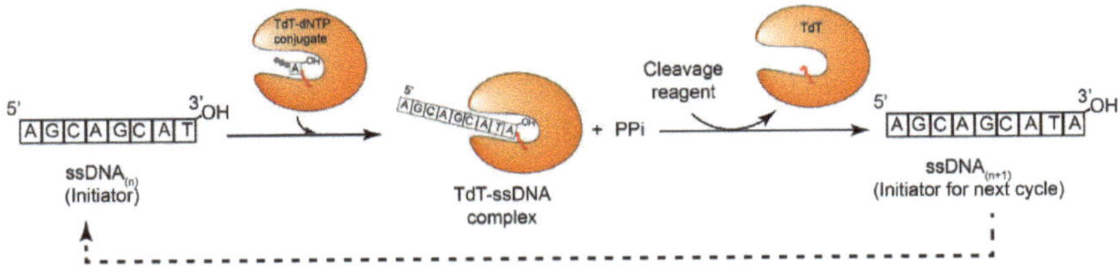

Fig. 3 Schematic for terminal-deoxynucleotidyl transferase (TdT) mediated synthesis of an oligonucleotide with user defined sequence (Palluk et al., 2018). The TdT with dNTP monomer conjugated at the active site *via* a cleavable linker is incubated with ssDNA initiator. After the extension event, treatment of the TdT-ssDNA complex with cleavage reagents (*e.g.*, DTT, peptidase, 365 nm light) releases the ssDNA and makes the 3′-OH available for next extension event. Depending on the intended sequence of oligonucleotide, the user can select the TdT-dNTP conjugate for the next cycle.

a primer and remains bound to the complex, blocking the 3′ end of the primer and inhibiting further addition. Excess unbound conjugate is removed and the 3′-OH is activated by cleaving the linker that connects dNTP to TdT. The oligonucleotide can then be subjected to the next round of extension with another TdT-dNTP conjugate to create the desired sequence (Palluk et al., 2018). As TdT can synthesize kilobase-length polynucleotides, this method can be used to overcome the limitations of sequence fidelity and length encountered in conventional chemical syntheses.

2.3 TdT

Terminal deoxynucleotidyl transferase was discovered in 1960 when it was purified from a calf thymus gland (Bollum, 1960). The full length 58 kDa protein undergoes proteolytic degradation, generating fragments some of which still show enzymatic activity (Chang, Plevani, & Bollum, 1982). However, this makes the purification of a homogenous, active TdT enzyme from calf thymus tedious. Growing interest in the structure-function relationship of TdT required producing pure enzyme in large quantities, thus other methods to obtain TdT have been explored, including expression of human TdT in human acute lymphoblastic leukemia cell lines or in baculovirus (Chang et al., 1988) and expression of avian TdT in *E.coli* (Yang, Gathy, & Coleman, 1995). Nevertheless, these systems were unable to generate sufficient quantities of active enzyme. Boulé et al. increased the expression level of murine TdT in *E.coli* by lowering the culture temperature and overexpressing a rare arginine tRNA (argU) (Boulé, Johnson, Rougeon, & Papanicolaou, 1998). In an attempt to improve the enzyme's

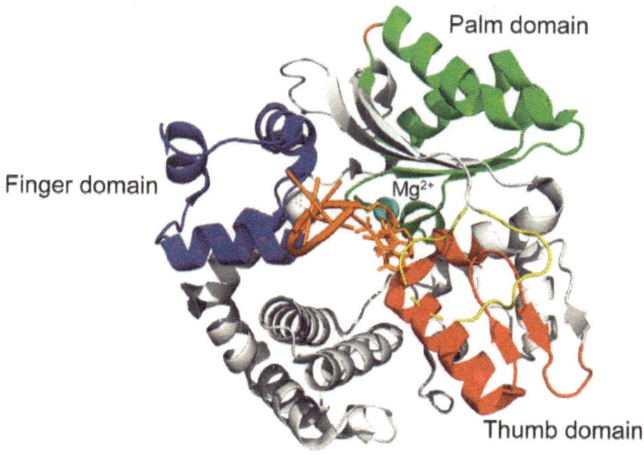

Fig. 4 A 2.6 Å resolution crystal structure of murine TdT ternary complex with ssDNA initiator and incoming nucleotide (PDB ID 4I27) (Gouge, Rosario, Romain, Beguin, & Delarue, 2013). The palm, thumb and finger domains are represented in green, red and blue, respectively. Access of a template strand to the active site is restricted by loop1 (yellow). Mg^{2+} ion: cyan; ssDNA and incoming nucleotide: orange.

activity for *in vitro* applications, a truncated version of calf thymus TdT with a 20–30 fold increase in activity was expressed in *E.coli* (Mueller et al., 2009).

In 2002, more than 40 years after the discovery of TdT, the first crystal structure of the catalytic core of murine TdT (130–510 amino acid residues) was solved (Delarue et al., 2002). Like in most DNA polymerases, TdT's catalytic core exhibits the typical "thumb," "palm," and "finger" domains (Fig. 4). However, unlike DNA pol β, and other DNA polymerases, TdT has an additional 16 residue loop (loop 1) that assumes a lariat-like conformation and is thought to hinder the accommodation of a template strand (Delarue et al., 2002). Thorough study of the initiator-TdT binary complex suggested that the last four nucleotides at the 3′-OH end interact with the polymerase, irrespective of the nature of the base. At the catalytic site metal ions coordinate with the conserved aspartate residues and the incoming dNTP, positioning its alpha phosphate (see Fig. 2) for nucleophilic attack by the 3′-OH. The absence of specific interactions of the nucleotide bases with TdT explains the lack of specificity toward addition of dNTP (Delarue et al., 2002; Gouge et al., 2013).

2.4 Metal ions as cofactors

Divalent cations coordinate catalytic polymerase residues with the incoming dNTP and thus play an important role as cofactors in the synthesis of polynucleotides by DNA polymerases (Steitz, 1993). Unlike other DNA polymerases, TdT can utilize several different metal ions including Mg^{2+}, Co^{2+}, Mn^{2+} and Zn^{2+}, though its preference for nucleotide incorporation depends on the metal ion used. For example, when the buffer is supplemented with Mg^{2+}, the purine nucleotide incorporation rate is 10-fold higher than that of pyrimidines (Douglas & Morgan, 1976). When the buffer is supplemented with Co^{2+}, the pyrimidine nucleotide incorporation rate is higher (Chang & Bollum, 1990). Interestingly, supplementing Mg^{2+} with submillimolar concentrations of Zn^{2+} increases TdT activity 2–4 fold and also rescues TdT inhibition by metal chelators (Chang & Bollum, 1990).

3. Applications

3.1 TdT catalyzed polynucleotide synthesis in solution

Tang et al. demonstrated that TcEP proceeds by a living chain-growth polycondensation mechanism, and thus the molecular weight of the product can be tuned by changing the feed ratio of nucleotide (monomer) to oligonucleotide (initiator) (M/I ratio) (Tang et al., 2017). Taking advantage of the template-free, promiscuous polymerization ability of TdT, single stranded polynucleotides with narrow molecular weight distribution and containing various unnatural nucleotides, including amines, alkynes, azides, and hydrophobes, can be synthesized (Gu et al., 2018). Thus it is easy to functionalize the ssDNA chains to produce stable nucleotide nanoparticles (Fig. 5) (Tang et al., 2014), to synthesize cytostatic polynucleotide drugs and to increase their nuclease resistance (Gu et al., 2018) which potentially helps to increase the half-life of the drug. Interestingly, a recent study by Que. et al. combined the template-independent property of TdT with the high amplification efficiency of RCA to develop a label-free, TdT-induced RCA based method to detect DNA Methyltransferase (MTase) (Que et al., 2019). This method has great potential for DNA MTase-related cancer diagnosis and disease therapeutics.

3.2 TdT catalyzed polynucleotide synthesis on surfaces

In addition to catalyzing nucleotide polymerization in solution, TdT can also be used for the *in situ* synthesis of polynucleotide brushes on surfaces by the "grafting-from" approach. This process was first introduced by Chow et al. who also coined the term "surface-initiated enzymatic polymerization" (SIEP) (Chow et al., 2005). Furthermore, SIEP can be used

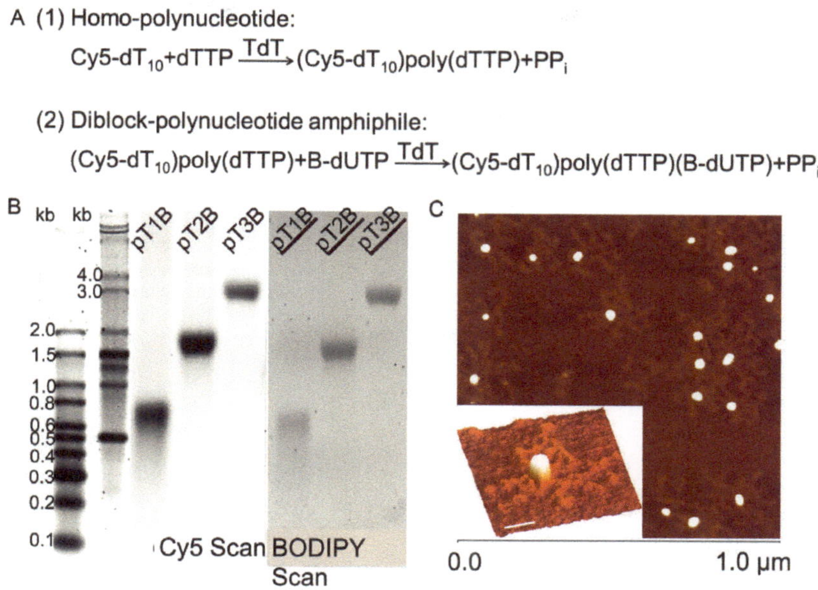

Fig. 5 (A) Schematic showing the enzyme-catalyzed synthesis of diblock polynucleotide amphiphiles. (B) Agarose gel electrophoresis results showing enzymatically synthesized poly(dTTP) with three different MWs (lanes pT1B-pT3B: M/I of 200:1, 500:1, and 1000:1), which were enzymatically end-functionalized with BODIPY-dUTP at a molar ratio of BODIPY-dUTP to poly(dTTP) 10:1. The presence of BODIPY-dUTP was verified by a BODIPY fluorescence scan. (C) AFM Tapping Mode height images of star like micelles formed by amphiphilic diblock polynucleotides. Inset: 3D image of a typical micelle (scale bars: 50nm XY, 1.5nm Z). *Adapted from Tang, L., Tjong, V., Li, N., Yingling, Y. G., Chilkoti, A., & Zauscher, S. (2014). Enzymatic polymerization of high molecular weight DNA amphiphiles that self-assemble into star-like micelles.* Advanced Materials 26 *(19), 3050–3054. doi:10.1002/adma.201306049.*

to functionalize nanoparticles, including DNA origami (Okholm, Aslan, Besenbacher, Dong, & Kjems, 2015), liposomes (Ruysschaert, Paquereau, Winterhalter, & Fournier, 2006) and gold nanoparticles (Wang et al., 2016), with ssDNA. The template-free SIEP is useful for the fabrication of sensors that can detect nucleases (Chen, Xu, Ji, & He, 2017), DNA (Tjong et al., 2011; Wan et al., 2013; Wan et al., 2014), RNA (Tjong, Yu, Hucknall, & Chilkoti, 2013), metal ions (Mei et al., 2016), tumor biomarkers (Wang et al., 2016) and proteins (Shi, Dou, Yang, Yuan, & Xiang, 2017). An early study by Tjong et al. demonstrated a sensing platform to detect ssDNA by first immobilizing a capture oligonucleotide on a glass surface, followed by SIEP at the 3′ end of the detected and hybridized ssDNA (Fig. 6) (Tjong et al., 2011). Efficient initiation ($\geq$50%) and narrow

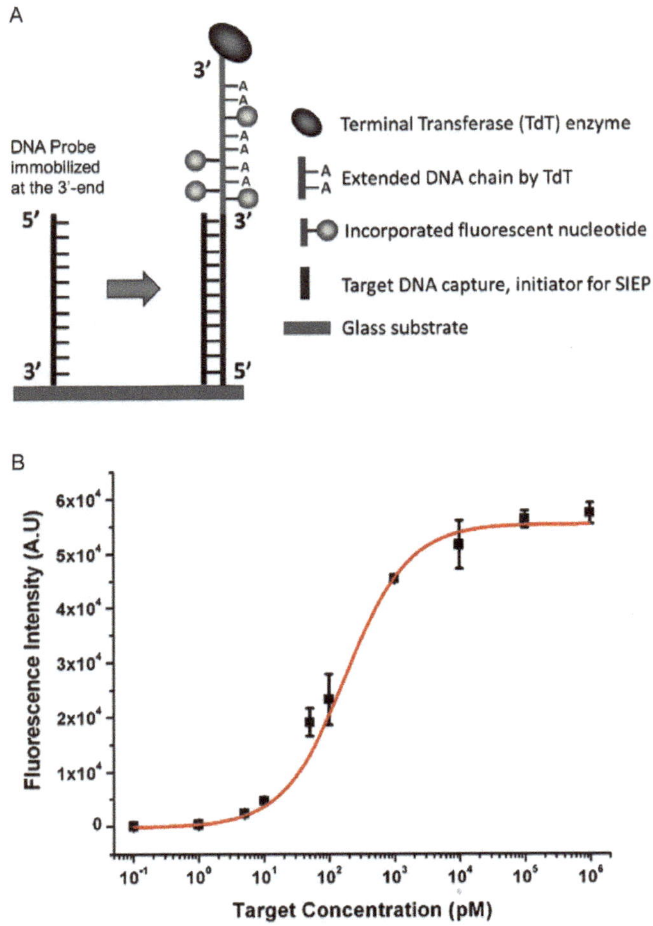

Fig. 6 (A) Schematic showing a single-step, on-chip detection and signal-amplification DNA microarray sensor. (B) Dose-response curve of the sensor shown in a). *Adapted with permission from Tjong, V., Yu, H., Hucknall, A., Rangarajan, S., & Chilkoti, A. (2011). Amplified on-chip fluorescence detection of DNA hybridization by surface-initiated enzymatic polymerization.* Analytical Chemistry 83*(13), 5153–5159. doi:10.1021/ac200946t. Copyright 2011 American Chemical Society.*

polydispersity of the extended product were observed when fluorophore-labeled dNTPs were incorporated. Up to ~50 Cy3-labeled dNTPs per kilobase could be incorporated into a ssDNA chain using TdT, which translated into a ~45-fold signal amplification compared to the incorporation of a single fluorophore. This led to a detection limit of about 1 pM, and a linear dynamic range of 2 logs.

Other studies used TdT to incorporate biotin-modified-dNTP to take advantage of the biotin-avidin interaction which increases sensor sensitivity.

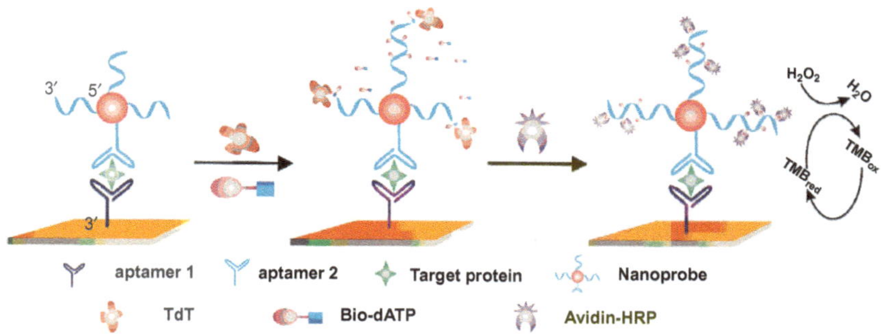

Fig. 7 Schematic showing the assembly of an electrochemical sensor for the highly sensitive detection of CEA. Aptamer 1 is immobilized on a gold electrode *via* its 3′-thiol modification, and aptamer 2 is immobilized on gold nanoparticles (GNPs) using its 5′-thiol modification to form a nanoprobe. If a target is present, it gets sandwiched between aptamer 1 and the aptamer 2-functionalized nanoprobe. After the incorporation of biotin-dATPs at the free 3′-OH groups of the GNPs using SIEP, numerous, long and biotin-labeled ssDNA strands are generated. This allows for the specific attachment of avidin-HRP complexes which aid in the electrochemical signal amplification. *Reprinted with permission from Wang, P., Wan, Y., Deng, S., Yang, S., Su, Y., Fan, C., … Zuo, X. (2016). Aptamer-initiated on-particle template-independent enzymatic polymerization (aptamer-OTEP) for electrochemical analysis of tumor biomarkers.* Biosensors & Bioelectronics 86, *536–541. doi:10.1016/j.bios.2016.07.025. Copyright 2016 Elsevier.*

For example, Wang et al. developed an electrochemical biosensor for the analysis of a cancer bio-marker, carcino-embryonic antigen (CEA), by incorporating a pair of aptamers which can specifically bind with CEA (Wang et al., 2016). One of the aptamers, immobilized on a gold electrode, serves as the capture probe, while the other aptamer is linked to a gold nanoparticle (GNP) probe. After the dual aptamer capture of CEA in "sandwich" format, the oligonucleotides on the GNPs are harnessed to initiate SIEP to incorporate biotin-dNTP, which are ultimately used for enhancing the signal (Fig. 7).

SIEP can also be used to produce complex biomolecular structures on flat surfaces with nanoscale resolution (Chow et al., 2005; Khan, Tjong, Chilkoti, & Zharnikov, 2012; Khan, Tjong, Chilkoti, & Zharnikov, 2013). Recently, SIEP was harnessed for the design of a digital information storage medium, in which information is stored in the transitions between non-identical nucleotides of ssDNA (Fig. 8) (Lee et al., 2019).

4. Materials

T_{50} and 5′ end-modified oligonucleotides can be procured from Integrated DNA Technology (IDT). Suppliers including Jena Biosciences,

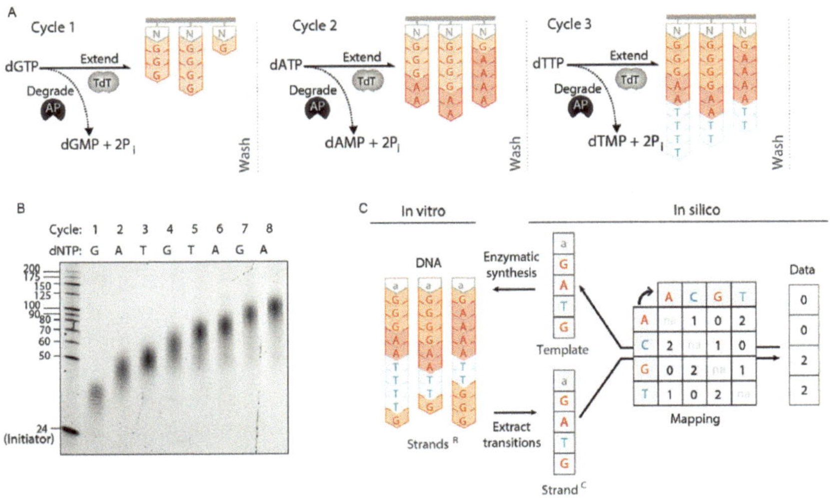

Fig. 8 An enzymatic synthesis strategy for storing information in ssDNA. (A) Schematic depiction of a series of enzymatic synthesis reactions consisting of oligonucleotide initiator (N), TdT and apyrase (AP). The initiator is tethered to a solid support or beads. (B) TBE-urea gel showing the DNA strands synthesized in each consecutive synthesis cycle. (C) Schematic showing the interconversion of DNA sequence to digital information. Raw strands (Strands^R) represent synthesized DNA, while a compressed strand (Strand^C) represents a sequence of transitions between non-identical nucleotides. Transitions between nucleotides are mapped from Strand^C to digital data in ternary digits. If a Strand^C is equivalent to the template sequence, all desired transitions are present and the information stored in DNA is retrieved. (Nucleotides A, T, G, C are colored as red, blue, orange, and dark blue, respectively.) *Adapted with permission from Lee, H. H., Kalhor, R., Goela, N., Bolot, J., & Church, G. M. (2019). Terminator-free template-independent enzymatic DNA synthesis for digital information storage.* Nature Communications 10 *(1), 2383. doi:10.1038/s41467-019-10258-1. Copyright 2019 Springer Nature.*

TriLink Biotechnologies, Sigma Aldrich and Thermo Fisher Scientific provide natural dNTPs as well as a range of unnatural dNTPs. Terminal deoxynucleotidyl transferase (TdT) and TdT buffer can be purchased from Promega. Ethylenediaminetetraacetic acid (EDTA) (0.5 M, pH 8) solution, sulfuric acid, nitric acid, sodium acetate, sodium chloride, sodium citrate, acetone, ethanol, 10 × tris-borate EDTA buffer (pH 8.3) and Tween 20 can be procured from Sigma Aldrich. 0.5 mL Microcon Centrifugal Filter Units can be purchased from Millipore, 5% TBE-PAGE gel from Biorad, Agarose (low EEO) from Thermo Scientific, glass slides from Solulink, and succinimidyl 4-formylbenzoate from VWR. Gold substrates can be fabricated by metal evaporation on a silicon wafer of first a ~10 nm titanium (or chromium) adhesion layer followed by a layer of ~100 nm gold.

5. Protocol for solution polymerization and characterization

5.1 Protocol

1.1 Prepare reaction solution containing 0.5 μM oligo initiator (*e.g.*, T_{50}), natural or unnatural dNTP (the final dNTP concentration depends on the desired length of the polynucleotide), and 1 × TdT reaction buffer (100 mM potassium cacodylate, 1 mM $CoCl_2$, and 0.1 mM DTT, pH 6.8). Then add 1.2 U/μL TdT and pipette to mix (DO NOT VORTEX!). If the amount of TdT used is too low, the reaction will yield polydisperse polynucleotides.

1.2 Incubate overnight at 37 °C.

1.3 Heat the reaction mixture at 90 °C for 3 min or at 70 °C for 10 min to inactivate the enzyme. If heat inactivation is unsuitable, add EDTA to a final concentration of 30 mM. Then add an equal volume of water and centrifuge at ≥16,000 rcf for 3 min to precipitate the enzyme. Collect supernatant by pipetting.

1.4 Purify 20–100 μL of the supernatant using a 0.5 mL Microcon Centrifugal Filter Unit (Millipore). Choose the molecular weight cut off (MWCO) of the filter according to the expected molecular weight of the polynucleotide. Perform three consecutive washes with Milli-Q grade water (>18.2 MΩ cm^{-1}) to remove all the free dNTP. After purification, concentrate the sample using the same filter, and recover the sample into a clean tube, supplied with the Microcon filter.

1.5 If a block copolymer is desired, repeat Steps 1–4 with a different dNTP.

1.6 Store the polymerization product at −20 °C. The products are stable for up to 2 month.

Tips

(a) Vortex mixing of enzymatic reactions may cause enzyme (protein) denaturation. Therefore, we recommend to mix the reaction by pipette and not by vortexing after addition of TdT.

(b) To obtain a constant polynucleotide molecular weight, maintain the relative ratio of initiator:dNTP:TdT in each reaction.

(c) It is easier to control the polynucleotide molecular weight by adjusting the ratio of dNTP monomer to initiator (M/I ratio) rather than the reaction time.

(d) If a random copolymer rather than a homo-polynucleotide is desired, use a mixture of dNTPs. However, note that nucleotide incorporation efficiency depends on the nucleotide and on the metal ion cofactor used (see Section 2.4).

(e) Use an initiator with a fluorescent tag to facilitate characterization using gel electrophoresis.

(f) Change the filter after 3–4 washes. Be sure to mix the sample on the centrifugal filter surface thoroughly by pipetting between each wash to prevent sample precipitation and loss on the filter. When using filters to remove fluorescent unnatural nucleotides, the number of wash steps is determined by the fluorescent intensity in the flow-through (wash the sample until no fluorescence is detected in the flow-through).

5.2 Characterization

2.1 The weight average and number average molecular weights ($\overline{M_w}$ and $\overline{M_n}$) and the polydispersity index ($PDI = \overline{M_w}/\overline{M_n}$) of the product can be determined by running the sample and the fractions of a molecular weight ladder on a 5% TBE-PAGE gel at 100 V for 40 min. The gel should be imaged with a high-resolution gel scanner to best capture the fluorescence intensity distribution. The image can subsequently be analyzed with ImageJ software (freely available at https://imagej.nih.gov/ij/) to obtain $\overline{M_w}$, $\overline{M_n}$ and the PDI of the product (Fig. 9), following the procedure developed by (Tang, 2016), pp.112–123.

2.2 A relatively accurate method to determine the molarity of the product is to fluorescently label the initiator and create a standard curve of the fluorophore using a fluorescence spectrometer. Another method to obtain the mass concentration of the product is to use a UV–vis photo-spectrometer (*e.g.*, Nanodrop 1000 UV–Vis spectrometer, Thermo Scientific), and determine the molarity using the molecular weight obtained from gel electrophoresis. Although HPLC is the most accurate way to determine the molecular weight and concentration, it is relatively difficult and time-consuming.

Tips:

(a) To obtain a more accurate molecular weight of a homo-polynucleotide, it is best to synthesize a ladder using the same initiator and oligonucleotide (*e.g.*, phosphorylated T50) used in polynucleotide synthesis, and concatenate these with T4 RNA-ligase in a buffer supplied with ATP.

(b) When using fluorescence to determine the molarity, the standard curve should be made with concentrations that yield a linear fluorescence dependence on concentration, and the sample concentration should be within the range of the standard curve.

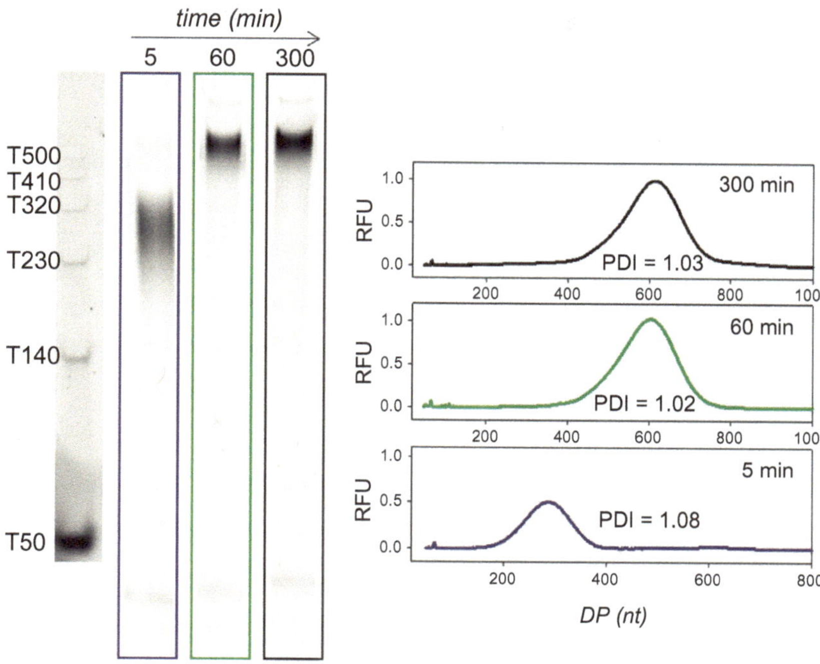

Fig. 9 Dependence of the molecular weight distribution on reaction time. Left: Cy5 scan of 5% TBE-PAGE gel images. Right: corresponding relative fluorescent intensity distributions plotted as a function of the degree of polymerization (DP). The resulting polydispersity indices (PDIs) are shown in the plots. *Adapted from Tang, L., Navarro, L. A., Jr., Chilkoti, A., & Zauscher, S. (2017). High-molecular-weight polynucleotides by transferase-catalyzed living chain-growth polycondensation.* Angewandte Chemie (International Ed. in English) 56 *(24), 6778–6782. doi:10.1002/anie.201700991.*

6. Protocol for surface-initiated enzymatic polymerization (SIEP) and characterization

6.1 Protocol for SIEP on flat gold substrates using TdT (Chow et al., 2005)

1.1 Clean the gold substrate by sequential sonication in Milli-Q water, acetone, and ethanol at 50 °C with intervening rinse steps. Dry the substrate in a stream of dry N_2 and store it in a clean, covered petri dish.

1.2 Prepare 10 μM 5′-thiol modified oligonucleotide in 1 × PBS. Drop cast the initiator solution on the gold surface (for reference: a 1 cm^2 gold surface needs ~100 μL solution for full coverage). Cover the petri-dish and incubate the substrate at room temperature overnight.

1.3 Remove the solution by rinsing with Milli-Q water. Sonicate the substrate first in 0.1% Tween 20, and then in water for 5 min at room temperature to remove any remaining free initiator. Dry the substrate in a stream of dry N_2.

1.4 Prepare a reaction mixture containing natural and/or unnatural dNTP and 1 × TdT reaction buffer. The choice of reactant concentration depends on the desired polynucleotide length. For random copolymers, a dNTP mixture with different nucleotide ratios can be used. After adding TdT to the solution to a final concentration of 0.5 U/μL, mix the solution by pipetting, and drop cast it immediately onto the initiator-modified substrate surface (as mentioned above, a 1 cm^2 gold surface needs about 100 μL solution for full coverage). Incubate the substrate in a covered petri dish at room temperature. Brush height (MW) is controlled by the length of the incubation time. However, do not exceed 4 h as solvent will evaporate.

1.5 Remove the reaction solution by rinsing thoroughly with Milli-Q water. Then sonicate the substrate first in 0.1% Tween 20 and then Milli-Q water for 5 min at room temperature. After sonication, rinse the surface with Milli-Q water and dry the substrate in a stream of dry N_2.

1.6 If a *block-co*-polynucleotide is desired, repeat Steps 4–5 with a different dNTP.

1.7 Store the DNA functionalized substrates under vacuum until further use.

Tips

(a) The surface needs to be meticulously clean to form an effective initiator monolayer and for subsequent polymerization reactions. If needed, a gold substrate can be first cleaned with plasma or piranha solution (Caruso, Serizawa, Furlong, & Okahata, 1995) before modification.

(b) A set of experiments with different reagent concentrations and reaction times is typically useful to determine optimal reaction conditions to achieve a desired brush height.

(c) Mixed thiol self-assembled monolayers of initiator- and (for example) PEG-modified thiols can be used to adjust the surface graft density of the polynucleotide brush.

6.2 Protocol for SIEP from Hylink™ glass slides using TdT (Tjong et al., 2011)

2.1 Conjugate 5′-amine-modified oligonucleotides to succinimidyl 4-formylbenzoate (SFB) linkers by reacting with a 20-fold molar excess of SFB in the modification buffer (100 mM sodium acetate, 150 mM NaCl, pH 7.2) overnight in the dark.

2.2 Remove unreacted SFB using Microcon YM-3 spin columns (Millipore). Dilute the concentrated SFB-modified-oligonucleotides in conjugation buffer (100 mM sodium citrate, 150 mM NaCl, pH 6.0), and spot onto the hydrazinonicotinamide glass substrates (Hylink Glass slides, Solulink, San Diego, CA) using a noncontact piezoelectric printer (*e.g.*, Piezorray, Perkin Elmer, Inc).

2.3 Incubate the spotted slides overnight in a humidified chamber to allow for the covalent immobilization of the oligonucleotides on the glass surface.

2.4 Remove unreacted oligonucleotides by rinsing in 1 × SSC buffer (15 mM sodium citrate, 150 mM NaCl, pH 7.0) containing 0.1% Tween 20, then rinse with Milli-Q water and spin-dry on a microarray slide centrifuge for 2 min.

2.5 Store the DNA functionalized slides under vacuum until further use.

6.3 Characterization

3.1 The average dry thickness of the polynucleotide brush can be determined by spectroscopic ellipsometry (Chow & Chilkoti, 2007) or AFM height measurements (Chow et al., 2005).

3.2 Surface patterns can be visualized by tapping mode atomic force microscopy (TM-AFM) (Chow et al., 2005; Chow & Chilkoti, 2007). For example, when ssDNA is grown on a gold surface modified with microfluidic channels (Chow & Chilkoti, 2007), a distinct pattern can be observed by AFM imaging (Fig. 10).

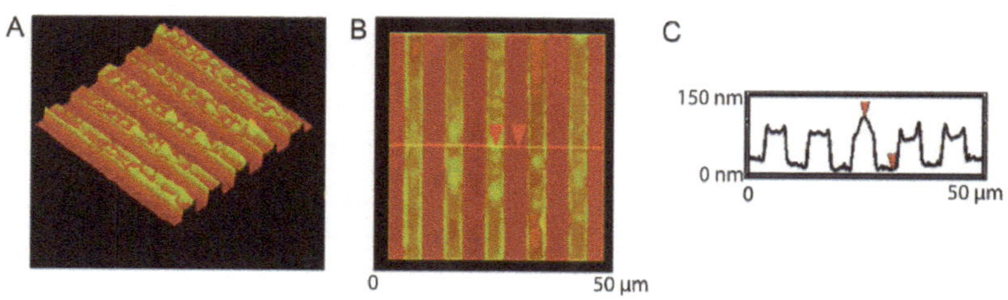

Fig. 10 (A) side and (B) top view TM-AFM height images in air of DNA brushes grown by surface-initiated enzymatic polymerization on a micropatterned initiator SAM (5′-SH $(CH_2)_6$-dTTP$_{25}$). (C) Height profile of the line indicated in (B). The dry brush height was about 60 nm. *Adapted with permission from Chow, D. C., & Chilkoti, A. (2007). Surface-initiated enzymatic polymerization of DNA.* Langmuir 23 *(23), 11712–11717. doi:10.1021/la701630g. Copyright 2007 American Chemical Society.*

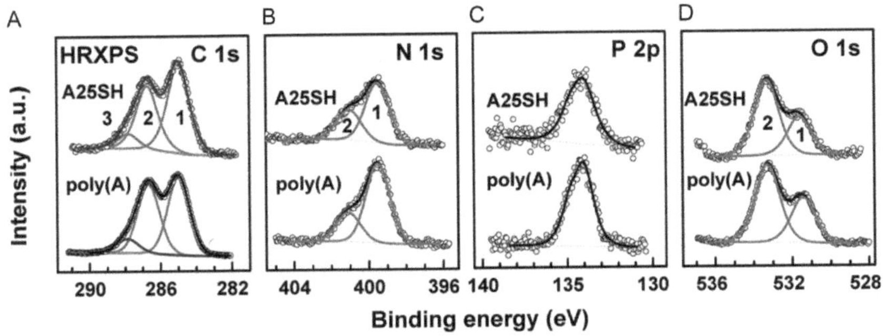

Fig. 11 (A) C 1 s, (B) N 1 s, (C) P 2p, and (D) O 1 s high resolution XPS spectra of the A25-SH SAM and poly(A) brush with a thickness of 25 nm (open circles). The C 1 s, N 1 s, and O 1 s spectra are deconvoluted (solid gray lines) into component peaks. The fitted spectral envelopes are drawn by solid black lines. *Adapted with permission from Khan, M. N., Tjong, V., Chilkoti, A., & Zharnikov, M. (2013). Spectroscopic study of a DNA brush synthesized in situ by surface initiated enzymatic polymerization.* The Journal of Physical Chemistry B 117 *(34), 9929–9938. doi:10.1021/jp404774x. Copyright 2013 American Chemical Society.*

3.3 The chemical integrity, purity, orientation, and ordering in the polynucleotide brushes can be characterized by using a combination of synchrotron-based X-ray photoelectron spectroscopy (XPS) and angular-resolved near-edge X-ray absorption fine structure (NEXAFS) spectroscopy (Khan et al., 2012, 2013). XPS provides information on the chemical composition of the samples based on the specific core level photoemissions and their chemical shifts (Fig. 11) (Khan et al., 2013). NEXAFS spectroscopy provides information by sampling of unoccupied molecular orbitals rather than the core levels probed by XPS. Furthermore, NEXAFS spectroscopy provides information about molecular orientation (Khan et al., 2013).

3.4 The kinetics of TdT-catalyzed DNA extension (Chow et al., 2005) and biosensor function (Chow & Chilkoti, 2007) can be measured in real time by surface plasmon resonance (SPR) spectroscopy. After forming a self-assembled monolayer on the surface, the surface can be mounted in an SPR spectrometer. A solution containing all SIEP reactants (described above) is injected into the spectrometer at a set flow rate (Chow et al., 2005). The SPR response curve shows the growth of ssDNA on the surface over time. After a certain time, the response curve reaches an asymptote, indicating the end of the reaction (Fig. 12).

3.5 The behavior of electrochemical sensors can be monitored by using cyclic voltammetry (CV) and differential pulse voltammetry (DPV) (Mei et al., 2016; Shi et al., 2017; Wan et al., 2013). CV measurements can also be used to characterize the stepwise modification process of the

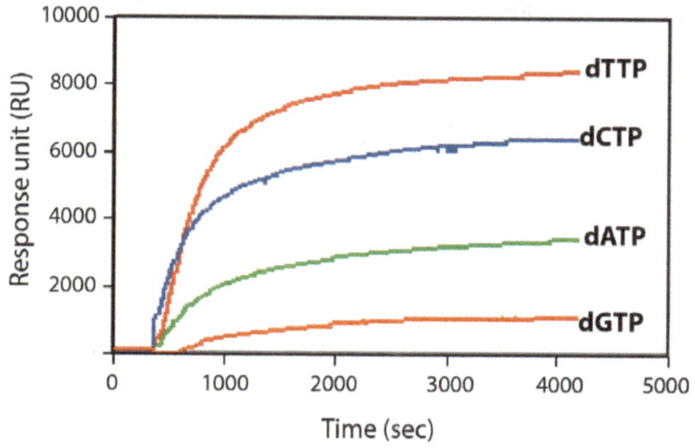

Fig. 12 Kinetics of TdT-catalyzed DNA extension using four different nucleotides measured by SPR. As a control, the same reaction was performed without the addition of nucleotides to account for the SPR signal generated by the buffer exchange and non-specific binding of the reaction components to the surface. *Adapted with permission from Chow, D. C., Lee, W. K., Zauscher, S., & Chilkoti, A. (2005). Enzymatic fabrication of DNA nanostructures: Extension of a self-assembled oligonucleotide monolayer on gold arrays.* Journal of the American Chemical Society 127 *(41), 14122–14123. doi:10.1021/ja052491z. Copyright 2005 American Chemical Society.*

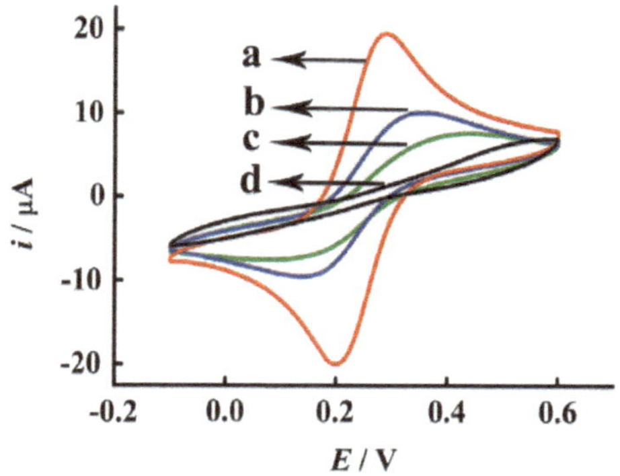

Fig. 13 CV characterizations of step-by-step modification of a gold electrode (AuE). (A) bare AuE, (B) T-HP/MCH/AuE, (C) (thrombin + S-HP)/T-HP/MCH/AuE and (D) TdT/(thrombin + S-HP)/MCH/T-HP/AuE. T-HP: Thiol-modified hairpin probe; MCH: 6-mercaptohexano; S-HP: signal hairpin probe. *Adapted with permission from Shi, K., Dou, B., Yang, J., Yuan, R., & Xiang, Y. (2017). Target-triggered catalytic hairpin assembly and TdT-catalyzed DNA polymerization for amplified electronic detection of thrombin in human serums.* Biosensors & Bioelectronics 87*, 495–500. doi:10.1016/j.bios.2016.08.056. Copyright 2017 Elsevier.*

sensing electrode. Because of the addition and the elongation of ssDNA sequences on the electrode, the negative charges and steric hindrance are increased, which inhibits the electron transfer and reduces the peak currents (Fig. 13) (Shi et al., 2017).

7. Summary

Template-independent terminal deoxynucleotidyl transferase catalyzed enzymatic polymerization (TcEP) provides an exciting alternative to conventional enzymatic and solid-phase DNA syntheses. Specifically, the tolerance of TdT for unnatural dNTPs can be harnessed to introduce a broad range of functional groups into polynucleotides and thus provides a means to synthesize functionalized polynucleotides for a broad range of applications. In this chapter we provided a brief background on TdT, and described the effect that different reactants used in TdT polymerization have on polynucleotide growth and functionality, with the goal to provide general guidelines for designing the polymerization reactions. Furthermore, we discussed some selected applications of TcEP in solution and on surfaces. Finally, we provided step-by-step protocols to carry out TcEP in solution and initiated from surfaces, paying special attention on how to control the TcEP reaction, and how to purify and characterize the reaction products.

Acknowledgments

This work was supported by NSF DMR-1411126, NSF DMR-1121107, and NIH 5 R21 EB 026590 (Trailblazer). We thank Loren Baugh for editorial comments.

References

Abolhasan, R., Mehdizadeh, A., Rashidi, M. R., Aghebati-Maleki, L., & Yousefi, M. (2019). Application of hairpin DNA-based biosensors with various signal amplification strategies in clinical diagnosis. *Biosensors and Bioelectronics*, *129*, 164–174. https://doi.org/10.1016/j.bios.2019.01.008.

Alberts, B., Johnson, A., Lewis, J., Martin, R., Roberts, K., & Walter, P. (2002). *The adaptive immune system. In Molecular biology of the cell.* (4th ed.). New York: Garland Science Chapter 24. Retrieved from https://www.ncbi.nlm.nih.gov/books/NBK21070/.

Bae, W., Kocabey, S., & Liedl, T. (2019). DNA nanostructures in vitro, in vivo and on membranes. *Nano Today*, *26*, 98–107. https://doi.org/10.1016/j.nantod.2019.03.001.

Berdis, A. J., & McCutcheon, D. (2007). The use of non-natural nucleotides to probe template-independent DNA synthesis. *Chembiochem*, *8*(12), 1399–1408. https://doi.org/10.1002/cbic.200700096.

Bergen, K., Steck, A.-L., Strütt, S., Baccaro, A., Welte, W., Diederichs, K., et al. (2012). Structures of KlenTaq DNA polymerase caught while incorporating C5-modified pyrimidine and C7-modified 7-deazapurine nucleoside triphosphates. *Journal of the American Chemical Society*, *134*(29), 11840–11843. https://doi.org/10.1021/ja3017889.

Bollum, F. J. (1960). Calf thymus polymerase. *The Journal of Biological Chemistry*, *235*, 2399–2403.

Bollum, F. J. (1974). 5. Terminal deoxynucleotidyl transferase. In *Protein synthesis DNA synthesis and repair RNA synthesis energy-linked ATPases synthetases* (pp. 145–171). Elsevier.

Bollum, F. J. (2006). Terminal deoxynucleotidyl transferase: Biological studies. In A. Meister (Ed.), *Advances in enzymology and related areas of molecular biology, vol. 47*. (pp. 347–374). John Wiley & Sons, Inc.

Boulé, J. B., Johnson, E., Rougeon, F., & Papanicolaou, C. (1998). High-level expression of murine terminal deoxynucleotidyl transferase in *Escherichia coli* grown at low temperature and overexpressingargU tRNA. *Molecular Biotechnology, 10*(3), 199–208.

Carey, M. F., Peterson, C. L., & Smale, S. T. (2013). The primer extension assay. *Cold Spring Harbor Protocols, 2013*(2), 164–173. https://doi.org/10.1101/pdb.prot071902.

Caruso, F., Serizawa, T., Furlong, D. N., & Okahata, Y. (1995). Quartz crystal microbalance and surface plasmon resonance study of surfactant adsorption onto gold and chromium oxide surfaces. *Langmuir, 11*(5), 1546–1552. https://doi.org/10.1021/la00005a023.

Chang, L. M. S., & Bollum, F. J. (1990). Multiple roles of divalent cation in the terminal deoxynucleotidyltransferase reactions. *The Journal of Biological Chemistry, 265*, 17436–17440.

Chang, L. M. S., Plevani, P., & Bollum, F. J. (1982). Proteolytic degradation of calf thymus terminal deoxynucleotidyl transferase. *The Journal of Biological Chemistry, 257*(10), 5700–5706.

Chang, L. M. S., Rafter, E., Rusquet-Valerius, R., Peterson, R. C., White, S. T., & Bollum, F. J. (1988). Expression and processing of recombinant human terminal transferase in the baculovirus system. *The Journal of Biological Chemistry, 263*(25), 12509–12513.

Chen, J., Xu, Y., Ji, X., & He, Z. (2017). Enzymatic polymerization-based formation of fluorescent copper nanoparticles for the nuclease assay. *Sensors and Actuators B: Chemical, 239*, 262–269. https://doi.org/10.1016/j.snb.2016.08.017.

Chow, D. C., & Chilkoti, A. (2007). Surface-initiated enzymatic polymerization of DNA. *Langmuir, 23*(23), 11712–11717. https://doi.org/10.1021/la701630g.

Chow, D. C., Lee, W. K., Zauscher, S., & Chilkoti, A. (2005). Enzymatic fabrication of DNA nanostructures: Extension of a self-assembled oligonucleotide monolayer on gold arrays. *Journal of the American Chemical Society, 127*(41), 14122–14123. https://doi.org/10.1021/ja052491z.

Dean, F. B., Nelson, J. R., Giesler, T. L., & Lasken, R. S. (2001). Rapid amplification of plasmid and phage DNA using Phi 29 DNA polymerase and multiply-primed rolling circle amplification. *Genome Research, 11*(6), 1095–1099. https://doi.org/10.1101/gr.180501.

Delarue, M., Boule, J. B., Lescar, J., Expert-Bezancon, N., Jourdan, N., Sukumar, N., et al. (2002). Crystal structures of a template-independent DNA polymerase: Murine terminal deoxynucleotidyltransferase. *The EMBO Journal, 21*(3), 427–439. https://doi.org/10.1093/emboj/21.3.427.

Douglas, J., & Morgan, R. A. (1976). The isolation of a high molecular weight terminal deoxynucleotidyl transferase from calf thymus. *Biochemical and Biophysical Research Communications, 72*(3), 840–849. https://doi.org/10.1016/s0006-291x(76)80209-4.

Flickinger, J. L., Gebeyehu, G., Buchman, G., Haces, A., & Rashtchian, A. (1992). Differential incorporation ofbiotinylated nucleotides by terminal deoxynucleotidyl transferase. *Nucleic Acids Research, 20*(9), 2382. https://doi.org/10.1093/nar/20.9.2382.

Fowler, J. D., & Suo, Z. (2006). Biochemical, structural, and physiological characterization of terminal deoxynucleotidyl transferase. *Chemical Reviews, 106*(6), 2092–2110. https://doi.org/10.1021/cr040445w.

Gavrieli, Y., Sherman, Y., & Ben-Sasson, S. A. (1992). Identification of programmed cell death in situ via specific labeling of nuclear DNA fragmentation. *Journal of Cell Biology, 119*(2), 493–501.

Gouge, J., Rosario, S., Romain, F., Beguin, P., & Delarue, M. (2013). Structures of intermediates along the catalytic cycle of terminal deoxynucleotidyltransferase: Dynamical aspects of the two-metal ion mechanism. *Journal of Molecular Biology, 425*(22), 4334–4352. https://doi.org/10.1016/j.jmb.2013.07.009.

Gu, R., Oweida, T., Yingling, Y. G., Chilkoti, A., & Zauscher, S. (2018). Enzymatic synthesis of nucleobase-modified single-stranded DNA offers tunable resistance to nuclease degradation. *Biomacromolecules*, *19*(8), 3525–3535. https://doi.org/10.1021/acs.biomac.8b00816.

Gu, R., Tang, L., Aritome, I., & Zauscher, S. (2017). DNA brushes: Advances in synthesis and applications. In *Polymer and biopolymer brushes* (pp. 627–654). John Wiley & Sons, Inc.

Guerra, C. E. (2006). Analysis of oligonucleotide microarrays by 3' end labeling using fluorescent nucleotides and terminal transferase. *Biotechniques*, *41*(1), 53–56. https://doi.org/10.2144/000112182.

Hansbury, E., Kerr, V. N., Mitchell, V. E., Ratliff, R. L., Smith, D. A., Williams, D. L., et al. (1970). Synthesis of polydeoxynucleotides using chemically modified subunits. *Biochimica et Biophysica Acta (BBA)—Nucleic Acids and Protein Synthesis*, *199*(2), 322–329. https://doi.org/10.1016/0005-2787(70)90075-4.

Hayes, F. N., Mitchell, V. E., Ratliff, R. L., Schwartz, A. W., & Williams, D. L. (1966). Incorporation efficiency of small oligo-5'-nucleotide initiators in the terminal deoxyribonucleotide transferase reaction. *Biochemistry*, *5*(11), 3625–3629. https://doi.org/10.1021/bi00875a035.

Houlihan, G., Arangundy-Franklin, S., & Holliger, P. (2017). Engineering and application of polymerases for synthetic genetics. *Current Opinion in Biotechnology*, *48*, 168–179. https://doi.org/10.1016/j.copbio.2017.04.004.

Igloi, G. L. (1998). Strategies for introducing non-radioactive labels during the automated sequence analysis of nucleic acids. *Electronic Journal of Biotechnology*, *1*(1), 23–30. https://doi.org/10.2225/vol1-issue1-fulltext-3.

Jung, D., & Alt, F. W. (2004). Unraveling V(D)J recombination: Insights into gene regulation. *Cell*, *116*(2), 299–311. https://doi.org/10.1016/S0092-8674(04)00039-X.

Kato, K.-I., Goncalves, J. M., Houts, G. E., & Bollum, F. J. (1967). Deoxynucleotide-polymerizing enzymes of calf thymus gland II. Properties of the terminal deoxynucleotidyl transferase. *The Journal of Biological Chemistry*, *242*, 2780–2789. June 10.

Khan, M. N., Tjong, V., Chilkoti, A., & Zharnikov, M. (2012). Fabrication of ssDNA/oligo(ethylene glycol) monolayers and complex nanostructures by an irradiation-promoted exchange reaction. *Angewandte Chemie (International Ed in English)*, *51*(41), 10303–10306. https://doi.org/10.1002/anie.201204245.

Khan, M. N., Tjong, V., Chilkoti, A., & Zharnikov, M. (2013). Spectroscopic study of a DNA brush synthesized in situ by surface initiated enzymatic polymerization. *The Journal of Physical Chemistry. B*, *117*(34), 9929–9938. https://doi.org/10.1021/jp404774x.

Kumar, A., Tchen, P., Roullet, F., & Cohen, J. (1988). Nonradioactive labeling of synthetic oligonucleotide probes with terminal deoxynucleotidyl transferase. *Analytical Biochemistry*, *169*(2), 376–382. https://doi.org/10.1016/0003-2697(88)90299-0.

Lee, H. H., Kalhor, R., Goela, N., Bolot, J., & Church, G. M. (2019). Terminator-free template-independent enzymatic DNA synthesis for digital information storage. *Nature Communications*, *10*(1), 2383. https://doi.org/10.1038/s41467-019-10258-1.

Liu, S., Jiang, Q., Wang, Y., & Ding, B. (2019). Biomedical applications of DNA-based molecular devices. *Advanced Healthcare Materials*, *8*(10), 1801658. https://doi.org/10.1002/adhm.201801658.

Mason, S. D., Tang, Y., Li, Y., Xie, X., & Li, F. (2018). Emerging bioanalytical applications of DNA walkers. *TrAC Trends in Analytical Chemistry*, *107*, 212–221. https://doi.org/10.1016/j.trac.2018.08.015.

Mathur, D., & Medintz, I. L. (2019). The growing development of DNA nanostructures for potential healthcare-related applications. *Advanced Healthcare Materials*, *8*(9), 1801546. https://doi.org/10.1002/adhm.201801546.

Matyašovský, J., Perlíková, P., Malnuit, V., Pohl, R., & Hocek, M. (2016). 2-substituted dATP derivatives as building blocks for polymerase-catalyzed synthesis of DNA modified in the minor groove. *Angewandte Chemie (International Ed. in English)*, *128*(51), 16088–16091. https://doi.org/10.1002/ange.201609007.

Mei, C., Lin, D., Fan, C., Liu, A., Wang, S., & Wang, J. (2016). Highly sensitive and selective electrochemical detection of Hg(2+) through surface-initiated enzymatic polymerization. *Biosensors & Bioelectronics*, *80*, 105–110. https://doi.org/10.1016/j.bios.2016.01.009.

Mueller, R., Pajatsch, M., Curdt, I., Sobek, H., Schmidt, M., Suppmann, B., et al. (2009). US Patent No. 7,494,797.

Nick-McElhinny, S. A., Havener, J. M., Garcia-Diaz, M., Juarez, R., Bebenek, K., Kee, B. L., et al. (2005). A gradient of template dependence defines distinct biological roles for family X polymerases in nonhomologous end joining. *Molecular Cell*, *19*(3), 357–366. https://doi.org/10.1016/j.molcel.2005.06.012.

Obeid, S., Baccaro, A., Welte, W., Diederichs, K., & Marx, A. (2010). Structural basis for the synthesis of nucleobase modified DNA by Thermus aquaticus DNA polymerase. *Proceedings of the National Academy of Sciences of the United States of America*, *107*(50), 21327–21331. https://doi.org/10.1073/pnas.1013804107.

Okholm, A. H., Aslan, H., Besenbacher, F., Dong, M., & Kjems, J. (2015). Monitoring patterned enzymatic polymerization on DNA origami at single-molecule level. *Nanoscale*, 7(25), 10970–10973. https://doi.org/10.1039/c5nr01945a.

Owen, J. A., Punt, J., Stranford, S. A., Jones, P. P., & Kuby, J. (2013). *Kuby immunology*. New York: W.H. Freeman.

Palluk, S., Arlow, D. H., de Rond, T., Barthel, S., Kang, J. S., Bector, R., et al. (2018). De novo DNA synthesis using polymerase-nucleotide conjugates. *Nature Biotechnology*, *36*(7), 645–650. https://doi.org/10.1038/nbt.4173.

Que, H., Yan, X., Guo, B., Ma, H., Wang, T., Liu, P., et al. (2019). Terminal deoxynucleotidyl transferase and rolling circle amplification induced G-triplex formation: A label-free fluorescent strategy for DNA methyltransferase activity assay. *Sensors and Actuators B: Chemical*, *291*, 394–400. https://doi.org/10.1016/j.snb.2019.04.091.

Rittié, L., & Perbal, B. (2008). Enzymes used in molecular biology: A useful guide. *Journal of Cell Communication and Signaling*, *2*(1), 25–45. https://doi.org/10.1007/s12079-008-0026-2.

Rothlisberger, P., & Hollenstein, M. (2018). Aptamer chemistry. *Advanced Drug Delivery Reviews*, *134*, 3–21. https://doi.org/10.1016/j.addr.2018.04.007.

Roychoudhury, R. (1972). Enzymatic synthesis of polynucleotides. *The Journal of Biological Chemistry*, *247*, 3010–3017, June 25.

Roychoudhury, R., Jay, E., & Wu, R. (1976). Terminal labeling and addition of homopolymer tracts to duplex DNA fragments by terminal deoxynucleotidyl transferase. *Nucleic Acids Research*, *3*(1), 101–116, (Republication: *Nucleic Acids Research*, *3* (4), 863–877).

Ruysschaert, T., Paquereau, L., Winterhalter, M., & Fournier, D. (2006). Stabilization of liposomes through enzymatic polymerization of DNA. *Nano Letters*, *6*(12), 2755–2757. https://doi.org/10.1021/nl061724x.

Schatz, D. G., Oettinger, M. A., & Schlissel, M. S. (1992). V(D)J recombination: Molecular biology and regulation. *Annual Review of Immunology*, *10*(1), 359–383. https://doi.org/10.1146/annurev.iy.10.040192.002043.

Schmitz, G. G., Walter, T., Seibl, R., & Kessler, C. (1991). Nonradioactive labeling of oligonucleotides in vitro with the hapten digoxigenin by tailing with terminal transferase. *Analytical Biochemistry*, *192*(1), 222–231. https://doi.org/10.1016/0003-2697(91)90212-C.

Shi, K., Dou, B., Yang, J., Yuan, R., & Xiang, Y. (2017). Target-triggered catalytic hairpin assembly and TdT-catalyzed DNA polymerization for amplified electronic detection of thrombin in human serums. *Biosensors & Bioelectronics*, *87*, 495–500. https://doi.org/10.1016/j.bios.2016.08.056.

Steitz, T. A. (1993). DNA- and RNA-dependent DNA polymerases. *Current Opinion in Structural Biology*, *3*(1), 31–38. https://doi.org/10.1016/0959-440x(93)90198-t.

Suo, Z., Chen, J., Hou, X., Hu, Z., Xing, F., & Feng, L. (2019). Growing prospects of DNA nanomaterials in novel biomedical applications. *RSC Advances*, *9*(29), 16479–16491. https://doi.org/10.1039/c9ra01261c.

Tang, L. (2016). *Enzymatic polymerization of high molecular weight DNA*. [Ph.D.]. Durham, NC, USA: Duke University. Retrieved from http://hdl.handle.net/10161/12909.

Tang, L., Navarro, L. A., Jr., Chilkoti, A., & Zauscher, S. (2017). High-molecular-weight polynucleotides by transferase-catalyzed living chain-growth polycondensation. *Angewandte Chemie (International Ed. in English)*, *56*(24), 6778–6782. https://doi.org/10.1002/anie.201700991.

Tang, L., Tjong, V., Li, N., Yingling, Y. G., Chilkoti, A., & Zauscher, S. (2014). Enzymatic polymerization of high molecular weight DNA amphiphiles that self-assemble into star-like micelles. *Advanced Materials*, *26*(19), 3050–3054. https://doi.org/10.1002/adma.201306049.

Tauraitė, D., Jakubovska, J., Dabužinskaitė, J., Bratchikov, M., & Meškys, R. (2017). Modified nucleotides as substrates of terminal deoxynucleotidyl transferase. *Molecules (Basel, Switzerland)*, *22*(4), 672. https://doi.org/10.3390/molecules22040672.

Terpe, K. (2013). Overview of thermostable DNA polymerases for classical PCR applications: From molecular and biochemical fundamentals to commercial systems. *Applied Microbiology and Biotechnology*, *97*, 10243–10254.

Tjong, V., Yu, H., Hucknall, A., & Chilkoti, A. (2013). Direct fluorescence detection of RNA on microarrays by surface-initiated enzymatic polymerization. *Analytical Chemistry*, *85*(1), 426–433. https://doi.org/10.1021/ac303132j.

Tjong, V., Yu, H., Hucknall, A., Rangarajan, S., & Chilkoti, A. (2011). Amplified on-chip fluorescence detection of DNA hybridization by surface-initiated enzymatic polymerization. *Analytical Chemistry*, *83*(13), 5153–5159. https://doi.org/10.1021/ac200946t.

Vosberg, H.-P. (1989). The polymerase chain reaction: An improved method for the analysis of nucleic acids. *Human Genetics*, *83*(1), 1–15. https://doi.org/10.1007/BF00274139.

Wan, Y., Wang, P., Su, Y., Zhu, X., Yang, S., Lu, J., et al. (2014). Ultrasensitive electrochemical DNA sensor based on the target induced structural switching and surface-initiated enzymatic polymerization. *Biosensors & Bioelectronics*, *55*, 231–236. https://doi.org/10.1016/j.bios.2013.11.077.

Wan, Y., Xu, H., Su, Y., Zhu, X., Song, S., & Fan, C. (2013). A surface-initiated enzymatic polymerization strategy for electrochemical DNA sensors. *Biosensors & Bioelectronics*, *41*, 526–531. https://doi.org/10.1016/j.bios.2012.09.017.

Wang, P., Wan, Y., Deng, S., Yang, S., Su, Y., Fan, C., et al. (2016). Aptamer-initiated on-particle template-independent enzymatic polymerization (aptamer-OTEP) for electrochemical analysis of tumor biomarkers. *Biosensors & Bioelectronics*, *86*, 536–541. https://doi.org/10.1016/j.bios.2016.07.025.

Winz, M. L., Linder, E. C., Andre, T., Becker, J., & Jaschke, A. (2015). Nucleotidyl transferase assisted DNA labeling with different click chemistries. *Nucleic Acids Research*, *43*(17). e110https://doi.org/10.1093/nar/gkv544.

Yang, B., Gathy, K. N., & Coleman, M. S. (1995). T-cell specific avian TdT: Characterization of the cDNA and recombinant enzyme. *Nucleic Acids Research*, *23*(11), 2041–2048. https://doi.org/10.1093/nar/23.11.2041.

Yin, H., Song, C. Q., Suresh, S., Wu, Q., Walsh, S., Rhym, L. H., et al. (2017). Structure-guided chemical modification of guide RNA enables potent non-viral in vivo genome editing. *Nature Biotechnology*, *35*(12), 1179–1187. https://doi.org/10.1038/nbt.4005.

Zhao, Y., Dai, X., Wang, F., Zhang, X., Fan, C., & Liu, X. (2019). Nanofabrication based on DNA nanotechnology. *Nano Today*, *26*, 123–148. https://doi.org/10.1016/j.nantod.2019.03.004.

CHAPTER SEVEN

Enzymatic synthesis of functional amylosic materials and amylose analog polysaccharides

Jun-ichi Kadokawa*
Graduate School of Science and Engineering, Kagoshima University, Kagoshima, Japan
*Corresponding author: e-mail address: kadokawa@eng.kagoshima-u.ac.jp

Contents

Abstract

Because polysaccharides have very complicated chemical structures constructed by a great diversity of monosaccharide residues and glycosidic linkages, enzymatic approaches have been identified as powerful tools to precisely synthesize polysaccharides as the reactions progress in highly controlled regio- and stereoarrangements.

Methods in Enzymology, Volume 627
ISSN 0076-6879
https://doi.org/10.1016/bs.mie.2019.06.009

α-Glucan phosphorylase (GP) is one of the enzymes that have acted as catalysts for the practical production of well-defined polysaccharides. GP can catalyze enzymatic polymerization of α-D-glucose 1-phosphate (Glc-1-P) as a monomer from a maltooligosaccharide primer to produce a pure amylose with well-defined structure via the formation of α(1→4)-glycosidic linkages. Here, the author presents methods which achieve the enzymatic synthesis of functional amylosic materials and amylose analog polysaccharides by GP-catalyzed enzymatic polymerization approaches. As the polymerization progresses at the non-reducing end of the primer, it can be conducted using polymeric primers that are modified at the reducing end and covalently attached on suitable polymeric chains. By using such polymeric primers, various amylose-grafted functional materials can be enzymatically synthesized. For example, the detailed protocol for the synthesis of amylose-grafted poly(γ-glutamic acid) is described. GP shows loose specificity for the recognition of substrates, which allows to recognize some monosaccharide 1-phosphates as analog substrates of Glc-1-P. Representatively, the experimental procedure of the GP-catalyzed enzymatic polymerization of α-D-glucosamine 1-phosphate as the analog substrate is presented to synthesize an α(1→4)-linked glucosamine polymer, that is called amylosamine. By means of a similar approach catalyzed by GP, several amylose analog polysaccharides have been obtained.

1. Introduction

Polysaccharides are widely present in nature, and accordingly, very important biomass resources (Schuerch, 1986). A variety of structures are present in natural polysaccharides, because they are constructed by a number of sugar units and several regio- and stereotypes in glycosidic linkages, that join repeating sugar residues (Yalpani, 1988). Because of such complicated structures of polysaccharides, their synthesis by general organic approaches exhibits difficulty in strictly controlled fashions, high yields, and exclusion of side-reactions and by-products (Paulsen, 1982; Schmidt, 1986; Toshima & Tatsuta, 1993). It has been well accepted that enzymatic approach is a very powerful tool to synthesize well-defined polysaccharides compared with the general organic reaction approach, because they progress in perfect controlled regio- and stereoarrangements under mild conditions, eliminating side-reactions and by-products (Shoda, 2001; Shoda, Izumi, & Fujita, 2003).

Of the six main classes of enzymes, hydrolases (glycosyl hydrolases) and transferases (glycosyl transferases) have been employed as catalysts to practically synthesize polysaccharides with well-defined structure (Blixt & Razi, 2008; Thiem & Thiem, 2008). α-Glucan phosphorylase (GP), which belongs to the GT35 family (EC 2.4.1.1), catalyzes enzymatic polymerization to produce a pure amylose (Fujii et al., 2003; Ohdan, Fujii, Yanase, Takaha, & Kuriki, 2006; Yanase, Takaha, & Kuriki, 2006; Ziegast & Pfannemüller, 1987). Amylose, composed of glucose repeating units linked through α(1 → 4)-glycosidic linkages, is a well-known natural polysaccharide as a component of starch (Schuerch, 1986). Besides the role as an energy resource in nature, amylose can be considered as a functional polymeric material owing to its left-handed helical conformation to construct hierarchically controlled higher-order structures (Putseys, Lamberts, & Delcour, 2010; Sarko & Zugenmaier, 1980). For example, enzymatically produced single amylose chains spontaneously form double helical assembly in aqueous media (Imberty, Chanzy, Perez, Buleon, & Tran, 1988; Imberty & Perez, 1988) and also bind appropriate guest polymers into cavity inside the helix during the GP-catalyzed polymerization to form supramolecular inclusion complexes (Hatanaka, Takemoto, Yamamoto, & Kadokawa, 2013; Kadokawa, 2014; Kadokawa, Arimura, Takemoto, & Yamamoto, 2012; Kaneko, Beppu, & Kadokawa, 2009; Kaneko & Kadokawa, 2005, 2006). However, a pure amylose sample is hardly provided from natural resources, principally because of difficulty in perfect separation of amylose from starch with the other component, amylopectin. Therefore, the GP-catalyzed enzymatic polymerization can be identified as a powerful tool to obtain the pure amylose sample, which has extensively been employed to precisely prepare amylose-based polymeric materials (Kadokawa, 2011b, 2016, 2017; Kadokawa & Kaneko, 2013; O'Neill & Field, 2015).

In nature, GP catalyzes phosphorolytic cleavage of α(1 → 4)-glucans, such as starch and glycogen, at the non-reducing end to produce α-D-glucose 1-phosphate (Glc-1-P) (Kitaoka & Hayashi, 2002; Nakai, Kitaoka, Svensson, & Ohtsubo, 2013) (Fig. 1A). Because a bond energy of the cleaved glycosidic linkage is comparable to that of the produced 1-phosphate linkage, GP-catalyzed reaction shows reversible nature. By means of the reversibility, the GP-catalyzed enzymatic α(1 → 4)-glycosidic bond formation, so-called glycosylation, can be conducted using Glc-1-P as a glycosyl donor under the appropriate conditions (Fig. 1B). For the α(1 → 4)-glycosidic bond formation, glycosyl acceptors having the α(1 → 4)-glucan non-reducing end structure, which are recognized by GP, should be used. The short α(1 → 4)-glucans, that is, maltooligosaccharides, are generally used as the glycosyl acceptors, where a Glc residue is transferred from the Glc-1-P donor to the non-reducing end of the maltooligosaccharide acceptor with the α(1 → 4)-glycosidic bond formation to produce the elongated maltooligosaccharide with a Glc unit. GP specifically has the smallest value

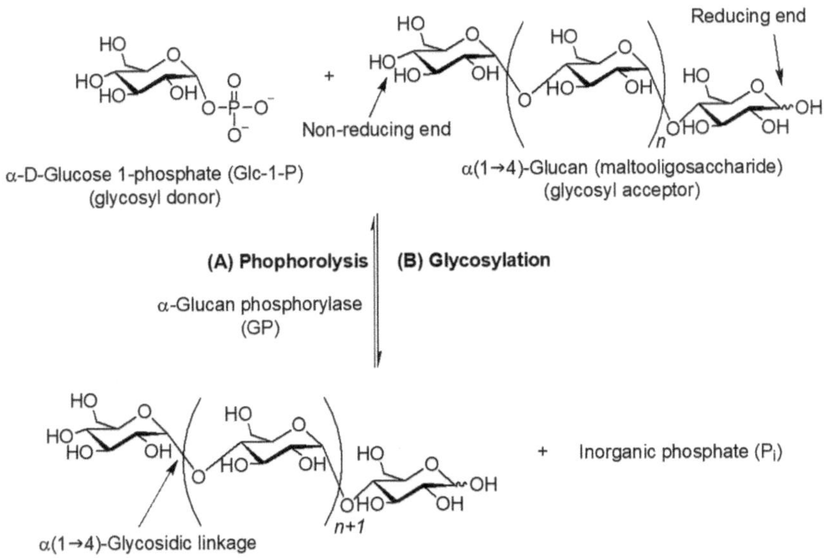

Fig. 1 GP-catalyzed enzymatic (A) phosphorolysis and (B) glycosylation.

of degree of polymerization (DP) in α(1→4)-linked glucose chains (glucans) to recognize, in which maltooligosaccharides with DPs higher than the smallest one can be employed as the glycosyl acceptor. The smallest glycosyl acceptor recognized by the most extensively studied GP, which is isolated from potato, is maltotetraose, while GP isolated from thermostable bacteria recognizes one smaller glycosyl acceptor, that is, maltotriose (Boeck & Schinzel, 1996; Takaha, Yanase, Takata, & Okada, 2001; Yanase, Takata, Fujii, Takaha, & Kuriki, 2005).

When an excess donor/acceptor molar ratio is employed, GP catalyzes consecutive glycosylations as the polymerization to produce the α(1→4)-glucan polymer, that is, amylose (Fig. 2). As the GP-catalyzed enzymatic polymerization of the monomer Glc-1-P progresses in the manner of chain-growth polymerization, the glycosyl acceptor is often defined as the "primer" of the polymerization. As the GP-catalyzed enzymatic polymerization is conceived analogously to a living polymerization without the occurrence of termination and chain-transfer reaction, the molecular weight of the produced amylose can be adjusted by the monomer/primer feed ratio, and polydispersity of polymeric product is typically narrow ($M_w/M_n < 1.2$) (Kitamura, 1996). As the reducing end of the maltooligosaccharide primer does not participate into the polymerization, it can be used as a reactive group to be covalently linked to other materials, such as polymeric chains, by

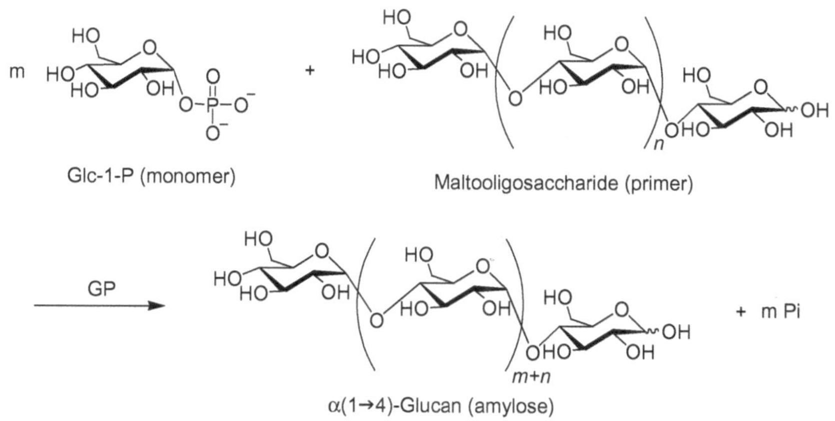

Fig. 2 GP-catalyzed enzymatic polymerization of Glc-1-P from maltooligosaccharide.

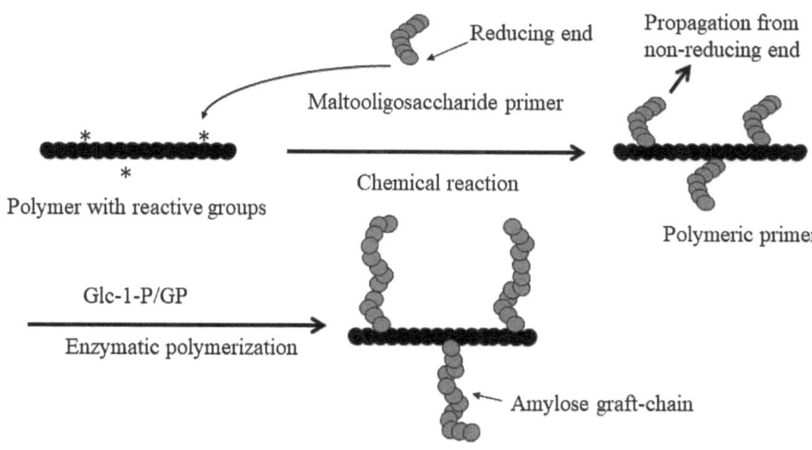

Fig. 3 Synthesis of amylose-grafted polymeric material by GP-catalyzed enzymatic polymerization using polymeric primer.

appropriate chemical reactions to obtain the polymeric primers, typically having plural α(1→4)-glucan non-reducing ends. The GP-catalyzed enzymatic polymerization then progresses from the non-reducing ends of the polymeric primers to produce amylose-grafted polymeric materials (Kadokawa, 2012, 2013a, 2014; Kaneko & Kadokawa, 2009; Nishimura & Akiyoshi, 2016) (Fig. 3).

Based on the fact that GPs recognize some analog substrates of Glc-1-P (1-phosphates of different monosaccharide residues) depending on their sources, the extension of the GP-catalyzed enzymatic polymerization has been investigated using several monosaccharide 1-phosphates as monomers to obtain non-natural (amylose analog) polysaccharides (Kadoakwa, 2015; Kadokawa, 2011a, 2013b).

In this chapter, the author first describes the synthesis and general characterization of the pure amylose sample by the GP-catalyzed enzymatic polymerization of Glc-1-P. As mentioned above, the polymeric primers can be used in the GP-catalyzed enzymatic polymerization to synthesize amylose-grafted polymeric materials. Based on weak specificity for the recognition of substrates, some amylose analog polysaccharides are efficiently synthesized by the GP-catalyzed enzymatic polymerization using different monosaccharide 1-phosphates as monomers under the specific conditions.

2. Synthesis of amylose by glucan phosphorylase-catalyzed polymerization

As abovementioned, the pure amylose can be synthesized by the GP-catalyzed enzymatic polymerization using Glc-1-P and maltooligosaccharide as the monomer and primer, respectively, in aqueous buffer. Detailed protocols for the synthesis of the general amylose are described below.

2.1 Enzymes

GP is extracted from potato, purified, and then used for the enzymatic polymerization (Kobayashi, Kamiya, & Enomoto, 1996). GPs have also been cloned and expressed from several mesophilic and thermophilic bacteria sources (Takaha et al., 2001). For example, hereafter, the protocols using GP from *Aquifex aeolicus* VF5 are described (Bhuiyan, Rus'd, Kitaoka, & Hayashi, 2003).

2.2 Preparation of chemicals

Glc-1-P as a sodium salt is commercially available and used without further purification. Maltooligosaccharides with different DPs are also purchased, but relatively expensive. Alternatively, maltoheptaose can be facilely prepared by selective cleavage of a glycosidic bond of β-cyclodextrin under acidic conditions according to literature procedure (von Braunmühl, Jonas, & Stadler, 1995). The product is purified by separation from the remaining β-cyclodextrin by means of its precipitation as an inclusion complex with *p*-xylene.

2.3 Measurements

^{1}H NMR spectra were recorded on JEOL ECA600 and ECX400 spectrometers. MALDI-TOF MS measurements were carried out by using a SHIMADZU Voyager Biospectrometry Workstation Ver.5.1 with 2.5-dihydroxybenzoic acid as matrix containing 0.05% trifluoroacetic acid under positive ion mode. UV–vis measurements were conducted using a Jasco V-650Q1 spectrometer. Powder X-ray diffraction (XRD) measurement was conducted using a PANalytical X'Pert Pro MPD with Ni-filtered CuKα radiation (λ=0.15418 nm).

2.4 Enzymatic polymerization

A solution of Glc-1-P (150 mg, 0.493 mmol) and maltotriose (5.5 mg, 0.0109 mmol) in 0.2 mol/L acetate buffer (pH 6.2, 5.0 mL) was incubated in the presence of thermostable GP (from *Aquifex aeolicus* VF5, 40 U) at 40 °C for 15 h (feed ratios of Glc-1-P/maltotriose = 50). After the reaction mixture was maintained at 100 °C for 1 h to deactivate the enzyme, the deactivated enzyme was separated by centrifugation. The supernatant was then dialyzed against water (molecular cut off: 1000) and lyophilized to give amylose (116 mg). ^{1}H NMR (DMSO-d_6, Fig. 4) δ 3.15–3.92 (m, H_2–H_6, overlapping with HOD), 4.56, 5.36, 5.46 (OH), 4.31–4.32 (d, $H_{1\beta}$ of reducing end), 4.90–4.92 (d, $H_{1\alpha}$ of reducing end), 5.10–5.11 (br, H_1).

2.5 Characterization of amylose

Degree of polymerization: The degree of polymerization (DP) of amylose was determined by UV–vis analysis of the complex with iodine as follows (Banks, Greenwood, & Khan, 1971; Kobayashi et al., 1996). A standard iodine-iodide solution was prepared by dissolving potassium iodide (26.3 mg, 0.158 mmol) and iodine (26.3 mg, 0.207 mmol) in water (50 mL). The amylose sample (1.0 mg) was dissolved in DMSO (0.20 mL) and the standard iodine-iodide solution (1.0 mL) was added to the solution. After the resulting solution was diluted with water (10 mL), the violet solution was then subjected to UV–vis spectroscopy to determine the DP value. The DP value was calculated using plots of the maximum wavelengths (λ_{max}) versus the DPs of amyloses provided in the previous literature (Banks et al., 1971; Kobayashi et al., 1996) to be 59.6, corresponding to the number-average molecular weight (M_n) = 9670. The X-ray diffraction (XRD) profile of the amylose sample showed two diffraction peaks at 2θ = ca. 17° and 22° (Fig. 5A), which was a typical pattern for double helix conformation of amylose.

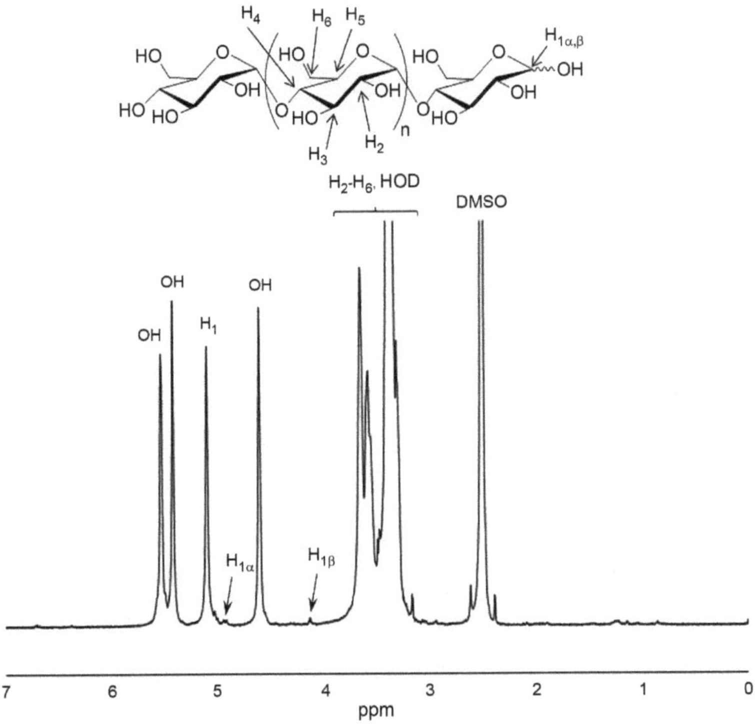

Fig. 4 ^{1}H NMR spectrum of enzymatically synthesized amylose by GP catalysis in DMSO-d_6.

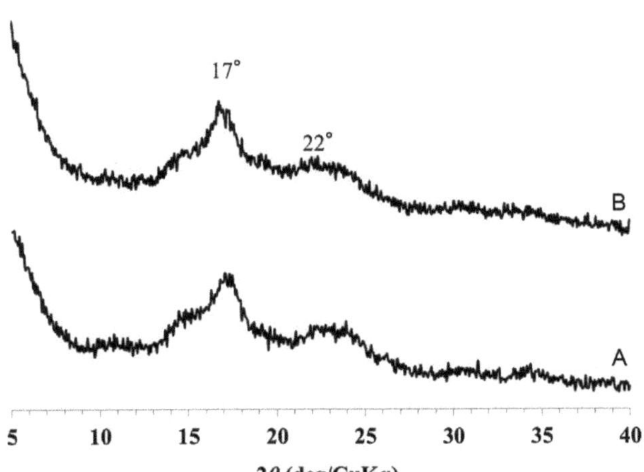

Fig. 5 XRD profiles of (A) enzymatically synthesized amylose and (B) amylose-grafted poly(γ-glutamic acid) (PGA).

2.6 Activity assay of GP

GP activities are evaluated for polymerization and phosphorolysis as follows (Bhuiyan et al., 2003; Takaha et al., 2001).

Preparation of phosphoric acid calibration curve: 0.1, 0.2, 0.3, 0.4, 0.5, 0.6 mmol/L potassium dihydrogen phosphate aqueous solutions were prepared. 800 μL of molybdate reagent (15 mmol/L ammonium molybdate and 100 mmol/L zinc acetate aqueous solution) and 200 μL of 0.57 mol/L sodium ascorbate aqueous solution were added to a mixture of 200 μL of distilled water and 200 μL of each phosphate solution. The mixtures were incubated at 30 °C for 20 min. Absorbance of the mixture was measured at 850 nm. A phosphoric acid calibration curve was prepared by plotting the concentration of phosphoric acid at Y-axis and the absorbance at 850 nm at X-axis.

Assay for polymerization: The production of phosphate (P_i) from Glc-1-P is measured according to the literature method (Bhuiyan et al., 2003; Takaha et al., 2001). Briefly, a reaction solution (200 μL) of Glc-1-P (monomer, 45 mmol/L), soluble starch (primer, 1% (*w/v*)) in acetate buffer (0.2 mol/L, pH 6.0), and thermostable GP (from *Aquifex aeolicus* VF5, 0.323 U) was incubated at temperature range from 22 °C to 100 °C for 30 min. The reaction was stopped by adding 20% (w/v) sodium dodecyl sulfate aqueous solution (10 μL). The amount of P_i released was measured by the method according to the literature (Saheki, Takeda, & Shimazu, 1985) with minor modification as follows. 800 μL of molybdate reagent (15 mmol/L ammonium molybdate and 100 mmol/L zinc acetate aqueous solution) and 200 μL of 0.57 mol/L sodium ascorbate aqueous solution were added to the reaction mixture. The mixture was incubated at 30 °C for 15 min. Absorbance of the mixture was measured at 850 nm. The amount of inorganic phosphate produced in the reaction was determined by using the abovementioned phosphoric acid calibration curve.

Assay for phosphorolysis: Activity was assayed by evaluating the amount of Glc-1-P produced from soluble starch and P_i by the ^{1}H NMR analysis. A reaction mixture (600 μL) containing sodium phosphate buffer (0.1 mol/L, pH 7.0), 1% (w/v) soluble starch, and thermostable GP (from *Aquifex aeolicus* VF5, 0.323 U) was incubated at range from 22 °C to 100 °C for 15 min. The reaction was stopped by adding 20% (w/v) sodium dodecyl sulfate aqueous solution (10 μL). The mixture was dissolved in 0.5 mol/L NaOD/D_2O for ^{1}H NMR measurement to determine the amount of Glc-1-P produced in the reaction.

The relative activities of the enzyme for polymerization and phosphorolysis are shown in Fig. 6.

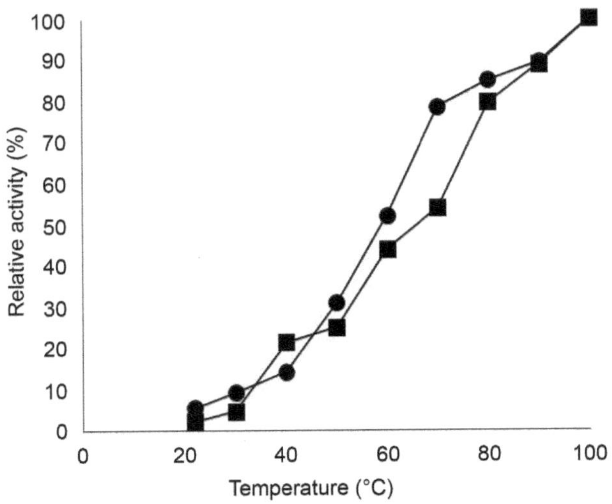

Fig. 6 Effect of temperature on the activity of thermostable GP (*Aquifex aeolicus* VF5) for polymerization (●) and phosphorolysis (■).

3. Enzymatic synthesis of functional amyloses

As mentioned above, the GP-catalyzed enzymatic polymerization can be performed from polymeric maltooligosaccharide primers which are covalently immobilized on other substances at their reducing ends, such as polymeric chains. By using the polymeric primers, various amylose-grafted polymeric materials have been synthesized, such as amylose-grafted polysaccharides, polypeptides, poly(vinyl alcohol), poly(dimethylsiloxane), polyacetylene, and polystyrene (Arimura, Omagari, Yamamoto, & Kadokawa, 2011; Hatanaka et al., 2013; Kadokawa et al., 2012; Kadokawa, Nakamura, Sasaki, Kaneko, & Nishikawa, 2008; Kamiya & Kobayashi, 1998; Kaneko, Matsuda, & Kadokawa, 2007, 2010; Kobayashi et al., 1996; Narumi et al., 2003; Omagari, Kaneko, & Kadokawa, 2010; Omagari, Matsuda, Kaneko, & Kadokawa, 2009; Sasaki, Kaneko, & Kadokawa, 2009; von Braunmühl et al., 1995). Amylose-grafted brushes were also fabricated by the GP-catalyzed enzymatic polymerization of Glc-1-P from maltoheptaose primers covalently immobilized on Au and Si surfaces (Mazzocchetti, Tsoufis, Rudolf, & Loos, 2014). In the amylose-grafted polypeptides, the detailed protocols to synthesize amylose-grafted poly (γ-glutamic acid) (PGA) are described below (Shouji, Yamamoto, & Kadokawa, 2017) (Fig. 7).

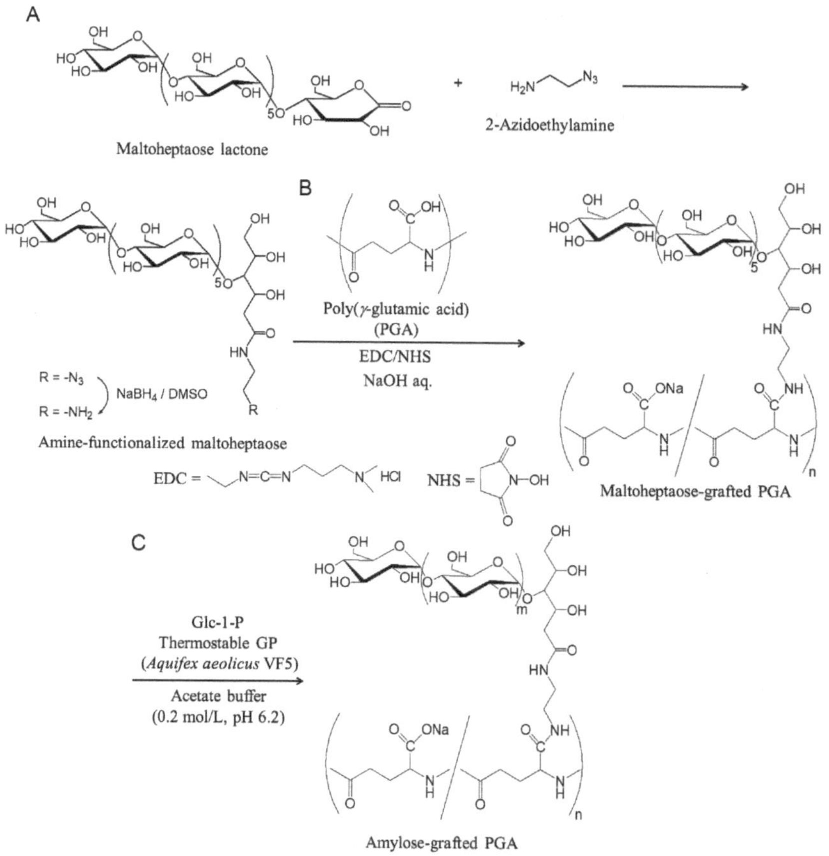

Fig. 7 Synthesis of (A) amine-functionalized maltoheptaose, (B) maltoheptaose-grafted PGA, and (C) amylose-grafted PGA.

3.1 Preparation of chemicals

Glc-1-P as a sodium salt is commercially available and used without further purification. Maltoheptaose can be facilely prepared by selective cleavage of a glycosidic bond of β-cyclodextrin under acidic conditions according to literature procedure (von Braunmühl et al., 1995). The product is purified by separation from the remaining β-cyclodextrin by means of its precipitation as an inclusion complex with *p*-xylene. PGA (M.W. $= 1.5–2.5 \times 10^6$) was purchased from Wako Pure Chemicals, Tokyo, Japan.

3.2 Measurements

^{1}H NMR spectra were recorded on JEOL ECA600 and ECX400 spectrometers. XRD measurement was conducted using a PANalytical X'Pert Pro MPD with Ni-filtered CuKα radiation ($\lambda = 0.15418$ nm).

3.3 Preparation of polymeric primers (maltoheptaose-grafted PGA)

As the polymeric primer, maltoheptaose-grafted PGA was prepared by condensation of carboxylate in PGA with an amine-functionalized maltoheptaose using a condensing agent.

Synthesis of the amine-functionalized maltoheptaose (Omagari et al., 2010) (Fig. 7A): Maltoheptaose was first converted into maltoheptaose lactone as follows. To a solution of maltoheptaose (7.70 g, 6.68 mmol) in water (10 mL), I_2/methanol (4.70 g/33 mL) was added and the mixture was stirred at 40 °C for 2 h. After KOH/methanol (5.10 g/102 mL) was added dropwise, the mixture was further stirred at room temperature for 1 h. The precipitate was separated by decantation and the supernatant was concentrated under reduced pressure. The residue was dissolved in water (25 mL) and the solution was treated with activated carbon for 30 min. The activated carbon was removed by filtration and the filtrate was concentrated under reduced pressure. The resulting powder was dissolved in water (20 mL) and the solution was treated with cation exchange resin (Amberlite IR-120B, H^+ form) for 1 h. The resin was removed by filtration and the filtrate was concentrated and dried under reduced pressure to give maltoheptaose lactone (6.10 g, 5.30 mmol, yield 79.3%). ^{1}H NMR (D_2O) δ 3.53–4.15 (br H_2–H_6), 4.05 (dd, terminal H_3, J=2.3 Hz, J=6.4 Hz), 4.25 (d, terminal H_2, J=2.3 Hz), 4.95 (d, terminal H_1, J=3.7 Hz), 5.19 (m, H_1).

2-Bromoethylamine (8.20 g, 40.0 mmol) was reacted with sodium azide (7.80 g, 120.0 mmol) in water (40 mL) at 80 °C for 15 h. After the reaction solution was concentrated to a half volume, potassium hydroxide (16 g) was added and extracted twice with diethyl ether (50 mL + 25 mL). The extracts were treated with potassium carbonate, filtered, evaporated, and dried under reduced pressure to produce 2-azidoethylamine (1.80 g, 20.1 mmol, yield 50.3%). ^{1}H NMR (D_2O) δ 2.78 (t, 2H, J=7.8 Hz), 3.39 (t, 2H, J=7.0 Hz).

Under argon, a mixture of the maltoheptaose lactone (1.39 g, 1.19 mmol) and 2-azidoethylamine (2.05 g, 23.8 mmol) was heated at 70 °C for 48 h. The resulting solution was poured into a large amount of acetone to precipitate a crude product, which was filtered, washed with acetone and dried under reduced pressure to give the azido-functionalized maltoheptaose (1.14 g, 0.91 mmol, yield 76.5%). ^{1}H NMR (D_2O) δ 3.45 (t, N_3-CH_2, J=3.66 Hz), 3.41–3.77 (m H_5, H_6, NH-CH_2), 3.87 (m H_2, H_3, J=9.6 Hz), 3.93 (m H_4, J=5.5 Hz), 4.20 (dd, terminal H_3, J=2.29 Hz, J=6.41 Hz), 4.33 (d, terminal H_2, J=2.29 Hz), 5.17 (d, terminal H_1, J=4.12 Hz), 5.41 (d, H_1, J=3.6 Hz).

Under argon, to a solution of the product (1.1 g, 0.878 mmol) in DMSO (3.0 mL) was added sodium borohydride (0.66 g, 17.56 mmol) and the mixture was stirred at 60 °C for 15 h. The reaction mixture was poured into a large amount of methanol to precipitate the product. The precipitate was isolated by filtration and dried under reduced pressure to give the amine-functionalized maltoheptaose (0.87 g, 0.706 mmol, yield 80.4%). ^1H NMR (D_2O) δ 3.19 (t, NH_2-CH_2, J=3.9 Hz), 3.41–3.77 (m H_5, H_6, NH-CH_2), 3.84 (m H_2, H_3, J=7.6 Hz), 3.95 (m H_4, J=9.7 Hz), 4.40 (m terminal H_2, H_3, J=2.8 Hz), 5.12 (d, terminal H_1, J=3.4 Hz), 5.39 (d, H_1, J=3.9 Hz).

Synthesis of maltoheptaose-grafted PGA (Fig. 7B): To a solution of PGA (0.065 g, 0.50 unit mmol) in 0.1 mol/L NaOH aq. (4.0 mL) was added the amine-functionalized maltoheptaose (2.45 g, 2.0 mmol), 1-ethyl-3-(3-dimethylaminopropyl)carbodiimide hydrochloride (EDC) (0.383 g, 2.0 mmol) and *N*-hydroxysuccinimide (NHS) (0.230 g, 2.0 mmol), and the mixture was stirred at 60 °C for 48 h. After the reaction solution was dialyzed against water (molecular cut off: 1000) overnight, the obtained material was purified further by precipitation into ethanol (300 mL). The precipitate was isolated by filtration and lyophilized to give the maltoheptaose-grafted PGA (0.257 g, 0.36 mmol, yield 71.8%). ^1H NMR (D_2O) δ 1.80–1.99, 1.99–2.18 (br, β-CH_2 of PGA), 2.25–2.58 (br, γ-CH_2 of PGA), 3.32–4.05 (br, sugar protons of H_2–H_6), 4.05–4.23 (br, α-CH of PGA), 5.19, 5.38 (br s, H_1 of maltoheptaose). The degree of grafting was determined by the integrated ratio of the H_1 signal of maltoheptaose to the γ-CH_2 signal of PGA to be 47.6%.

3.4 Enzymatic polymerization to synthesize amylose-grafted PGA (Fig. 7C)

Maltoheptaose-grafted PGA (0.0025 g, 0.0037 unit mmol) was dissolved in 0.2 mol/L acetate buffer (pH 6.2, 2.0 mL) and Glc-1-P disodium salt (0.16 g, 0.51 mmol, 300 equiv. with primer) was added to the solution. Thermostable GP (from *Aquifex aeolicus* VF5, 43 U) was added to this solution, which was then maintained at 45 °C for 20 h. The reaction mixture was left standing in refrigerator for 6 h for hydrogelation. The resulting hydrogel was soaked and washed with water for purification (1.35 g). The purified hydrogel was lyophilized to give a cryogel (71 mg). ^1H NMR (NaOD/D_2O) δ 1.81–1.99, 1.99–2.16 (br, β-CH_2 of PGA), 2.24–2.48 (br, γ-CH_2 of PGA), 3.35–4.06 (br, H_2–H_6 of amylose), 4.06–4.18 (br, α-CH of PGA, 5.25 (br s, H_1 of amylose).

3.5 Characterization

From the weight of the lyophilized sample from the hydrogel, the following data were calculated; monomer conversion; 83%, degree of polymerization (molecular weight of amylose); 249 (40300), amylose content; 97%, water content; 95.0%. The XRD profile of the lyophilized sample showed two diffraction peaks at 2θ = ca. 17° and 22°, which was a typical pattern for double helix conformation of amylose (Fig. 5B).

4. Enzymatic synthesis of amylose analog polysaccharides

Potato GP has been reported to recognize some monosaccharide 1-phosphates as the analog substrates of Glc-1-P, such as 2-deoxy-α-D-glucose, α-D-mannose (Man), α-D-xylose, α-D-glucosamine (GlcN), and *N*-formyl-α-D-glucosamine 1-phosphates. Accordingly, the previous studies have performed potato GP-catalyzed glycosylations using such glycosyl donors using maltotetraose as a glycosyl acceptor to produce non-natural pentasaccharides having the respective monosaccharide residues at the non-reducing end (Evers, Mischnick, & Thiem, 1994; Evers & Thiem, 1997; Kawazoe, Izawa, Nawaji, Kaneko, & Kadokawa, 2010; Nawaji, Izawa, Kaneko, & Kadokawa, 2008a, 2008b; Percival & Withers, 1988). Thermostable GP from *Aquifex aeolicus* VF5 shows more tolerance in the recognition specificities for monosaccharide 1-phosphate than that of potato GP. For example, α-D-glucuronic acid 1-phosphate (GlcA-1-P) is not recognized by potato GP, while thermostable GP shows ability for the recognition of GlcA-1-P in addition to the above analog glycosyl donors (Umegatani et al., 2012). When the thermostable GP-catalyzed glycosylation of the maltotriose acceptor with GlcN-1-P or Man-1-P was conducted, consecutive glycosylations occurred to obtain non-natural hetero-oligosaccharides composed of $\alpha(1 \rightarrow 4)$-linked GlcN/Man chains at the non-reducing ends (Shimohigoshi, Takemoto, Yamamoto, & Kadokawa, 2013). An attempt to accelerate the consecutive glucosaminylations was made by removing P_i as an ammonium magnesium phosphate precipitate using ammonium buffer (0.5 mol/L, pH 8.6) containing $MgCl_2$ as a reaction solvent system. As a result, the thermostable GP-catalyzed enzymatic polymerization of the GlcN-1-P monomer using the maltotriose primer (30:1) in such buffer system smoothly took place to produce $\alpha(1 \rightarrow 4)$-linked non-natural (amylose analog) aminopolysaccharide with a DP of the GlcN units of ~20 corresponding to the structure of a chitosan stereoisomer,

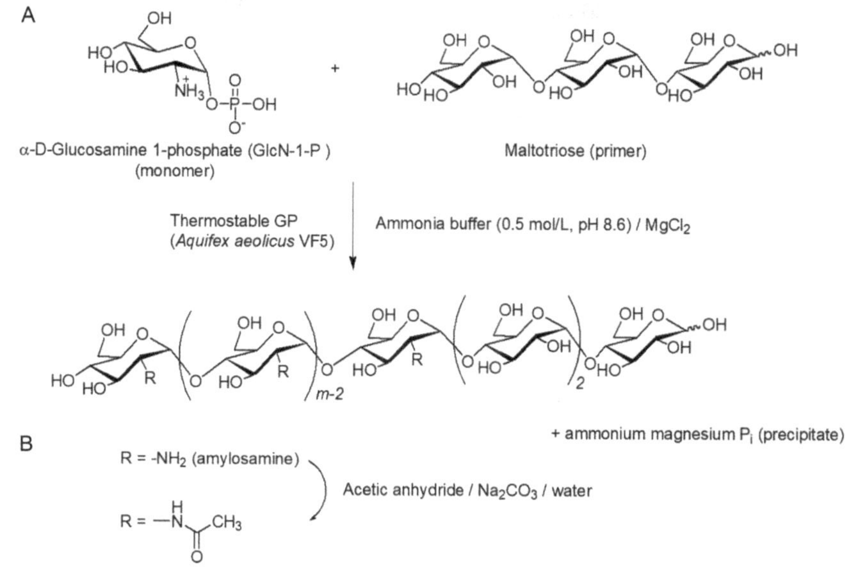

Fig. 8 (A) Thermostable GP (*Aquifex aeolicus* VF5)-catalyzed enzymatic polymerization of GlcN-1-P to synthesize amylosamine and (B) its *N*-acetylation.

which be named "amylosamine" (Kadokawa, Shimohigoshi, Yamashita, & Yamamoto, 2015) (Fig. 8A). Amylosamine was further converted into a chitin stereoisomer by *N*-acetylation (Fig. 8B). The thermostable GP-catalyzed enzymatic copolymerizations of Glc-1-P with GlcN-1-P and with Man-1-P have successfully occurred under the conditions of removal of P_i in ammonia/$MgCl_2$ buffer to obtain non-natural glucosaminoglucan comprising Glc/GlcN units and mannoglucan comprising Glc/Man units (Baba, Yamamoto, & Kadokawa, 2016; Yamashita, Yamamoto, & Kadoakwa, 2015).

4.1 Preparation of chemicals

Dibenzyl *N*,*N*-diethylphosphoramidite is synthesized according to literature procedure (Perich & Johns, 1987; Yu & Fraser-Reid, 1988).

4.2 Measurements

^{1}H NMR spectra were recorded on JEOL ECA600 and ECX400 spectrometers. MALDI-TOF MS measurements were carried out by using a SHIMADZU Voyager Biospectrometry Workstation Ver.5.1 with 2.5-dihydroxybenzoic acid as matrix containing 0.05% trifluoroacetic acid under positive ion mode.

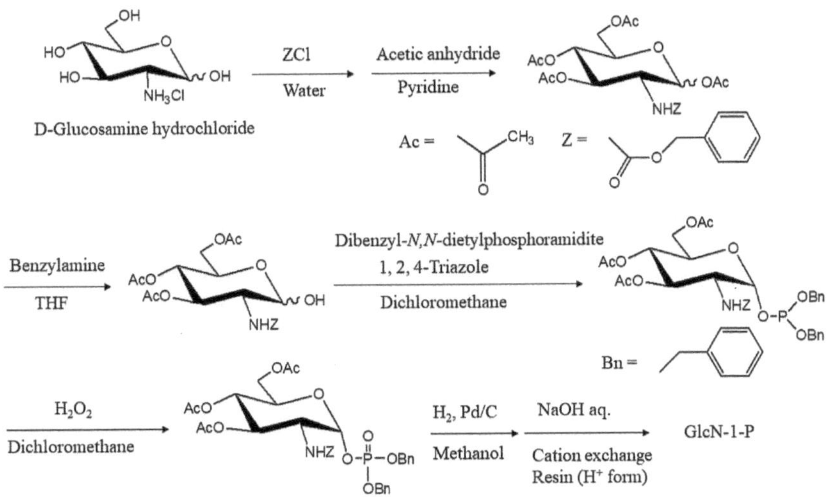

Fig. 9 Synthesis of GlcN-1-P.

4.3 Synthesis of monosaccharide 1-phosphates (GlcN-1-P)

As a typical example for synthesis of monosaccharide 1-phosphates, the procedure to obtain GlcN-1-P is described (Takata, Shimohigoshi, Yamamoto, & Kadokawa, 2014) (Fig. 9). To a solution of D-glucosamine hydrochloride (Nacalai Tesque, Inc., Japan, 10.0 g, 46.6 mmol) and $NaHCO_3$ (9.0 g, 107.1 mmol) in water (50 mL) was added benzyl chloroformate (ZCl, 12.0 g, 70.6 mmol) at room temperature and the mixture was stirred at that temperature for 1 h. After the mixture was left in refrigerator overnight, dichloromethane (120 mL) was added and the precipitate was isolated by filtration, washed with water, and dried under reduced pressure. Then, the material and acetic anhydride (50 mL, 0.48 mol) was mixed with pyridine (75 mL) at 0 °C and the mixture was stirred at room temperature for 12 h. After the reaction mixture was concentrated and diluted with dichloromethane (30 mL), the solution was washed successively with 1 mol/L hydrochloric acid aq., saturated $NaHCO_3$ aq., and saturated NaCl aq. Then, the organic layer was dried over anhydrous Na_2SO_4, filtered, and concentrated to give 1,3,4,6-tetra-*O*-acetyl-*N*-benzyloxycarbonyl-2-deoxy-D-glucopyranose (12.6 g, 26.2 mmol) in 56.2% yield.

Under argon, a solution of the product (5.0 g, 10.4 mmol) and benzylamine (2.2 mL, 20.2 mmol) in THF (30 mL) was stirred at room temperature for 12 h. After cold water (50 mL) was added to the reaction solution, the mixture was extracted three times with dichloromethane. After the extracts were combined, dried over anhydrous Na_2SO_4, filtered, and concentrated, the residue was subjected to column chromatography on silica gel (hexane/ethyl acetate = 2:1 (*v*/v)) to give 3,4,6-tri-*O*-acetyl-*N*-benzyloxycarbonyl-2-deoxy-D-glucopyranose (2.79 g, 6.35 mmol) in 60.9% yield.

Under argon, to a solution of the product (2.79 g, 6.35 mmol) in dichloromethane (20 mL) were added 1,2,4-triazole (1.72 g, 24.9 mmol) and dibenzyl *N,N*-diethylphosphramidite (4.44 g, 15.6 mmol) at room temperature and the mixture was stirred at that temperature overnight. After the reaction mixture was diluted with dichloromethane, the solution was washed successively with saturated $NaHCO_3$ aq., water, and saturated NaCl aq., dried over anhydrous Na_2SO_4, and concentrated. The residue was subjected to column chromatography on silica gel (hexane/ethyl acetate/triethylamine = 4:1:0.01 (v/v/v)) to give a crude dibenzyl 3,4,6-tri-*O*-acetyl-*N*-benzyloxycarbonyl-2-deoxy-D-glucopyranosyl phosphite. The product was dissolved with dichloromethane (10 mL) and H_2O_2 (30%, 10 mL) was added to the solution at 0 °C. After the mixture was stirred at room temperature for 1 h, it was washed with saturated NaCl aq. The organic layer was dried over anhydrous Na_2SO_4 and concentrated. The residue was subjected to column chromatography on silica gel (hexane/ethyl acetate/triethylamine = 3:1:0.01 (v/v/v)) and the fractions were combined, concentrated, and poured into hexane (100 mL) to precipitate dibenzyl 3,4,6-tri-*O*-acetyl-*N*-benzyloxycarbonyl-2-deoxy-D-glucopyranosyl phosphate. Then, the product was dissolved with methanol (10 mL) and the solution was stirred in the presence of 10% Pd/C (0.10 g) at room temperature for 6 h under hydrogel atmosphere. After the mixture was filtered and the filtrate was concentrated, 1 mol/L NaOH aq. (6.6 mL) was added to the residue and the resulting mixture was stirred at room temperature for 1 h and neutralized by treatment with cation-exchange resin (Amberlite IRC 76, H^+ form). After the resin was filtered off, the filtrate was concentrated and poured into methanol to precipitate the product. The precipitate was isolated by filtration and lyophilized to give GlcN-1-P (0.103 g, 0.40 mmol, yield 6.3%). ^{1}H NMR (D_2O) δ 3.32 (dd, H_2, $J_{1,2}=3.0$ Hz, $J_{2,3}=10.2$ Hz), 3.51 (t, H4, $J_{3,4}=J_{4,5}=9.9$ Hz), 3.81 (dd, H_{6a}, $J_{5,6a}=12.6$ Hz, $J_{6a,6b}=4.8$ Hz), 3.86–3.96 (m, H_3, H_5, H_{6b}), 5.64 (dd, H_1, $J_{1,2}=3.0$ Hz, $J_{1,P}=7.2$ Hz).

4.4 Enzymatic polymerization to synthesize amylosamine

A mixture of GlcN-1-P (37.0 mg, 0.143 mmol) and maltotriose (2.4 mg, 4.76 μmol) in 0.5 mol/L ammonia buffer (pH 8.6, 300 μL, 0.150 mmol) dissolving $MgCl_2$ (14.0 mg, 0.147 mmol) was incubated in the presence of thermostable GP (from *Aquifex aeolicus* VF5, 20 U) at 40 °C for 7 days. The precipitated material was then separated by centrifugation, and the supernatant was maintained at 100 °C for 1 h to deactivate the enzyme. The deactivated enzyme was removed by centrifugation, and the supernatant was lyophilized. After the produced crude products were dissolved in water (1.0 mL), the resulting solution was treated with an anion-exchange resin (Amberlite IRA-400J Cl^- form) for 1 h, filtered, and concentrated. Methanol was subsequently added to the concentrated solution to precipitate the product, which was isolated by filtration to give amylosamie (15.4 mg, yield 61.8% based on the total GlcN and Glc residues present in the reaction system). ^{1}H NMR (D_2O + acetic acid-d_4, Fig. 10) δ 3.45–4.14 ($H_{2,\ 3,\ 4,\ 5,\ 6}$), 4.64 ($H_{1\beta}$), 5.22 ($H_{1\alpha}$), 5.37 (Glc-H_1), and 5.73 (GlcN-H_1). From the integrated ratio of the reducing anomeric signals at δ 4.46 and 5.22 to the GlcN-H_1 signal at δ 5.73, the average DP value of the GlcN chain in

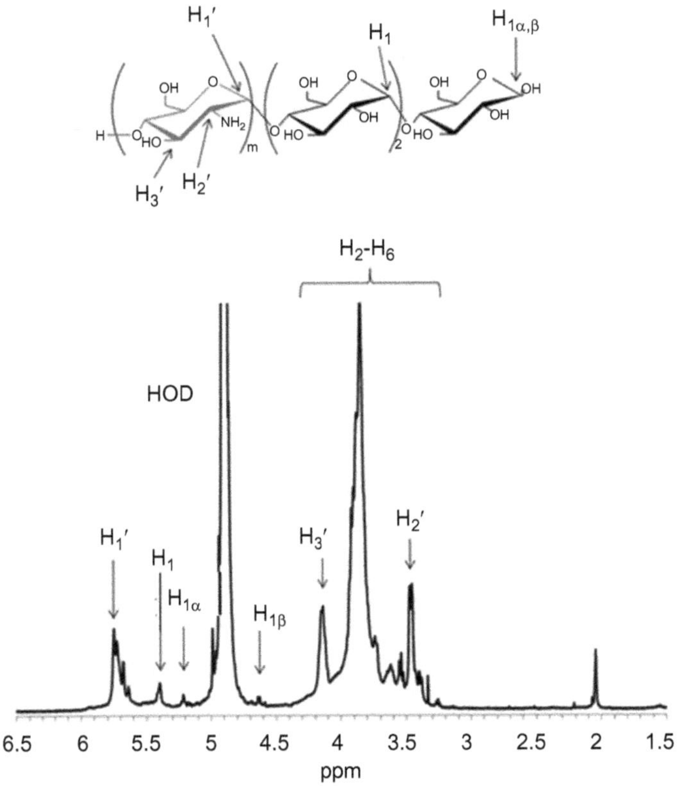

Fig. 10 ^{1}H NMR spectrum of amylosamine in D_2O/acetic acid-d_4.

the polysaccharide product was calculated to be 20.8. Based on this DP value, the average M_n value of the produced polysaccharide including the maltotriose chain was calculated to be 3850.

4.5 *N*-acetylation of amylosamine

To amylosamine prepared above (15.7 mg) and Na_2CO_3 (9.50 mg, 0.0896 mmol) in water (1.0 mL) was added acetic anhydride (0.0086 mL, 0.0860 mmol), and the mixture was maintained at room temperature for 10 min with stirring. The reaction mixture was then treated successively with cation-exchange resin (Amberlite IR-120(plus) H^+ form) and anion-exchange resin (Amberlite IRA-400J Cl^- form) for 5 min each, and the resulting solution was concentrated. Ethanol was then added to the concentrated solution to precipitate the product, which was isolated by filtration, and lyophilized to give the chitin stereoisomer (15.7 mg, yield 81.6%). ^1H NMR (D_2O) δ 2.07 (CH_3), 3.75–3.99 ($H_{2,\ 3,\ 4,\ 5,\ 6}$), 4.66 ($H_{1\beta}$), 5.24 ($H_{1\alpha}$), and 5.40 (Glc-H_1 and GlcN-H_1).

4.6 Characterization of amylosamine

MALDI-TOF MS measurement: A mixture of GlcN-1-P (12.7 mg, 0.0490 mmol) and maltotriose (2.4 mg, 4.76 μmol) in 0.5 mol/L ammonia buffer (pH 8.6, 100 μL, 0.050 mmol) dissolving $MgCl_2$ (4.8 mg, 0.0505 mmol) was incubated in the presence of thermostable GP (from *Aquifex aeolicus* VF5, 20 U) at 40 °C for 7 days. After the precipitate was removed by centrifugation, supernatant was maintained at 100 °C for 1 h in order to deactivate the enzyme. The deactivated enzyme was removed by centrifugation and the supernatant was lyophilized and subjected to MALDI-TOF MS measurement. The spectrum in Fig. 11 displays signals separated by m/z 161, corresponding to the molecular masses of polysaccharides containing up to 12 GlcN units.

^1H-^1H COSY NMR measurement of fully acetylated amylosamine: A mixture of the *N*-acetylated amylosamine (8.5 mg), acetic anhydride (0.50 mL, 5.00 mmol), and *N*,*N*-dimethylaminopyridine (0.50 mg, 4.09 μmol) in pyridine (1.0 mL) was stirred at 80 °C for 6 h. After concentration of the reaction mixture, diethyl ether was added to the residue. The resulting precipitate was isolated by filtration, and dried under reduced pressure to give the fully acetylated amylosamine (7.50 mg, yield 60.1%). ^1H NMR (DMSO-d_6) δ 1.80–2.13 (CH_3), 3.87 (H_5), 3.93 (H_4), 4.13 (H_2), 4.30, 4.36 ($H_{6a,\ 6b}$), 5.07 (H_1), 5.14 (H_3), and 6.56 (NH). ^1H-^1H COSY

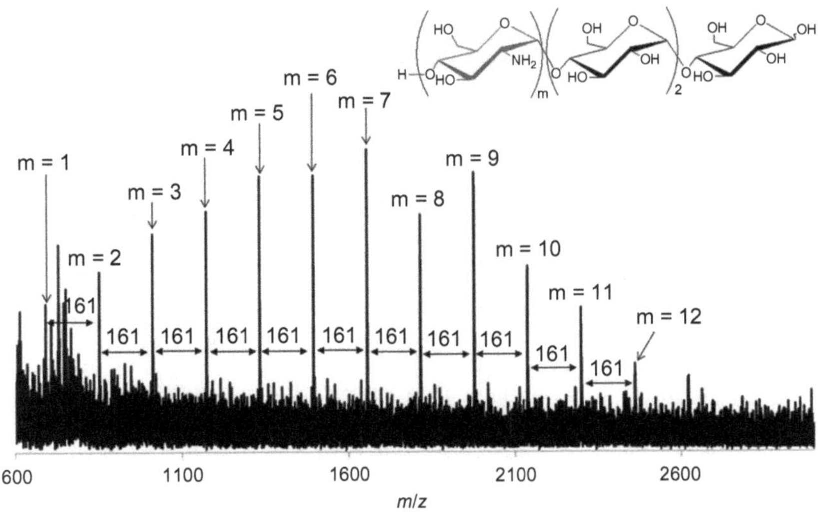

Fig. 11 MALDI-TOF MS of amylosamine (Kadokawa et al., 2015). *Reproduced and adapted by permission of The Royal Society of Chemistry.*

NMR data indicated that hydroxy groups at the C-2, 3, and 6 positions were acetylated, whereas the acetylation at the C-1, 4, and 5 positions did not occur because of the absence of hydroxy groups. It could therefore be concluded that formation of the glycosidic linkages between C-1 and C-4 in the GlcN residues was successful using thermostable GP, to give an $\alpha(1 \rightarrow 4)$-linked GlcN chain, that is, amylosamine.

5. Conclusion

The GP-catalyzed enzymatic polymerization has been identified as a powerful tool to precisely synthesize amylose-based functional materials and amylose analog polysaccharides. For example, the pure amylose is produced by the GP-catalyzed enzymatic polymerization of a native monomer, Glc-1-P. The molecular weights of the products can be controlled by monomer/primer feed ratios. As the polymerization progresses from one of two primer terminuses, that is the non-reducing end, the immobilization of the other terminus, i.e., the reducing end, by covalent linkage does not suppress the progress of the polymerization. Accordingly, the polymeric primers with plural primer chains, where the reducing ends are covalently attached on some polymeric materials can be used for the GP-catalyzed enzymatic polymerization to produce amylose-grafted functional polymeric materials. As GP shows weak specificity for recognition of the substrates, analog glycosyl donors, such as GlcN-1-P and Man-1-P can be used to occur consecutive glycosylations, giving rise to α(1 → 4)-linked GlcN and Man chains. Particularly, under the conditions of removal of P_i, the glycosylations are accelerated for the occurrence of the polymerization of analog monomers to give several amylose analog polysaccharides.

Acknowledgments

The author is indebted to the co-workers, whose names are found in references from his papers, for their enthusiastic collaborations.

References

Arimura, T., Omagari, Y., Yamamoto, K., & Kadokawa, J. (2011). Chemoenzymatic synthesis and hydrogelation of amylose-grafted xanthan gums. *International Journal of Biological Macromolecules*, *49*(4), 498–503. https://doi.org/10.1016/j.ijbiomac.2011.06.003.

Baba, R., Yamamoto, K., & Kadokawa, J. (2016). Synthesis of α(1 → 4)-linked non-natural mannoglucans by alpha-glucan phosphorylase-catalyzed enzymatic copolymerization. *Carbohydrate Polymers*, *151*, 1034–1039. https://doi.org/10.1016/j.carbpol.2016.06.057.

Banks, W., Greenwood, C. T., & Khan, K. M. (1971). The interaction of linear, amylose oligomers with iodine. *Carbohydrate Research*, *17*(1), 25–33. https://doi.org/10.1016/S0008-6215(00)81539-6.

Bhuiyan, S. H., Rus'd, A. A., Kitaoka, M., & Hayashi, K. (2003). Characterization of a hyperthermostable glycogen phosphorylase from *Aquifex aeolicus* expressed in *Escherichia coli*. *Journal of Molecular Catalysis B: Enzymatic*, *22*(3–4), 173–180. https://doi.org/10.1016/s1381-1177(03)00029-8.

Blixt, O., & Razi, N. (2008). Enzymatic glycosylation by transferases. In B. O. Fraser-Reid, K. Tatsuta, & J. Thiem (Eds.), *Glycoscience chemistry and chemical biology* (2nd ed., pp. 1361–1385). Cham; New York: Springer.

Boeck, B., & Schinzel, R. (1996). Purification and characterisation of an α-glucan phosphorylase from the thermophilic bacterium *Thermus thermophilus*. *European Journal of Biochemistry*, *239*(1), 150–155. https://doi.org/10.1111/j.1432-1033.1996.0150u.x.

Evers, B., Mischnick, P., & Thiem, J. (1994). Synthesis of 2-deoxy-α-d-arabino-hexopyranosyl phosphate and 2-deoxy-maltooligosaccharides with phosphorylase. *Carbohydrate Research*, *262*(2), 335–341. https://doi.org/10.1016/0008-6215(94)84189-6.

Evers, B., & Thiem, J. (1997). Further syntheses employing phosphorylase. *Bioorganic & Medicinal Chemistry*, *5*(5), 857–863. https://doi.org/10.1016/S0968-0896(97)00031-X.

Fujii, K., Takata, H., Yanase, M., Terada, Y., Ohdan, K., Takaha, T., et al. (2003). Bioengineering and application of novel glucose polymers. *Biocatalysis and Biotransformation*, *21*(4–5), 167–172. https://doi.org/10.1080/10242420310001614379.

Hatanaka, D., Takemoto, Y., Yamamoto, K., & Kadokawa, J. (2013). Hierarchically self-assembled nanofiber films from amylose-grafted carboxymethyl cellulose. *Fibers*, *2*, 34–44.

Imberty, A., Chanzy, H., Perez, S., Buleon, A., & Tran, V. (1988). The double-helical nature of the crystalline part of a-starch. *Journal of Molecular Biology*, *201*(2), 365–378. https://doi.org/10.1016/0022-2836(88)90144-1.

Imberty, A., & Perez, S. (1988). A revisit to the three-dimensional structure of b-type starch. *Biopolymers*, *27*(8), 1205–1221. https://doi.org/10.1002/bip.360270803.

Kadoakwa, J. (2015). Enzymatic synthesis of non-natural oligo- and polysaccharides by phosphorylase-catalyzed glycosylations using analogue substrates. In H. N. Cheng, R. A. Gross, & P. B. Smith (Eds.), *ACS symposium series: Vol. 1192. Green polymer chemistry: Biobased materials and biocatalysis* (pp. 87–99). Washington, DC: American Chemical Society.

Kadokawa, J. (2011a). Facile synthesis of unnatural oligosaccharides by phosphorylase-catalyzed enzymatic glycosylations using new glycosyl donors. In N. S. Gordon (Ed.), *Oligosaccharides: Sources, properties and applications* (pp. 269–281). Hauppauge, NY: Nova Science Publishers, Inc.

Kadokawa, J. (2011b). Precision polysaccharide synthesis catalyzed by enzymes. *Chemical Reviews*, *111*(7), 4308–4345. https://doi.org/10.1021/Cr100285v.

Kadokawa, J. (2012). Synthesis of amylose-grafted polysaccharide materials by phosphorylase-catalyzed enzymatic polymerization. In P. B. Smith & R. A. Gross (Eds.), *ACS symposium series: Vol. 1105. Biobased monomers, polymers, and materials* (pp. 237–255). Washington, DC: American Chemical Society.

Kadokawa, J. (2013a). Synthesis of new polysaccharide materials by phosphorylase-catalyzed enzymatic α-glycosylations using polymeric glycosyl acceptors. In H. N. Cheng, R. A. Gross, & P. B. Smith (Eds.), *ACS symposium series: Vol. 1144. Green polymer chemistry: Biocatalysis and materials II* (pp. 141–161). Washington, DC: American Chemical Society.

Kadokawa, J. (2013b). Synthesis of non-natural oligosaccharides by α-glucan phosphorylase-catalyzed enzymatic glycosylations using analogue substrates of α-d-glucose 1-phosphate. *Trends in Glycoscience and Glycotechnology*, *25*(142), 57–69. https://doi.org/10.4052/Tigg.25.57.

Kadokawa, J. (2014). Chemoenzymatic synthesis of functional amylosic materials. *Pure and Applied Chemistry*, *86*(5), 701–709. https://doi.org/10.1515/pac-2013-1116.

Kadokawa, J. (2016). Precision synthesis of functional polysaccharide materials by phosphorylase-catalyzed enzymatic reactions. *Polymers*, *8*(4), 138. https://doi.org/10.3390/polym8040138.

Kadokawa, J. (2017). A-glucan phosphorylase: A useful catalyst for precision enzymatic synthesis of oligo- and polysaccharides. *Current Organic Chemistry*, *21*(13), 1192–1204. https://doi.org/10.2174/1385272821666170222124526.

Kadokawa, J., Arimura, T., Takemoto, Y., & Yamamoto, K. (2012). Self-assembly of amylose-grafted carboxymethyl cellulose. *Carbohydrate Polymers*, *90*(3), 1371–1377. https://doi.org/10.1016/j.carbpol.2012.07.006.

Kadokawa, J., & Kaneko, Y. (2013). *Engineering of polysaccharide materials—By phosphorylase-catalyzed enzymatic chain-elongation*. Singapore: Pan Stanford Publishing Pte Ltd.

Kadokawa, J., Nakamura, Y., Sasaki, Y., Kaneko, Y., & Nishikawa, T. (2008). Chemoenzymatic synthesis of amylose-grafted polyacetylenes. *Polymer Bulletin*, *60*(1), 57–68. https://doi.org/10.1007/s00289-007-0842-1.

Kadokawa, J., Shimohigoshi, R., Yamashita, K., & Yamamoto, K. (2015). Synthesis of chitin and chitosan stereoisomers by thermostable α-glucan phosphorylase-catalyzed enzymatic polymerization of α-d-glucosamine 1-phosphate. *Organic & Biomolecular Chemistry*, *13*, 4336–4343. https://doi.org/10.1039/C5OB00167F.

Kamiya, S., & Kobayashi, K. (1998). Synthesis and helix formation of saccharide-poly (l-glutamic acid) conjugates. *Macromolecular Chemistry and Physics*, *199*(8), 1589–1596. https://doi.org/10.1002/(Sici)1521-3935(19980801)199:8<1589::Aid-Macp1589>3.0.Co;2-V.

Kaneko, Y., Beppu, K., & Kadokawa, J. (2009). Amylose selectively includes a specific range of molecular weights in poly(tetrahydrofuran)s in vine-twining polymerization. *Polymer Journal*, *41*(9), 792–796. https://doi.org/10.1295/polymj.PJ2009104.

Kaneko, Y., & Kadokawa, J. (2005). Vine-twining polymerization: A new preparation method for well-defined supramolecules composed of amylose and synthetic polymers. *Chemical Record*, *5*(1), 36–46. https://doi.org/10.1002/Tcr.20031.

Kaneko, Y., & Kadokawa, J. (2006). Synthesis of nanostructured bio-related materials by hybridization of synthetic polymers with polysaccharides or saccharide residues. *Journal of Biomaterials Science Polymer Edition*, *17*(11), 1269–1284. https://doi.org/10.1163/156856206778667479.

Kaneko, Y., & Kadokawa, J. (2009). Chemoenzymatic synthesis of amylose-grafted polymers. In R. Ito & Y. Matsuo (Eds.), *Handbook of carbohydrate polymers: Development, properties and applications* (pp. 671–691). Hauppauge, NY: Nova Science Publishers, Inc.

Kaneko, Y., Matsuda, S., & Kadokawa, J. (2007). Chemoenzymatic syntheses of amylose-grafted chitin and chitosan. *Biomacromolecules*, *8*(12), 3959–3964. https://doi.org/10.1021/Bm701000t.

Kaneko, Y., Matsuda, S., & Kadokawa, J. (2010). Chemoenzymatic synthesis of amylose-grafted poly(vinyl alcohol). *Polymer Chemistry*, *1*(2), 193–197. https://doi.org/10.1039/B9py00215d.

Kawazoe, S., Izawa, H., Nawaji, M., Kaneko, Y., & Kadokawa, J. (2010). Phosphorylase-catalyzed *n*-formyl-α-glucosaminylation of maltooligosaccharides. *Carbohydrate Research*, *345*(5), 631–636. https://doi.org/10.1016/j.carres.2010.01.001.

Kitamura, S. (1996). Starch polymers, natural and synthetic. In C. Salamone (Ed.), *Vol. 10. The polymeric materials encyclopedia, synthesis, properties and applications* (pp. 7915–7922). New York: CRC Press.

Kitaoka, M., & Hayashi, K. (2002). Carbohydrate-processing phosphorolytic enzymes. *Trends in Glycoscience and Glycotechnology*, *14*(75), 35–50.

Kobayashi, K., Kamiya, S., & Enomoto, N. (1996). Amylose-carrying styrene macromonomer and its homo- and copolymers: Synthesis via enzyme-catalyzed polymerization and complex formation with iodine. *Macromolecules*, *29*(27), 8670–8676. https://doi.org/10.1021/Ma9603443.

Mazzocchetti, L., Tsoufis, T., Rudolf, P., & Loos, K. (2014). Enzymatic synthesis of amylose brushes revisited: Details from x-ray photoelectron spectroscopy and spectroscopic ellipsometry. *Macromolecular Bioscience*, *14*(2), 186–194. https://doi.org/10.1002/mabi.201300273.

Nakai, H., Kitaoka, M., Svensson, B., & Ohtsubo, K. (2013). Recent development of phosphorylases possessing large potential for oligosaccharide synthesis. *Current Opinion in Chemical Biology*, *17*(2), 301–309. https://doi.org/10.1016/j.cbpa.2013.01.006.

Narumi, A., Kawasaki, K., Kaga, H., Satoh, T., Sugimoto, N., & Kakuchi, T. (2003). Glycoconjugated polymer 6. Synthesis of poly[styrene-*block*-(styrene-*graft*-amylose)] via potato phosphorylase-catalyzed polymerization. *Polymer Bulletin*, *49*(6), 405–410. https://doi.org/10.1007/s00289-003-0126-3.

Nawaji, M., Izawa, H., Kaneko, Y., & Kadokawa, J. (2008a). Enzymatic α-glucosaminylation of maltooligosaccharides catalyzed by phosphorylase. *Carbohydrate Research*, *343*(15), 2692–2696. https://doi.org/10.1016/j.carres.2008.08.013.

Nawaji, M., Izawa, H., Kaneko, Y., & Kadokawa, J. (2008b). Enzymatic synthesis of α-d-xylosylated maltooligosaccharides by phosphorylase-catalyzed xylosylation. *Journal of Carbohydrate Chemistry*, *27*(4), 214–222. https://doi.org/10.1080/07328300802105340.

Nishimura, T., & Akiyoshi, K. (2016). Amylose engineering: Phosphorylase-catalyzed polymerization of functional saccharide primers for glycobiomaterials. *Wiley Interdisciplinary Reviews. Nanomedicine and Nanobiotechnology*, *9*, e1423. https://doi.org/10.1002/wnan.1423.

Ohdan, K., Fujii, K., Yanase, M., Takaha, T., & Kuriki, T. (2006). Enzymatic synthesis of amylose. *Biocatalysis and Biotransformation*, *24*(1–2), 77–81. https://doi.org/10.1080/10242420600598152.

Omagari, Y., Kaneko, Y., & Kadokawa, J. (2010). Chemoenzymatic synthesis of amylose-grafted alginate and its formation of enzymatic disintegratable beads. *Carbohydrate Polymers*, *82*(2), 394–400. https://doi.org/10.1016/j.carbpol.2010.04.078.

Omagari, Y., Matsuda, S., Kaneko, Y., & Kadokawa, J. (2009). Chemoenzymatic synthesis of amylose-grafted cellulose. *Macromolecular Bioscience*, *9*(5), 450–455. https://doi.org/10.1002/mabi.200800237.

O'Neill, E. C., & Field, R. A. (2015). Enzymatic synthesis using glycoside phosphorylases. *Carbohydrate Research, 403*, 23–37. https://doi.org/10.1016/j.carres.2014.06.010.

Paulsen, H. (1982). Advances in selective chemical syntheses of complex oligosaccharides. *Angewandte Chemie (International Ed. in English), 21*(3), 155–173.

Percival, M. D., & Withers, S. G. (1988). Applications of enzymes in the synthesis and hydrolytic study of 2-deoxy-α-d-glucopyranosyl phosphate. *Canadian Journal of Chemistry—Revue Canadienne De Chimie, 66*(8), 1970–1972. https://doi.org/10.1139/V88-317.

Perich, J. W., & Johns, R. B. (1987). A new, convenient and efficient general procedure for the conversion of alcohols into their dibenzyl phosphorotriesters using N,N-diethyl dibenzyl phosphoramidite. *Tetrahedron Letters, 28*(1), 101–102. https://doi.org/10.1016/S0040-4039(00)95660-0.

Putseys, J. A., Lamberts, L., & Delcour, J. A. (2010). Amylose-inclusion complexes: Formation, identity and physico-chemical properties. *Journal of Cereal Science, 51*(3), 238–247. https://doi.org/10.1016/j.jcs.2010.01.011.

Saheki, S., Takeda, A., & Shimazu, T. (1985). Assay of inorganic phosphate in the mild pH range, suitable for measurement of glycogen phosphorylase activity. *Analytical Biochemistry, 148*(2), 277–281. https://doi.org/10.1016/0003-2697(85)90229-5.

Sarko, A., & Zugenmaier, P. (1980). Crystal structures of amylose and its derivatives. In A. D. French & K. H. Gardner (Eds.), *ACS symposium series: Vol. 141. Fiber diffraction methods* (pp. 459–482). Washington, DC: American Chemical Society.

Sasaki, Y., Kaneko, Y., & Kadokawa, J. (2009). Chemoenzymatic synthesis of amylose-grafted polyacetylene by polymer reaction manner and its conversion into organogel with DMSO by cross-linking. *Polymer Bulletin, 62*(3), 291–303. https://doi.org/10.1007/s00289-008-0025-8.

Schmidt, R. R. (1986). New methods for the synthesis of glycosides and oligosaccharides—Are there alternatives to the Königs-Knorr method? *Angewandte Chemie (International Ed. in English), 25*(3), 212–235.

Schuerch, C. (1986). Polysaccharides. In H. F. Mark, N. Bilkales, & C. G. Overberger (Eds.), *Vol. 13. Encyclopedia of polymer science and engineering* (2nd ed., pp. 87–162). New York: John Wiley & Sons.

Shimohigoshi, R., Takemoto, Y., Yamamoto, K., & Kadokawa, J. (2013). Thermostable α-glucan phosphorylase-catalyzed successive α-mannosylations. *Chemistry Letters, 42*(8), 822–824. https://doi.org/10.1246/Cl.130286.

Shoda, S. (2001). Enzymatic glycosylation. In B. O. Fraser-Reid, K. Tatsuta, & J. Thiem (Eds.), *Vol. II. Glycoscience chemistry and chemical biology* (pp. 1465–1496). Berlin, Heidelberg, New York: Springer-Verlag.

Shoda, S., Izumi, R., & Fujita, M. (2003). Green process in glycotechnology. *Bulletin of the Chemical Society of Japan, 76*(1), 1–13.

Shouji, T., Yamamoto, K., & Kadokawa, J. (2017). Chemoenzyamtic synthesis and self-assembling gelation behavior of amylose-grafted poly(gamma-glutamic acid). *International Journal of Biological Macromolecules, 97*, 99–105. https://doi.org/10.1016/j.ijbiomac.2017.01.001.

Takaha, T., Yanase, M., Takata, H., & Okada, S. (2001). Structure and properties of *Thermus aquaticus* α-glucan phosphorylase expressed in *Escherichichia coli. Journal of Applied Glycoscience, 48*(1), 71–78.

Takata, Y., Shimohigoshi, R., Yamamoto, K., & Kadokawa, J. (2014). Enzymatic synthesis of dendritic amphoteric α-glucans by thermostable phosphorylase catalysis. *Macromolecular Bioscience, 14*(10), 1437–1443. https://doi.org/10.1002/mabi.201400204.

Thiem, J., & Thiem, J. (2008). Enzymatic glycosylation by glycohydrolases and glycosynthases. In B. O. Fraser-Reid, K. Tatsuta, & J. Thiem (Eds.), *Glycoscience chemistry and chemical biology* (2nd ed., pp. 1387–1409). Cham; New York: Springer.

Toshima, K., & Tatsuta, K. (1993). Recent progress in *o*-glycosylation methods and its application to natural products synthesis. *Chemical Reviews*, *93*(4), 1503–1531.

Umegatani, Y., Izawa, H., Nawaji, M., Yamamoto, K., Kubo, A., Yanase, M., et al. (2012). Enzymatic α-glucuronylation of maltooligosaccharides using α-glucuronic acid 1-phosphate as glycosyl donor catalyzed by a thermostable phosphorylase from *Aquifex aeolicus* VF5. *Carbohydrate Research*, *350*, 81–85. https://doi.org/10.1016/j.carres.2011.12.022.

von Braunmühl, V., Jonas, G., & Stadler, R. (1995). Enzymatic grafting of amylose from poly(dimethylsiloxanes). *Macromolecules*, *28*(1), 17–24.

Yalpani, M. (1988). *Polysaccharides: Syntheses, modifications, and structure/property relations*. Amsterdam; New York: Elsevier.

Yamashita, K., Yamamoto, K., & Kadoakwa, J. (2015). Synthesis of non-natural heteroaminopolysaccharides by α-glucan phosphorylase-catalyzed enzymatic copolymerization: A(1→4)-linked glucosaminoglucans. *Biomacromolecules*, *16*, 3989–3994.

Yanase, M., Takaha, T., & Kuriki, T. (2006). α-Glucan phosphorylase and its use in carbohydrate engineering. *Journal of the Science of Food and Agriculture*, *86*(11), 1631–1635. https://doi.org/10.1002/Jsfa.2513.

Yanase, M., Takata, H., Fujii, K., Takaha, T., & Kuriki, T. (2005). Cumulative effect of amino acid replacements results in enhanced thermostability of potato type l α-glucan phosphorylase. *Applied and Environmental Microbiology*, *71*(9), 5433–5439. https://doi.org/10.1128/Aem.71.9.5433-5439.2005.

Yu, K.-L., & Fraser-Reid, B. (1988). A novel reagent for the synthesis of myo-inositol phosphates: N,N-diisopropyl dibenzyl phosphoramidite. *Tetrahedron Letters*, *29*(9), 979–982. https://doi.org/10.1016/0040-4039(88)85313-9.

Ziegast, G., & Pfannemüller, B. (1987). Linear and star-shaped hybrid polymers. Phosphorolytic syntheses with di-functional, oligo-functional and multifunctional primers. *Carbohydrate Research*, *160*, 185–204. https://doi.org/10.1016/0008-6215(87)80311-7.

CHAPTER EIGHT

Synthesis of glycomonomers via biocatalytic methods

Azis Adharis, Katja Loos*
Macromolecular Chemistry and New Polymeric Materials, Zernike Institute for Advanced Materials, University of Groningen, Groningen, The Netherlands
*Corresponding author: e-mail address: k.u.loos@rug.nl

Contents

Abstract

The demand to develop sustainable polymers has increased in the last decades. Sustainable polymers contain some or all components that are derived from renewable resources. Carbohydrates, or saccharides, are one of the most abundant renewable biomass produced annually. Harnessing carbohydrates for building blocks of polymers can generate novel functional water-soluble materials. One of the examples is called glycopolymers, that are synthetic polymers having pendant saccharide moieties. In this report, we present the synthesis of glycomonomers, the precursors of glycopolymers, by

Methods in Enzymology, Volume 627
ISSN 0076-6879
https://doi.org/10.1016/bs.mie.2019.04.015

enzymatic approaches. Some enzymes, like lipases, proteases, glycosidases, and glycosyltransferases, are exploited for this purpose. These enzymes catalyze the formation of ester or glycosidic bonds through transesterification reaction, reverse hydrolysis reaction, transglycosylation reaction, or glycosyl transfer from donors to acceptors.

1. Introduction

1.1 Carbohydrates

Carbohydrates are one of the most abundant natural organic substances on Earth and they mainly serve as structural and functional elements of cellular materials. In general, carbohydrates are defined as biomolecules consisting of carbon (C), hydrogen (H), and oxygen (O) atoms with the molecular formula $C_m(H_2O)_n$ also known as hydrates of carbon (Alén, 2018). According to the IUPAC definition, another term for carbohydrates is saccharides; which includes monosaccharides, oligosaccharides, and polysaccharides as well as their derivative components (Moss, Smith, & Tavernier, 1995).

Monosaccharides are the simplest carbohydrates and they typically contain three to nine carbon atoms. Fig. 1 shows some examples of monosaccharide structures in the chair conformation. Carbohydrates with two or more monosaccharide units linked together by glycosidic bonds are classified as oligo- and polysaccharides (Mcnaught, 1996). Oligosaccharides consist of a small number of repeating monosaccharides, generally between 2 and

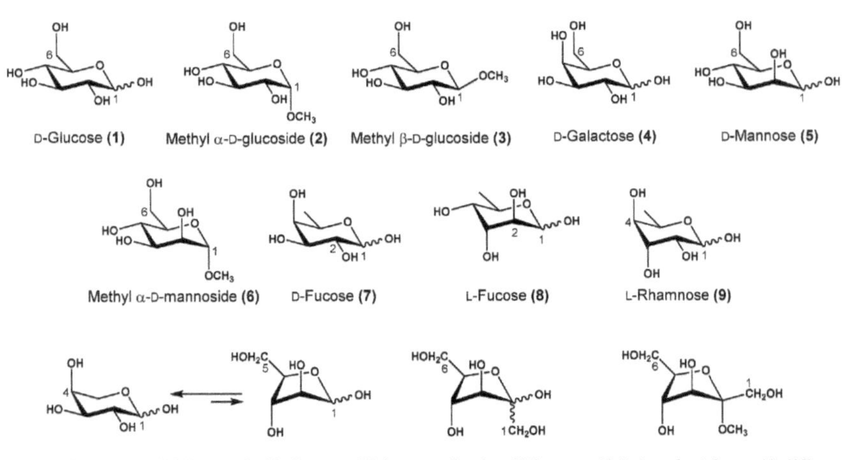

Fig. 1 Monosaccharide structures in the chair conformation. The numbers within the structures indicate the position of functionalization by vinyl groups as discussed in Sections 2–4.

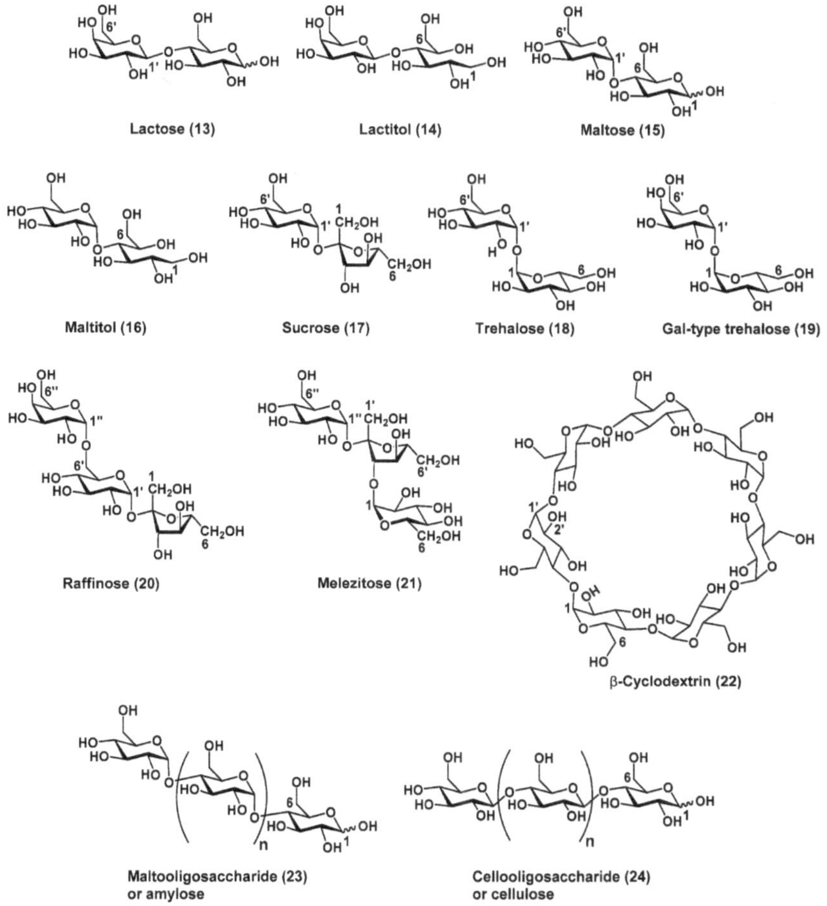

Fig. 2 Oligo- and polysaccharide structures in the chair conformation. The numbers within the structures indicate the position of functionalization by vinyl groups as discussed in Sections 2–4.

10 units. In the case of two or three monosaccharides combined, specific terms are used, namely, disaccharides or trisaccharides. Fig. 2 presents some examples of oligo- and polysaccharides.

1.2 Enzymes as biocatalysts in synthetic organic chemistry

Enzymes are catalysts that perform biosynthetic reactions in the cells at a remarkably high rate under mild reaction conditions (temperature, pH, and pressure) (Blanco & Blanco, 2017). The enzymatic reaction rates are about 10^6- to 10^{12}-fold of the uncatalyzed reaction, even up to 10^{20} times

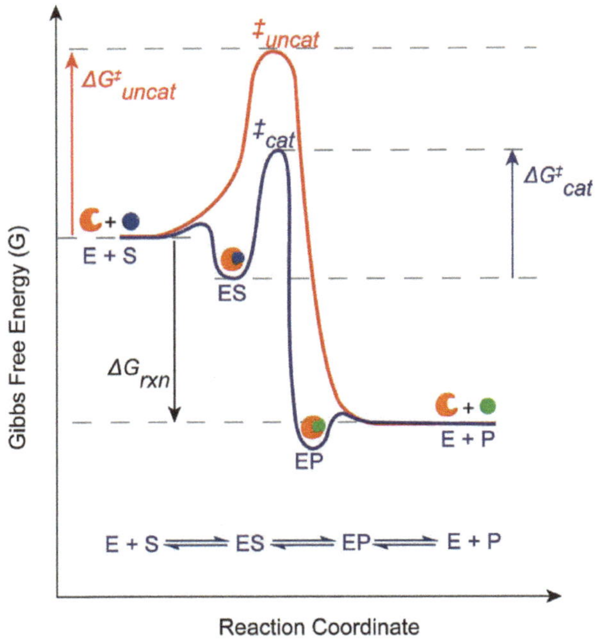

Fig. 3 Energy diagram of a chemical reaction with and without enzyme.

in some special cases (Borman, 2004). The energy diagram of a typical chemical reaction with enzymes (*in vivo*) and without enzymes is presented in Fig. 3. In principle, the role of enzymes as biocatalysts is identical to chemical catalysts in terms of speeding up the reaction rate without changing the total free energy from reactants to products. This is achieved by providing an alternative reaction mechanism, which has a lower activation energy than the non-catalyzed reaction. Besides, enzymes show amazing efficiency and specificity as well as the fact that they are non-toxic, natural, and renewable materials in comparison with conventional catalysts (Loos, 2010). As a result, *in vitro* enzymatic reactions have been developed in order to provide chemical synthesis in a green approach.

Probably the first *in vitro* enzymatic reaction was reported in the 1930s for the synthesis of esters catalyzed by lipase in organic solvents (Sym, 1933, 1936). However, it began to gain widespread interest after Zaks and Klibanov published their articles in the 1980s dealing with different lipases as catalysts for various organic reactions in anhydrous solvents (Zaks & Klibanov, 1984, 1985). Since then, biocatalysts in synthetic organic chemistry have been one of the most studied topics with a constantly growing reputation.

Many enzymes are found to be able to work not only with their native substrates but also with unnatural substrates which results in broad variety of materials (Shoda, Uyama, Kadokawa, Kimura, & Kobayashi, 2016).

1.3 Enzymatic synthesis of glycomonomers

The demand to develop sustainable materials and production methods, especially in the field of polymers, has increased in the last decades (Gandini & Lacerda, 2015; Jiang & Loos, 2016). This is reasonable considering that the main components for polymer production are based on fossil resources which have been predicted to be exhausted in the next several hundred years. In addition, the manufacturing of fossil resources often creates environmental issues like global warming, pollution, and dangerous waste. In order to significantly decrease the disadvantages created by the use of fossil resources, the European Union and the United Nations, through their policies and legislation, support the useof eco-friendly materials and simultaneously limit the consumption of environmentally hazardous substances in industry (Satyanarayana, Arizaga, & Wypych, 2009; Zhu, Romain, & Williams, 2016).

Sustainable polymers contain some or all portions of building blocks that are derived from renewable feedstocks (Gandini & Lacerda, 2015; Jiang & Loos, 2016; Zhu et al., 2016). One of the most promising examples of renewable feedstocks is biomass and its derivatives since these resources are produced by nature via photosynthetic pathways. Especially carbohydrates, one of the most abundant biomass, are interesting materials for the production of durable polymers. However, the chemical industry is still only using a small percentage of the accessible carbohydrates (Lichtenthaler & Peters, 2004; Nakajima, Dijkstra, & Loos, 2017), although carbohydrates give rise to a new family of functional water-soluble monomers and polymers. One example for this is glycopolymers, synthetic polymers composed of saccharides as the pendant group (Adharis, Ketelaar, Komarudin, & Loos, 2019; Becer & Hartmann, 2015).

The potential applications of glycopolymers have attracted much research in the last decades starting in the 1970s when synthetic glycopolymers with polyacrylamide backbones could interact with carbohydrate-binding proteins called lectins (Hořoejší, Chaloupecká, & Kocourek, 1978; Schnaar & Lee, 1975). Glycopolymers exhibit a high affinity and specificity to lectins, thus mimicking glycolipids and glycoproteins at the cell surface

(Kiessling & Grim, 2013; Miura, Hoshino, & Seto, 2016). Moreover, glycopolymers have been used as inhibitors that prevent the adhesion of pathogens to the cell surface during an infection (Bhatia, Camacho, & Haag, 2016). Other applications of glycopolymers cover, for example, drug delivery (Appelhans, Klajnert-Maculewicz, Janaszewska, Lazniewska, & Voit, 2015; Sunasee, Adokoh, Darkwa, & Narain, 2014; Utama, Jiang, Zetterlund, & Stenzel, 2015), biosensors (Hushegyi, Klukova, Bertok, & Tkac, 2015; Zeng, Qu, & Rehman, 2016), and diseases inhibitors (Bhatia et al., 2016; Branson & Turnbull, 2013).

Glycomonomers, the precursors of glycopolymers, contain saccharide units that are linked to some polymerizable groups. Currently, vinyl groups are the most exploited ones. The preparation of glycomonomers from saccharides and vinyl containing molecules can be carried out using chemical reactions (Ambrosi, Cameron, Davis, & Stolnik, 2005; Schuster, Mortell, Hegeman, & Kiessling, 1997; Wulff, Schmid, & Venhoff, 1996). However, this approach is time-consuming given that it involves multiple protection steps of hydroxyl groups of saccharides in the course of the reactions. An elegant way to synthesize glycomonomers is by using enzymes as the biocatalysts (Adharis, Vesper, Koning, & Loos, 2018; Albertin, Stenzel, Barner-Kowollik, Foster, & Davis, 2005; Gill & Valivety, 2000; Kloosterman, Roest, Priatna, Stavila, & Loos, 2014). As already mentioned in the previous section, enzymes exhibit an excellent reaction specificity compared to otherwise used chemical catalysts. Therefore, in the enzymatic synthesis of glycomonomers, the vinyl group can specifically be attached to a particular position, among multiple possible positions, of saccharides without involving the tedious protection steps. Some vinyl groups of the synthesized glycomonomers are presented in Fig. 4.

To the best of our knowledge, the first work on the synthesis of vinyl monomers containing saccharide moieties by biocatalytic approaches was published by Klibanov and coworkers in 1986 (Therisod & Klibanov, 1986, 1987). In this work, lipases were used to catalyze the acylation of various monosaccharides. The following sections discuss some of the latest results found in the last two decades. The early findings in this area were summarized in several excellent review articles published in the early 2000 (Gross, Kumar, & Kalra, 2001; Kadereit & Waldmann, 2001; La Ferla, 2002; Wulff et al., 1996). Enzymes from the class of hydrolases (EC 3), such as lipases, proteases, and glycosidases, were mainly used for the bio-catalytically synthesized glycomonomers since they are able to catalyze the formation of ester-, amide-, and glycosidic bonds as well as reverse hydrolysis reactions.

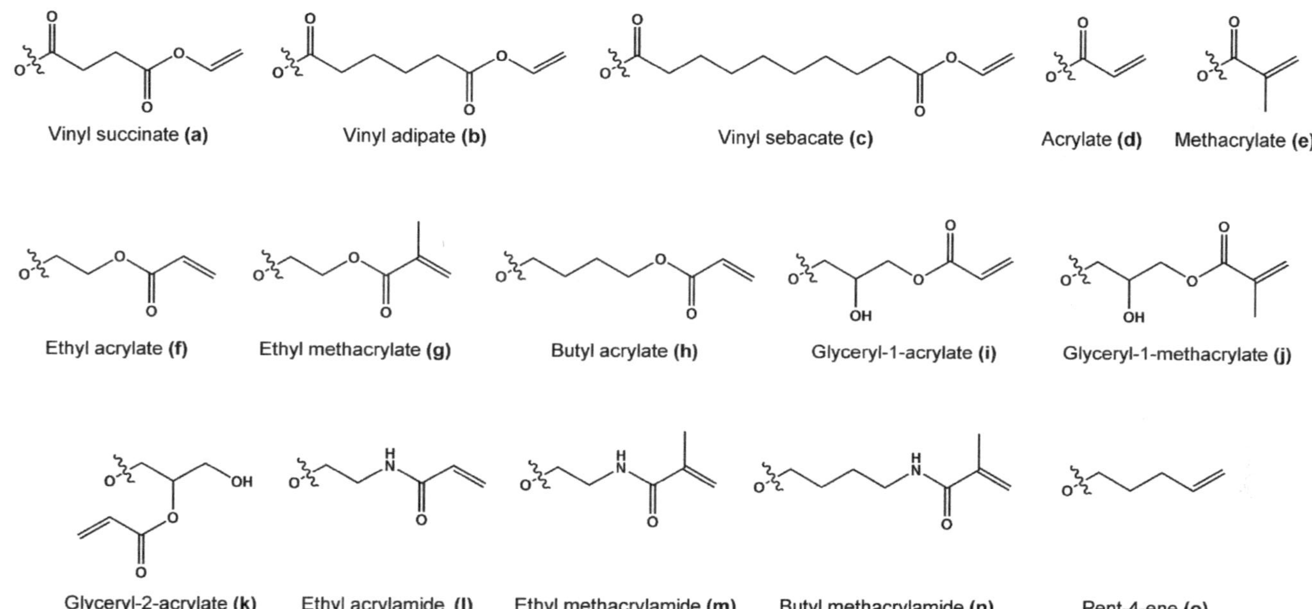

Fig. 4 Some examples of vinyl functionalities of glycomonomers.

2. Lipase-catalyzed synthesis of glycomonomers

Lipases are a subclass of esterases that, in nature, catalyze the hydrolysis reaction of ester bonds of long-chain triglycerides to diglycerides, monoglycerides, glycerols, and fatty acids (Rosales, Pazos, & Ángeles Sanromán, 2018). Lipases can be isolated from various origins such as bacteria, fungi, and mammals. *In vitro*, lipases can perform the reverse reaction of its natural functions, meaning that lipases create ester bonds from alcohols and carboxylic acids of substrates in dry organic solvents.

For the case of glycomonomer synthesis, lipases catalyze the transesterification of hydroxyl groups of saccharides with divinyl esters. The transesterification reaction occurs in an equilibrium state. Therefore, in order to shift the equilibrium reaction to the products side, divinyl esters are favored over carboxylic acids as substrates. The esters generate vinyl alcohol as a side product which immediately converts, via tautomerization, to volatile acetaldehyde. Fig. 5 exemplarily shows the lipase-catalyzed synthesis of glycomonomers from glucose and divinyl adipate.

Table 1 summarizes some examples of glycomonomers that were synthesized by lipases. Lipase from *Candida antarctica* was the most frequently used biocatalyst; however, other sources of lipases from *Alcaligenes* sp., *Pseudomonas* sp., and *Thermomyces* sp. were also utilized. These lipases show a good activity in commonly available organic solvents like acetonitrile, pyridine, acetone, DMF, *t*-butanol, and butanone. Nevertheless, working with a particular solvent may be more preferred than the others. For instance, Cameron et al. reported that by replacing the solvent from acetonitrile to butanone for the synthesis of methyl 6-O-methacryloyl-α-D-glucoside, the glycomonomer yield was improved from 70% to 85% and the reaction time was decreased from 5 to 2 days (Cameron et al., 2008). In general, the lipase-catalyzed transesterification reaction was performed between 1 and 5 days at 30–60 °C to obtain a product yield up to 85%.

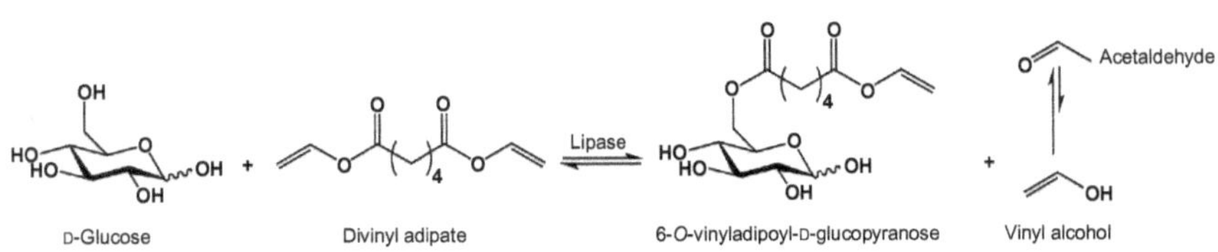

Fig. 5 Reaction scheme of the lipase-catalyzed synthesis of glycomonomers (Albertin, Kohlert, Stenzel, Foster, & Davis, 2004).

Table 1 Lipase-catalyzed synthesis of glycomonomers.

Saccharides	Vinyl group	Vinyl position	Enzyme sources	Reaction condition	Yield (%)	References
1	**b**	C-6	*Candida antarctica*	Acetonitrile, 50 °C, 1 day	50	Albertin, Kohlert, et al. (2004)
1	**c**	C-6	*Alcaligenes* sp.	Pyridine, 30 °C, 7 days	31	Kitagawa and Tokiwa (1998)
1	**e**	C-6	*Candida antarctica*	*t*-Butanol, 50 °C, 1 day	34	Perin and Felisberti (2018)
2	**d**	C-6	*Candida antarctica*	Acetone, 60 °C, 18 h	82	Li and Rethwisch (2002)
2	**e**	C-6	*Candida antarctica*	Acetonitrile, 50 °C, 5 days	70	Albertin, Stenzel, et al. (2004) and Albertin et al. (2005)
2	**e**	C-6	*Candida antarctica*	Butanone, 60 °C, 2 days	85	Cameron et al. (2008)
3	**d, e**	C-6	*Candida antarctica*	*t*-Butanol, 50 °C, 1 day	100[a]	Kim et al. (2004) and Park, Haam, Lee, Kim, and Kim (2004)
5	**e**	C-6	*Candida antarctica*	Acetone, 50 °C, 5 days	74	Granville, Quémener, Davis, Barner-Kowollik, and Stenzel (2007)
6	**e**	C-6	*Candida antarctica*	Acetonitrile, 50 °C, 4 days	48	Albertin et al. (2005)
9	**b**	C-4	*Pseudomonas* sp.	Pyridine, 30 °C, 7 days	11	Raku and Tokiwa (2003)
11	**e**	C-1, C-6, C-1,6	*Candida antarctica*	*t*-Butanol, 50 °C, 1 day	95[a]	Perin and Felisberti (2018)
14	**b, c**	C-6	*Candida antarctica*	Pyridine, 50 °C, 3 days	70[a], 99[a]	Miura, Ikeda, and Kobayashi (2003) and Miura, Ikeda, Wada, Sato, and Kobayashi (2003)

Continued

Table 1 Lipase-catalyzed synthesis of glycomonomers.—cont'd

Saccharides	Vinyl group	Vinyl position	Enzyme sources	Reaction condition	Yield (%)	References
16	**b, c**	C-6	*Candida antarctica*	Pyridine, 50 °C, 3 days	7[a], 44[a]	Miura, Ikeda, and Kobayashi (2003) and Miura, Ikeda, Wada, et al. (2003)
17	**b**	C-6,6′	*Candida antarctica*	Acetone, 50 °C, 2 days	21	Park, Kim, and Dordick (2000)
18	**b**	C-6,6′	*Candida antarctica*	Acetone, 50 °C, 2 days	20	Park et al. (2000)
18	**c**	C-6′	*Pseudomonas* sp.	Pyridine, 60 °C, 3 days	19	Miura, Wada, Nishida, Mori, and Kobayashi (2004)
19	**c**	C-6′	*Pseudomonas* sp.	Pyridine, 60 °C, 3 days	44	Miura et al. (2004)
20	**b**	C-6″	*Candida antarctica*	Pyridine, 60 °C, 3 days	42	Pérez-Victoria and Morales (2007)
20	**b**	C-6″	*Thermomyces lanuginosus*	Pyridine, 60 °C, 1 day	75	Pérez-Victoria and Morales (2007)
21	**b**	C-6	*Candida antarctica*	Pyridine, 60 °C, 3 days	38	Pérez-Victoria and Morales (2007)
21	**b**	C-6″	*Thermomyces lanuginosus*	Pyridine, 60 °C, 1 day	50	Pérez-Victoria and Morales (2007)

[a]Substrate conversion (%), yield data are not available.
The structure of saccharides and vinyl groups correspond to Figs. 1, 2, and 4.

Lipase from *Candida antarctica* selectively catalyzes the monosubstitution of vinyl groups toward a primary alcohol at the C-6 position of monosaccharides with pyranose structures (Albertin, Stenzel, et al., 2004; Granville et al., 2007; Kim et al., 2004; Li & Rethwisch, 2002; Park et al., 2004), at the C-6 position of reduced disaccharides (Miura, Ikeda, & Kobayashi, 2003; Miura, Ikeda, Wada, et al., 2003), and at C-6 or C-6″ positions of trisaccharides (Pérez-Victoria & Morales, 2007). Mono- and disubstitution of vinyl groups was observed when fructose, sucrose, and trehalose were used as the substrates (Park et al., 2000; Perin & Felisberti, 2018). This happened probably because they possess two primary alcohols on the β-position allowing the enzymes to perform two substitutions of vinyl groups in a saccharide unit. Also, this enzyme shows a reaction selectivity toward divinyl esters with different alkyl chain lengths. For example, Miura et al. reported that divinyl sebacate (Fig. 4C) yielded a better substrate conversion than divinyl adipate (Fig. 4B) during the synthesis of glycomonomers (Miura, Ikeda, Wada, et al., 2003). A similar pattern was observed during the enzymatic polymerization of polyesters where this lipase exhibits a preference toward more hydrophobic diols and diesters (Pellis et al., 2018).

A functionalization at the C-4 position of monosaccharide rhamnose with divinyl adipate was reported by Raku and Tokiwa using lipase from *Pseudomonas* species (Raku & Tokiwa, 2003). In that case, a transesterification occurred at the secondary alcohol since no primary alcohol was present in the rhamnose structure. If the disaccharide trehalose—bearing a primary alcohol at the C-6′ position—was used as the substrate for that lipase, monosubstitution of vinyl groups selectively occurred in the position of that alcohol (Miura et al., 2004). Moreover, a monosubstitution of trisaccharides with divinyl adipate was observed when lipase from *Thermomyces* sp. catalyzed the reaction (Pérez-Victoria & Morales, 2007).

The following procedure is based on the study by Blanazs and coworkers that generates the highest yield of the synthesized glycomonomers (Cameron et al., 2008). The reaction was catalyzed by lipase from *Candida antarctica* whereas methyl-α-D-glucoside served as the substrate.

2.1 Equipment

- Analytical balance
- Common glass laboratory equipments such as erlenmeyer flask, measuring cylinder, beaker glass, funnel, round-bottom flask, and dropping pippetes
- Temperature controlled shaker

- Rotary evaporator
- Hot plate
- Automated flash chromatography on a Biotage SP1 system with automatic fraction collector and on-board UV detector
- Glass-backed silica-gel plates for thin layer chromatography (TLC)
- Electrospray ionization mass spectrometry (ESI-MS) on a Micromass LCT spectrometer
- Nuclear magnetic resonance (NMR) spectroscopy on a Varian Inova 500 spectrometer

2.2 Materials

- Methyl-α-D-glucoside from Sigma-Aldrich
- Lipase (Novozym 435) from Novozymes A/S
- Vinyl methacrylate from Sigma-Aldrich
- Solvents (butanone, methanol, chloroform, ethanol) from Sigma-Aldrich
- Sulfuric acid from Sigma-Aldrich
- Deuterated methanol (CD_3OD) from Sigma-Aldrich

2.3 Procedure

1. Prepare an erlenmeyer flask containing methyl-α-D-glucoside (8.0 g, 0.041 mol), lipase (4.0 g), vinyl methacrylate (4.4 g, 0.039 mol), and butanone (40 mL).
2. Seal the flask with a rubber septum and shake at 200 rpm and 60 °C for 2 days.
3. Stop the reaction by separating the enzyme from the reaction mixture.
4. Wash the filtrate with methanol (100 mL).
5. Evaporate the solvent of the collected organic phases by rotary evaporator until a yellow-brown syrup is obtained.
6. Purify the product by flash chromatography using the eluent of chloroform/methanol (8:2).
7. Identify the product in the collected fractions by TLC using the same eluent as in the previous step. The TLC plates is immersed in a 20% H_2SO_4/ethanol solution and heated for spots detection. The product is detected at retention faction of 0.39.
8. Evaporate the pooled fractions containing the products by rotary evaporator until a clear syrup is obtained.
9. Characterize the product with ESI-MS, ^1H and ^{13}C NMR spectroscopy. CD_3OD is used as the solvent for NMR measurement.

3. Protease-catalyzed synthesis of glycomonomers

Proteases, or proteinases/peptidases, naturally hydrolyze peptide bonds of amino acids during the metabolism of proteins in all organisms (Clark & Pazdernik, 2016). Depending on the type of proteases, the hydrolysis reaction may be specific at particular sites of a protein structure or non-specific resulting in individual amino acids. For the *in vitro* reaction, proteases work in a similar approach like lipase; that is catalyzing transesterification of saccharides with divinyl esters during the synthesis of glycomonomers in organic solvents. For example, Lin and coworkers used galactose and several divinyl esters as the substrate for protease (see Fig. 6) (Wu, Wang, Xiao, Lu, & Lin, 2004).

Unlike lipases, proteases are able to produce the highest yield of glycomonomers when monosaccharides are combined with divinyl esters bearing a few methylene groups (Wu et al., 2004), meaning that this enzyme has a good selectivity toward more hydrophilic substrates. Comparable results were reported for dissaccharide (Wang, Wu, Wang, & Lin, 2005; Wu et al., 2004) and trisaccharide (Qi, Deshui, Zhichun, Guohua, & Xianfu, 2003; Wu, Chen, Lu, & Lin, 2006) substrates. However, an exception was found when glucose reacted with divinyl adipate (bearing four methylene groups). This resulted in an increased yield compared to the reaction with divinyl succinate (bearing two methylene groups) and divinyl sebacate (bearing eight methylene groups) (Wu et al., 2001). It is interesting to note that some lipases were successfully commercialized in the immobilized state whereas proteases are only available in the liquid or granulate form, which limits their reusability drastically.

Numerous examples of glycomonomer syntheses by proteases are outlined in Table 2 in which the enzymes were mainly obtained from *Bacillus* species, especially *Bacillus subtilis*. Like lipases, proteases also work well in organic solvents especially pyridine and DMF. Protease-catalyzed synthetic reactions are typically performed in 1–7 days with a temperature between 30 °C and 55 °C. The highest yield of the isolated product was reported to be up to 85%.

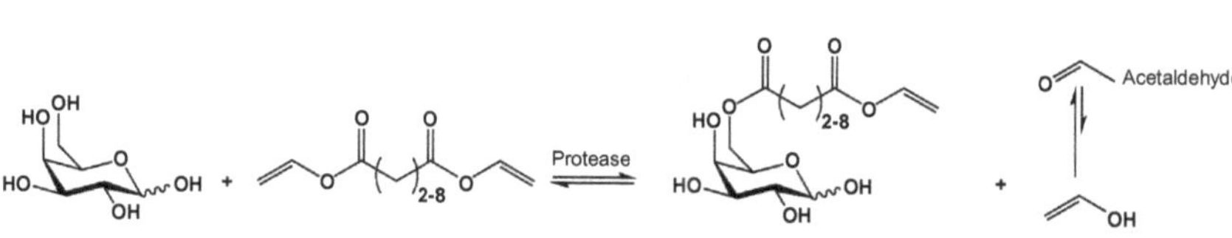

Fig. 6 Reaction scheme of the protease-catalyzed synthesis of glycomonomers.

Table 2 Protease-catalyzed synthesis of glycomonomers.

Saccharides	Vinyl group	Vinyl position	Enzyme sources	Reaction condition	Yield (%)	References
1	**a, b, c**	C-6	*Bacillus subtilis*	Pyridine, 55 °C, 3 days	30, 53, 35	Wu et al. (2001)
1	**b**	C-6	*Bacillus subtilis*	DMF, 50 °C, 5 days	80[a]	Borges, dos Santos, Vieira, and Balaban (2009)
1	**e**	C-6	*Bacillus subtilis*	DMF, 45 °C, 1 day	36	Perin and Felisberti (2018)
2	**e**	C-6	*Bacillus subtilis*	DMF, 45 °C, 7 days	33	Potier, Bouchu, Gagnaire, and Queneau (2001)
4	**a, b, c**	C-6	*Bacillus subtilis*	Pyridine, 50 °C, 0.5/3/5 days	70, 47, 35	Wu et al. (2004)
5	**a, b, c**	C-6	*Bacillus subtilis*	Pyridine, 50 °C, 0.5/4/5 days	85, 69, 43	Wu et al. (2004)
7	**b**	C-2	*Bacillus subtilis*	DMF, 30 °C, 7 days	55	Raku and Tokiwa (2003)
8	**b**	C-2	*Bacillus subtilis*	DMF, 30 °C, 7 days	32	Raku and Tokiwa (2003)
10	**b**	C-4 (**10y**), C-5 (**10z**)	*Bacillus subtilis*	DMF, 50 °C, 7 days	66	Borges and Balaban (2014)
10	**b**	C-5 (**10z**)	*Streptomyces* sp.	DMF, 30 °C, 7 days	46	Tokiwa et al. (1999)
11	**e**	C-1, C-6	*Bacillus subtilis*	DMF, 45 °C, 1 day	60[a]	Perin and Felisberti (2018)
12	**e**	C-1, C-6	*Bacillus subtilis*	DMF, 45 °C, 3 days	33	Potier et al. (2001)

13	**a, b, c**	C-6′	*Bacillus subtilis*	Pyridine, 50 °C, 5 days	62, 46, 31	Wang et al. (2005) and Wu et al. (2004)
15	**a, b, c**	C-6′	*Bacillus subtilis*	Pyridine, 50 °C, 5 days	53, 42, 34	Wu et al. (2004)
17	**b**	C-1	*Bacillus subtilis*	Pyridine, 50 °C, 5 days	55	Wang et al. (2005) and Wu et al. (2004)
17	**b**	C-1	*Bacillus subtilis*	DMF, 50 °C, 5 days	90[a]	Borges and Balaban (2007)
17	**e**	C-1	*Bacillus subtilis*	DMF, 45 °C, 3 days	70	de Oliveira and Felisberti (2013) and Potier, Bouchu, Descotes, and Queneau (2000)
17	**b**	C-1	*Bacillus licheniformis*	Pyridine, 45 °C, 2 days	13	Park et al. (2000)
18	**e**	C-6′	*Bacillus subtilis*	DMF, 45 °C, 3 days	25	Potier et al. (2001)
18	**b**	C-6′	*Bacillus licheniformis*	Pyridine, 45 °C, 2 days	22	Park et al. (2000)
20	**a, b, c**	C-1	*Bacillus subtilis*	Pyridine, 50 °C, 6 days	55, 48, 30	Qi et al. (2003) and Wu et al. (2006)
20	**b**	C-1	*Bacillus licheniformis*	Pyridine, 40 °C, 1 day	65	Pérez-Victoria and Morales (2007)
21	**b**	C-6′	*Bacillus licheniformis*	Pyridine, 40 °C, 1 day	55	Pérez-Victoria and Morales (2007)
22	**a, b, c**	C-2′	*Bacillus subtilis*	DMF, 50 °C, 5 days	5, 10, 13	Xiao, Wu, Wang, and Lin (2004)

[a]Substrate conversion (%), yield data are not available.
The structure of saccharides and vinyl groups correspond to Figs. 1, 2, and 4.

Protease from *Bacillus subtilis* selectively catalyzes the monosubstitution of vinyl groups to primary alcohols of saccharides at various positions. The substitution was located at the C-6 position of monosaccharides with pyranose structures (Borges et al., 2009; Perin & Felisberti, 2018; Potier et al., 2001; Wu et al., 2001, 2004), at the C-1 or C-6 position of monosaccharides with furanose structures (Perin & Felisberti, 2018; Potier et al., 2001), at the C-6′ position of the disaccharides lactose/maltose/trehalose (Potier et al., 2001; Wang et al., 2005; Wu et al., 2004), as well as at the C-1 position of the disaccharide sucrose and trisaccharide raffinose (Borges & Balaban, 2007; de Oliveira & Felisberti, 2013; Potier et al., 2000; Wang et al., 2005; Wu et al., 2004). Besides that, monosubstitution was also found at the secondary alcohol (at the C-2 position) when fucose and β-cyclodextrin served as the substrates (Raku & Tokiwa, 2003; Xiao et al., 2004).

Proteases from two resources can provide different reaction selectivity and this was observed when L-arabinose was used as the substrate. In solution, L-arabinose possesses two isomers that exhibit different structures, either furan- or pyran-like molecules. Protease from *Bacillus subtilis* catalyzed the transesterification of L-arabinose on both isomers (Borges & Balaban, 2014) while protease from *Streptomyces* sp. only worked with the furanose isomer (Tokiwa et al., 1999) under similar reaction condition.

Wu et al. reported the synthesis of a mannose-based glycomonomer using protease from *Bacillus subtilis* (Wu et al., 2004). The isolated product yield was remarkably high, up to 85%. The following section describes the experimental procedure for preparation of 6-*O*-vinylbutanedioyl-D-mannopyranose according to literature.

3.1 Equipment

- Analytical balance
- Common glass laboratory equipments such as conical flask, measuring cylinder, beaker glass, funnel, round-bottom flask, and dropping pippetes
- Temperature controlled shaker
- Rotary evaporator
- Glass column loaded with silica gel
- Glass-backed silica-gel plates for thin layer chromatography (TLC)
- Electrospray ionization mass spectrometry (ESI-MS) on a Bruker Esquire-LC

- Infrared (IR) spectroscopy on a Nicolet Nexus FTIR 670 spectrophotometer
- Nuclear magnetic resonance (NMR) spectroscopy on a Bruker Avance DMX500 spectrometer

3.2 Materials

- Protease from Wuxi Enzyme Co. Ltd. (Wuxi, PR China)
- D-Mannose from Sigma-Aldrich
- Divinyl butanedioate from Sigma-Aldrich
- Solvents (pyridine, methanol, ethyl acetate) from Sigma-Aldrich
- Sulfuric acid from Sigma-Aldrich
- DMSO-d_6 from Sigma-Aldrich

3.3 Procedure

1. Prepare a conical flask containing D-mannose (1.4 g, 7.8 mmol), protease (1 g, 25 mg mL^{-1}), divinyl butanedioate (5.3 g, 4 equiv), and pyridine (40 mL).
2. Seal the flask with a rubber septum and shake at 250 rpm and 50 °C for 12 h.
3. Stop the reaction by filtering off the enzyme.
4. Evaporate the pyridine by rotary evaporator.
5. Isolate the product by column chromatography using the eluent of ethyl acetate/methanol/H_2O (150:10:1).
6. Analyze the product formation by TLC using the same eluent as in the previous step. The TLC plates is sprayed by 10% H_2SO_4/methanol solution and heated for spots detection.
7. Evaporate the pooled fractions containing the products by rotary evaporator until tan amorphous solid is formed.
8. Characterize the product with ESI-MS, IR, ^{1}H and ^{13}C NMR spectroscopy. DMSO-d_6 is used as the solvent for NMR measurement.

4. Glycosidase-catalyzed synthesis of glycomonomers

Glycosidases, or glycosyl hydrolases, are carbohydrate-active enzymes that *in vivo* catalyze the hydrolytic degradation of glycosidic bonds in carbohydrates (Kwan, 2017). Considering their versatility, glycosidases have been utilized extensively in industrial processes. For instance, glycosidases are applied as biocatalysts during the degradation of biomass (cellulose and other plant-derived polysaccharides) yielding mono- and oligosaccharides for biofuels and biobased chemical manufacturing (Kumar, Barrett, Delwiche, & Stroeve, 2009; Olofsson, Bertilsson, & Lidén, 2008; Pérez, Muñoz-Dorado, De La Rubia, & Martínez, 2002).

For the synthesis of glycomonomers, glycosidases catalyze the monosubstitution of vinyl groups at the anomeric carbon (the C-1 position) of saccharides. Depending on the enzyme type, the linkage of the vinyl group may be on the α or β-position of the anomeric carbon which results in anomerically pure products. Glycosidases provide a different location for the attachment of vinyl groups compared to lipases and proteases which usually functionalize the saccharides at their primary alcohols. It was reported that functionalization at the primary alcohol of monosaccharides could be a disadvantage because the resulted glycopolymers were found to lose their activity with saccharide-binding proteins (Granville et al., 2007).

Harnessing glycosidases for the glycomonomer synthesis is done via two approaches, namely, thermodynamically controlled reversed hydrolysis and kinetically controlled transglycosylation. Both methodologies were exploited in the past with their advantages and limitations summarized elsewhere (Van Rantwijk, Van Woudenberg-, & Sheldon, 1999). The following paragraphs discuss briefly the differences between these two methods for the synthesis of saccharide-vinyl monomers listed in Table 3.

4.1 Thermodynamically controlled reverse hydrolysis

Hydrolysis reactions of saccharides by glycosidases are performed under equilibrium conditions. In principle, reverse hydrolysis reactions are used for the synthesis of glycomonomers by reacting monosaccharides with vinyl-based alcohols. Nevertheless, a small amount of water is still required to maintain the enzyme activity and to increase the solubility of the monosaccharides in the reaction mixture. Consequently, this method is limited to monosaccharide substrates since oligo- and polysaccharides can be hydrolyzed by glycosidases in the presence of water.

Fig. 7 shows an example of the thermodynamically controlled reverse hydrolysis of glucose and 2-hydroxyethyl acrylate catalyzed by β-glucosidase derived from almonds (Millqvist-Fureby et al., 1998). Some vinyl-based alcohols (Fig. 4F–K) were tested as well, but the obtained glycomonomer yields were rather low (up to 21%) although highly concentrated glucose solutions were used to shift the equilibrium reaction to the product side. In order to improve the product yield, Kloosterman et al. optimized the

Table 3 Glycosidase-catalyzed synthesis of glycomonomers.

Saccharides	Vinyl group	Vinyl position	Enzyme	Reaction condition	Yield (%)	References
1	**f, g, h, i, j, k**	C-1β	β-Glucosidase from almonds	Citrate buffer pH 5.2, 37 °C, 1–3 days	16, 3, 3, 21, 18, 14	Millqvist-Fureby, MacManus, Davies, and Vulfson (1998)
1	**f, g, h**	C-1β	β-Glucosidase from almonds	Water/dioxane, 50 °C, 1 day	41, 43, 21	Kloosterman, Roest, et al. (2014)
1	**f**	C-1β	Cellulase from *Trichoderma reesei*	Citrate buffer pH 5.0, 50 °C, 3 days	5	Kloosterman, Brouwer, and Loos (2014)
1	**f**	C-1α	Amyloglucosidase from *Aspergillus niger*	Acetate buffer pH 5.0, 55 °C, 5 days	N/A	Kloosterman, Jovanovic, Brouwer, and Loos (2014)
1	**l, m, n**	C-1β	β-Glucosidase from almonds	Water, 50 °C, 1 h	16, 12, 18	Adharis, Vesper, et al. (2018)
4	**g, o**	C-1β	β-Galactosidase from *Escherichia coli*	Phosphate buffer pH 7, 40 °C, 1.5 days	58, 39	Gill and Valivety (2000)
4	**g**	C-1β	β-Galactosidase from *Aspergillus niger*	Phosphate buffer pH 7, 40 °C, 1.5 days	43	Gill and Valivety (2000)
4	**o**	C-1β	β-Galactosidase from *Aspergillus oryzae*	Phosphate buffer pH 7, 40 °C, 1.5 days	60	Gill and Valivety (2000)
4	**g**	C-1β	β-Galactosidase from *Sulfolobus solfataricus*	Phosphate buffer pH 6.5, 75 °C, 1.5 h	75–80	Santin, Rosso, Sada, and Peluso (1996)
15	**f, g, h**	C-1α	Amylase from *Bacillus stearothermophilus*	Acetate buffer pH 5, 55 °C, 4 days	5	Kloosterman, Jovanovic, et al. (2014)

N/A = Not available, the product was not isolated.
The structure of saccharides and vinyl groups correspond to Figs. 1, 2, and 4.

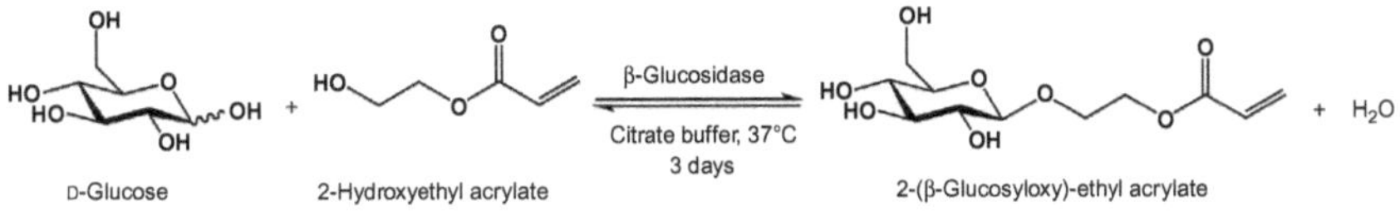

Fig. 7 Thermodynamically controlled reverse hydrolysis reaction catalyzed by β-glucosidase.

reaction time, the composition of alcohols and water, as well as adding an organic solvent into the reaction mixtures (Kloosterman, Roest, et al., 2014). As a result, the synthesized glycomonomers were obtained in a higher yield (up to 41%) than in the previous report.

In addition, less product yield was generated when 4-hydroxybutyl acrylate was used rather than 2-hydroxyethyl acrylate as the substrate in both studies, revealing the enzyme has preferences toward hydrophilic alcohols. To overcome the issue with hydrophobic alcohols, Gill and Valivety proposed to conduct the enzymatic reaction in plasticized glass phases using β-galactosidase from *E. coli*, *A. niger*, and *A. oryzae* (Gill & Valivety, 2000). In their study, the highest yield up to 71% was achieved by mixing saccharides/alkyl glycoside/dopant with a small amount of buffer, followed by melting the reaction mixture and finally adding alcohols and enzymes.

The synthetic procedure reported by Kloosterman et al. is suitable for both hydrophilic and hydrophobic vinyl-containing alcohols which result in moderate yield of glycomonomers (Kloosterman, Roest, et al., 2014). In addition, they utilize cheap starting materials as the substrates for β-glucosidase in a simple experimental method. Therefore, the mentioned procedure in the next paragraphs refers to that reference.

4.1.1 Equipment

- Analytical balance
- Common glass laboratory equipments such as round-bottom flasks, measuring cylinder, beaker glass, funnel, and dropping pippetes
- Stir/hot plate
- Rotary evaporator
- Glass column loaded with silica gel
- Aluminum sheet silica gel 60/kieselguhr for thin layer chromatography (TLC)
- Electrospray ionization mass spectrometry (ESI-MS) on a Thermo Scientific LTQ Orbitrap mass spectrometer
- Nuclear magnetic resonance (NMR) spectroscopy on a 400 MHz Varian VXR spectrometer

4.1.2 Materials

- D-Glucose from Sigma-Aldrich
- 2-Hydroxyethyl acrylate and 2-hydroxyethyl methacrylate from Sigma-Aldrich
- 4-Hydroxybutyl acrylate from ABCR Chemicals
- β-Glucosidase from Sigma-Aldrich
- Dioxane from Acros Organics
- Other solvents (chloroform, methanol, ethanol, sulfuric acid) from Lab-Scan
- Deuterated water (D_2O) from Sigma-Aldrich

4.1.3 Procedure

1. In a round-bottom flask, prepare D-glucose solution (0.7 g of glucose in 1 mL water).
2. Put the flask in a hotplate equipped with waterbath.
3. Add 2-hydroxyethyl acrylate/2-hydroxyethyl methacrylate/4-hydroxybutyl acrylate (12 mL) and dioxane (1 mL).
4. Start the reaction by adding the enzyme solution (70 mg of β-glucosidase in 1 mL water).
5. Close the flask and stir the reaction mixture at 200 rpm and 50 °C for 1 day.
6. Identify the product formation by TLC using the eluent of chloroform/methanol (4/1).
7. Spray the TLC plates by 5% H_2SO_4/ethanol solution and heated for spots detection. 2-(β-Glucosyloxy)-ethyl acrylate, 2-(β-glucosyloxy)-ethyl methacrylate, and 4-(β-glucosyloxy)-butyl acrylate are detected at retardation factor of 0.55, 0.59, and 0.69, respectively.
8. Isolate the product by column chromatography using the same eluent as in the previous step.
9. Evaporate the pooled fractions containing the products by rotary evaporator until a yellowish syrup is obtained.
10. Characterize the product with ESI-MS, ^{1}H and ^{13}C NMR spectroscopy. D_2O is used as the solvent for NMR measurement.

4.2 Kinetically controlled transglycosylation

In the synthesis of glycomonomers by kinetically controlled transglycosylation, glycosidases catalyze the reaction between saccharides working as glycosyl donors, and vinyl-based alcohols serving as glycosyl acceptors. The enzymatic reaction is carried out in aqueous solutions. As shown in Fig. 8, water acts not only as the solvent, but also as a competing reactant

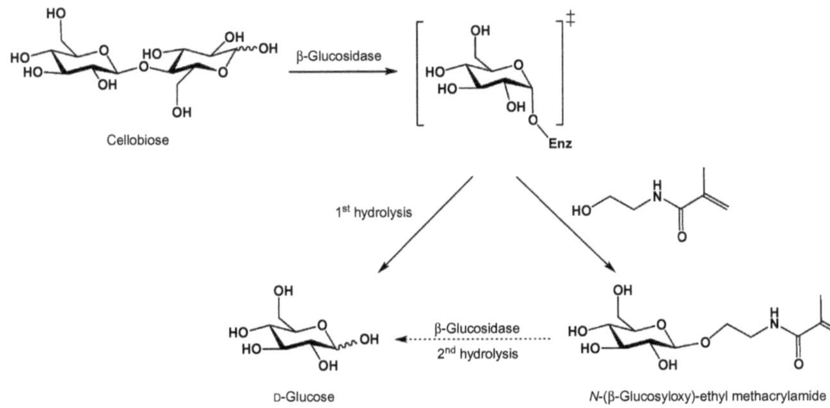

Fig. 8 Competing transglycosylation and hydrolysis catalyzed by β-glucosidase.

to glycosyl acceptors in the transglycosylation reaction. Therefore, it is crucial to optimize the time required for the reaction, as short reaction times may provide a low substrate conversion while hydrolysis of the synthesized glycomonomers is likely to occur during the longer reaction times.

In contrast to the previous approach, oligo- and polysaccharides can be used as the substrate of glycosidases in this method. For example, cellulase from *Trichoderma reesei* catalyzed the formation of 2-(β-glucosyloxy)-ethyl acrylate from cellulose (Kloosterman, Brouwer, et al., 2014). In addition, the same glycomonomer with the acrylate unit linked at the α-position was prepared from starch using amyloglucosidase from *Aspergillus niger* (Kloosterman, Jovanovic, et al., 2014). Moreover, maltogenic amylase from *Bacillus stearothermophilus* was able to catalyze the synthesis of maltose-based glycomonomers from starch (Kloosterman, Jovanovic, et al., 2014). A disadvantage was the low glycomonomers yield of less than 5%.

In order to improve the product yield, activated saccharides can be used as the glycosyl donors rather than native saccharides. Activated saccharides usually contain aryl or vinyl groups which are much better leaving groups in transglycosylation reaction than glycosyl units of natural sugars (Chiffoleau-Giraud, Spangenberg, & Rabiller, 1997; van Woudenberg-, Van Belle, Van Rantwijk, & Sheldon, 1998). For instance, Santin et al. reported *p*-nitrophenyl β-D-galactopyranoside as the substrate of β-galactosidase from *Sulfolobus solfataricus* with an isolated glycomonomer yield of about 75% (Santin et al., 1996). Another way to improve the product yield is by adding cosolvents like organic solvents (Bridiau, Issaoui, & Maugard, 2010;

Giacomini, Irazoqui, Gonzalez, Batista-viera, & Brena, 2002; Zeuner, Riisager, Mikkelsen, & Meyer, 2014) or ionic liquids (Bayón, Cortés, Berenguer, & Hernáiz, 2013; Kaftzik, Wasserscheid, & Kragl, 2002; Lang, Kamrat, & Nidetzky, 2006) to increase the enzyme-substrate interactions. For example, the yield of glucosyl-(meth)acrylamide monomers was improved from 16% to 68% upon addition of the ionic liquid $BMIMPF_6$ and replacing the native cellobiose with activated glucose (Adharis, Vesper, et al., 2018). This report comprehensively studied the synthesis of glycomonomers by providing either native or activated saccharides as the substrate for the enzyme. Furthermore, different solvent systems, either single or mixture solvents, were utilized and its influence into the product yields were investigated. Hence, the synthetic procedure mentioned in the following section is based on this literature.

4.2.1 Equipment

- Analytical balance
- Common glass laboratory equipments such as round-bottom flasks, measuring cylinder, beaker glass, funnel, and dropping pippetes
- Stir/hot plate
- Falcon tube
- Centrifuges on a Thermo Scientific Heraeus Labofuge 400 R
- Rotary evaporator
- Glass column loaded with silica gel
- Aluminum sheet silica gel 60/kieselguhr for thin layer chromatography (TLC)
- Electrospray ionization mass spectrometry (ESI-MS) on a Thermo Scientific LTQ Orbitrap mass spectrometer
- Nuclear magnetic resonance (NMR) spectroscopy on a 400 MHz Varian VXR spectrometer

4.2.2 Materials

- Cellobiose from Acros Organics
- *N*-Hydroxyethyl acrylamide from Sigma-Aldrich
- *N*-Hydroxyethyl methacrylamide and *N*-hydroxybutyl methacrylamide were synthesized according to literature (Rijcken et al., 2005)
- β-Glucosidase from Sigma-Aldrich
- *p*-Nitrophenyl β-D-glucopyranoside (*p*-NPG) from Alfa Aesar

- 1-*N*-Butyl-3-methylimidazolium hexafluorophosphate ($BMIMPF_6$) from Alfa Aesar
- Other solvents (chloroform, methanol, ethanol, sulfuric acid) from Avantor
- Deuterated water (D_2O) from Sigma-Aldrich

4.2.3 Procedure

1. In a round-bottom flask, prepare cellobiose solution (0.5 g of cellobiose in 2.25 mL water) or *p*-NPG solution (0.50 g in 4.50 mL water).
2. Put the flask in a hotplate equipped with waterbath.
3. Add *N*-hydroxyethyl acrylamide (0.65 g, 5.65 mmol), *N*-hydroxyethyl methacrylamide (0.75 g, 5.80 mmol), or *N*-hydroxybutyl methacrylamide (0.9 g, 5.72 mmol).
4. Start the reaction by adding the enzyme solution (50 mg of β-glucosidase in 0.25 mL water).
5. Close the flask and stir the reaction mixture at 200 rpm and 50 °C for 1 h.
6. Identify the product formation by TLC using the eluent of chloroform/methanol (4/1).
7. Spray the TLC plates by 5% H_2SO_4/ethanol solution and heated for spots detection. *N*-(β-Glycosyloxy)-ethyl acrylamide, *N*-(β-glycosyloxy)-ethyl methacrylamide, and *N*-(β-glycosyloxy)-butyl methacrylamide are detected at retardation factor of 0.30, 0.35, and 0.37, respectively.
8. Evaporate the water in the reaction mixture by rotary evaporator.
9. Precipitate the unreacted cellobiose and side products by addition of methanol into the reaction mixture.
10. Separate the supernatant by centrifugation (4500 rpm, 10 min, 4 °C).
11. Concentrate the supernatant by evaporating the solvent in rotary evaporator.
12. Isolate the product by column chromatography using the same eluent as in the previous step.
13. Evaporate the pooled fractions containing the products by rotary evaporator until a yellowish syrup is obtained.
14. Characterize the product with ESI-MS, ^{1}H and ^{13}C NMR spectroscopy. D_2O is used as the solvent for NMR measurement.

Note: In the case of using the ionic liquid $BMIMPF_6$ as cosolvent (30 v%), centrifugation of the reaction mixture is required after the reaction is stopped. The aim of this step is to separate the ionic liquid phase and the aqueous phase that contains the product.

5. Glycosyltransferase-catalyzed synthesis of glycomonomers

Glycosyltransferases are enzymes that catalyze the formation of glycosidic linkages by transferring glycosyl units from glycosyl donors to glycosyl acceptors (Gloster, 2014; Lairson, Henrissat, Davies, & Withers, 2008). In the preparation of saccharide-vinyl macromonomers (Table 4), various glucose-based monomers were utilized as the glycosyl acceptors, and either natural saccharides or glucosyl phosphates were used as the glycosyl donors. However, the number of glycosyltransferases used for this purpose was limited, probably because the isolation process of these enzymes on the practical scale is rather challenging (Shoda et al., 2016).

Cyclodextrin glycosyltransferase from *Bacillus macerans* catalyzes the synthesis of 2-(β-maltooligooxy)-ethyl (meth)acrylate macromonomers (Kloosterman, Van Dijk, et al., 2014). In that reaction, α-cyclodextrin and 2-(β-glucosyloxy)-ethyl (meth)acrylate are served as the glycosyl donor and acceptor, respectively. The glycomonomers contained 1–15 glucosyl units that are linked through α-(1,4)-glycosidic bonds while the β-linkage of the vinyl group remained intact after the coupling reaction.

In another example, a cellooligosaccharide-based glycomonomer was prepared from 2-(β-glucosyloxy)-ethyl methacrylate and α-D-glucose 1-phosphate (Wang et al., 2017). Cellodextrin phosphorylase from *Clostridium thermocellum* (CtCdP) was utilized as the biocatalyst. The synthesized

Table 4 Glycosyltransferase-catalyzed synthesis of glycomonomers.

Saccharides	Vinyl group	Vinyl position	Enzyme	Reaction condition	Yield (%)	References
23	**f, g**	C-1β	Cyclodextrin glycosyltransferase from *Bacillus macerans*	Tris buffer pH 6, 60°C, 1.5h	38	Kloosterman, Van Dijk, and Loos (2014)
24	**f, g, h, l, m**	C-1β	Cellodextrin phosphorylase from *Clostridium thermocellum*	HEPES buffer pH 7.5, 45°C, 3 days	65, 67, 58, 63, 70	Adharis, Petrović, Özdamar, Woortman, and Loos (2018) and Wang, Niu, Sawada, Shao, and Serizawa (2017)

The structure of the saccharides and vinyl groups correspond to Figs. 2 and 4.

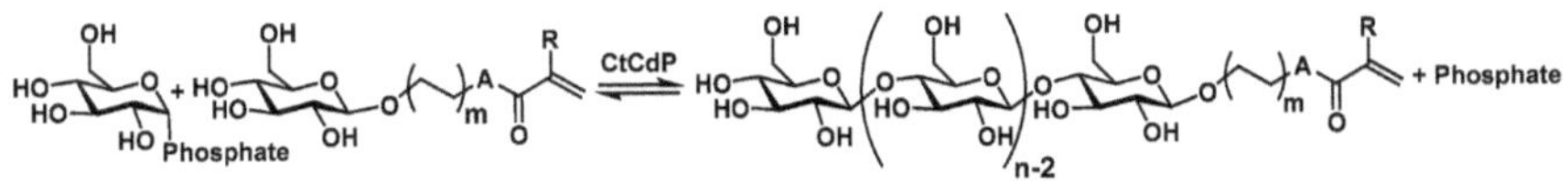

Fig. 9 Enzymatic synthesis of cellooligosaccharide-based glycomonomers catalyzed by CtCdP (m=1 or 2, A=O or NH, and R=H or CH_3).

macromonomer possesses an average number of nine glucosyl units which are coupled by β-(1,4)-glycosidic bonds. A disadvantage is that the solubility issue of the product hinders the enzyme ability to extend the number of glucosyl units further. Recently, the versatility of this enzyme with some unnatural substrates was reported by us (Adharis, Petrović, et al., 2018). In this work, several types of vinyl glucosides bearing (meth)acrylate/(meth)acrylamide functionalities were applied as the glucosyl acceptors (Fig. 9) and the isolated yields were about 65%. Interestingly, these vinyl glucosides were synthesized from renewable materials following an enzymatic approach in aqueous solutions, thus, providing an environmentally friendly process. The following procedure is described on the basis to this report.

5.1 Equipment

- Analytical balance
- Falcon tube
- Common glass laboratory equipments such as measuring cylinder, beaker glass, and dropping pippetes
- Eppendorf thermomixer comfort
- Centrifuges on a Thermo Scientific Heraeus Labofuge 400 R
- Aluminum sheet silica gel 60/kieselguhr for thin layer chromatography (TLC)
- Matrix-assisted laser desorption/ionization time-of-flight mass spectrometry (MALDI-ToF MS) on a Voyager DE-PRO instrument from Applied Biosystems
- Electrospray ionization mass spectrometry (ESI-MS) on a Thermo Scientific LTQ Orbitrap mass spectrometer
- Nuclear magnetic resonance (NMR) spectroscopy on a 400 MHz Varian VXR spectrometer
- Size exclusion chromatography (SEC) on an Agilent Technologies 1260 Infinity from PSS
- Wide-angle X-ray diffraction (WAXD) on a Bruker D8 Advance diffractometer

5.2 Materials

- α-D-Glucose 1-phosphate disodium salt hydrate from Sigma-Aldrich
- Cellobiose from Acros Organics
- Solvents (ethanol, isopropyl alcohol, *n*-butanol) from Avantor
- Sulfuric acid from Sigma-Aldrich
- Deuterated water (D_2O) and sodium deuteoxide (NaOD) from Sigma-Aldrich
- 2-(β-Glucosyloxy)-ethyl acrylate (G-EA), 2-(β-glucosyloxy)-ethyl methacrylate (G-EMA), 4-(β-glucosyloxy)-butyl acrylate (G-BA), *N*-(β-glucosyloxy)-ethyl acrylamide (G-EAAm), and *N*-(β-glucosyloxy)-ethyl methacrylamide (G-EMAAm) were synthesized following the literature (Adharis, Vesper, et al., 2018; Kloosterman, Roest, et al., 2014)
- Cellodextrin phosphorylase from Clostridium thermocellum (CtCdP) was isolated according to the literature (Petrović, Kok, Woortman, Ćirić, & Loos, 2015)

5.3 Procedure

1. Prepare a falcon tube containing α-Glc1P (1.82 g, 6.0 mmol) and 30 mL HEPES buffer 500 mM, pH 7.5.
2. Add G-EA (83.4 mg, 0.3 mmol), G-EMA (87.6 mg, 0.3 mmol), G-BA (96 mg, 0.3 mmol), G-EAAm (83.1 mg, 0.3 mmol), G-EMAAm (87.3 mg, 0.3 mmol), or cellobiose (102.7 mg, 0.3 mmol) into the reaction mixture.
3. Add the CtCdP solution (250 μL).
4. Put the tube on a temperature controlled shaker at 600 rpm and 45 °C for 72 h.
5. Observe the formation of the products after few hours as indicated from changing a transparent solution to a white turbid solution.
6. Identify the product in the reaction mixture by TLC using eluent of *n*-butanol/isopropyl alcohol/H_2O (1/2.5/1.5).
7. Spray the TLC plates by 5% H_2SO_4/ethanol solution and heated for spots detection. The product is detected at retardation factor of 0.63.
8. Isolate the product by centrifugation at 4500 rpm and 4 °C for 20 min.
9. Wash the precipitates at least three times with Milli-Q water.
10. Lyophilize the product in a freeze-drier at −45 °C and 0.01 mbar for overnight.
11. Characterize the product with MALDI-Tof MS, ^1H NMR spectroscopy, SEC, and WAXD. 4% NaOD in D_2O is used as the solvent for NMR measurement.

6. Summary

Various types of saccharide-vinyl monomers have been succesfully synthesized using enzymes as the biocatalyst. In organic solvents, lipases and proteases catalyzed transesterification of hydroxyl groups of saccharides with divinyl esters. Depending on the saccharide types and enzyme sources, mono- and disubstitution of vinyl groups were observed at several locations in a saccharide unit. Furthermore, an excellent selectivity was displayed by glycosidases in which the synthesized glycomonomers were characterized to be anomerically pure and monofunctional. Both thermodynamically controlled reversed hydrolysis and kinetically controlled transglycosylation approaches were used to generate these glycomonomers. Moreover, glycosyltransferases catalyzed the synthesis of saccharide-vinyl macromonomers by transferring the glycosyl units from donors to acceptors.

A well-defined structure of glycomonomers provides a good starting point for the investigation of the interaction of glycopolymers with sugar-binding proteins (Kiessling & Grim, 2013; Yilmaz & Becer, 2015). Therefore, further experiments will be aimed to prepare glycopolymers by controlled radical polymerization techniques such as atom-transfer radical polymerization, reversible addition-fragmentation chain transfer polymerization, nitroxide-mediated radical polymerization, and cyanoxyl-mediated free radical polymerization.

Acknowledgments

A.A. greatly acknowledges the scholarship from the Indonesia Endowment Fund for Education (Lembaga Pengelola Dana Pendidikan Republik Indonesia/LPDP RI). We would also like to thank Dr. Judith Schöbel and Dr. Peter Dijkstra for valuable suggestions as well as Dina Maniar for creating the energy diagram illustration in Fig. 3.

References

Adharis, A., Ketelaar, T., Komarudin, A. G., & Loos, K. (2019). Synthesis and self-assembly of double-hydrophilic and amphiphilic block glycopolymers. *Biomacromolecules*, *20*(3), 1325–1333.

Adharis, A., Petrović, D. M., Özdamar, I., Woortman, A. J. J., & Loos, K. (2018). Environmentally friendly pathways towards the synthesis of vinyl-based oligocelluloses. *Carbohydrate Polymers*, *193*, 196–204.

Adharis, A., Vesper, D., Koning, N., & Loos, K. (2018). Synthesis of (meth)acrylamide-based glycomonomers using renewable resources and their polymerization in aqueous systems. *Green Chemistry*, *20*(2), 476–484.

Albertin, L., Kohlert, C., Stenzel, M., Foster, J. R., & Davis, T. P. (2004). Chemoenzymatic synthesis of narrow-polydispersity glycopolymers: Poly(6-O-vinyladipolyl-D-glucopyranose). *Biomacromolecules, 5*(2), 255–260.

Albertin, L., Stenzel, M., Barner-Kowollik, C., Foster, L. J. R., & Davis, T. P. (2004). Well-defined glycopolymers from RAFT polymerization: Poly(methyl 6-O-methacryloyl-α-d-glucoside) and its block copolymer with 2-hydroxyethyl methacrylate. *Macromolecules, 37*(20), 7530–7537.

Albertin, L., Stenzel, M. H., Barner-Kowollik, C., Foster, L. J. R., & Davis, T. P. (2005). Well-defined diblock glycopolymers from RAFT polymerization in homogeneous aqueous medium. *Macromolecules, 38*(22), 9075–9084.

Alén, R. (2018). *Carbohydrate chemistry—Fundamentals and applications* (1st ed.). Singapore: World Scientific.

Ambrosi, M., Cameron, N. R., Davis, B. G., & Stolnik, S. (2005). Investigation of the interaction between peanut agglutinin and synthetic glycopolymeric multivalent ligands. *Organic & Biomolecular Chemistry, 3*(8), 1476–1480.

Appelhans, D., Klajnert-Maculewicz, B., Janaszewska, A., Lazniewska, J., & Voit, B. (2015). Dendritic glycopolymers based on dendritic polyamine scaffolds: View on their synthetic approaches, characteristics and potential for biomedical applications. *Chemical Society Reviews, 44*(12), 3968–3996.

Bayón, C., Cortés, Á., Berenguer, J., & Hernáiz, M. J. (2013). Highly efficient enzymatic synthesis of Galβ-(1→3)-GalNAc and Galβ-(1→3)-GlcNAc in ionic liquids. *Tetrahedron, 69*(24), 4973–4978.

Becer, C. R., & Hartmann, L. (Eds.), (2015). *Glycopolymer code: Synthesis of glycopolymers and their applications* (1st ed.): The Royal Society of Chemistry.

Bhatia, S., Camacho, L. C., & Haag, R. (2016). Pathogen inhibition by multivalent ligand architectures. *Journal of the American Chemical Society, 138*(28), 8654–8666.

Blanco, A., & Blanco, G. (2017). Enzymes. In A. Blanco & G. Blanco (Eds.), *Medical biochemistry* (1st ed., pp. 153–175). London: Academic Press.

Borges, M. R., & Balaban, R. (2007). Sucrose-branched polymer synthesized by protease from *Bacillus subtilis*. *Macromolecular Symposia, 258*(1), 25–29.

Borges, M. R., & Balaban, R. C. (2014). L-Arabinose (pyranose and furanose rings)-branched poly(vinyl alcohol): Enzymatic synthesis of the sugar esters followed by free radical polymerization. *Journal of Biotechnology, 192*, 42–49.

Borges, M. R., dos Santos, J. A., Vieira, M., & Balaban, R. (2009). Polymerization of a water soluble glucose vinyl ester monomer with tensoactive properties synthesized by enzymatic catalyst. *Materials Science and Engineering: C, 29*(2), 519–523.

Borman, S. (2004). Much ado about enzyme mechanisms. *Chemical & Engineering News, 82*(8), 35–39.

Branson, T. R., & Turnbull, W. B. (2013). Bacterial toxin inhibitors based on multivalent scaffolds. *Chemical Society Reviews, 42*(11), 4613–4622.

Bridiau, N., Issaoui, N., & Maugard, T. (2010). The effects of organic solvents on the efficiency and regioselectivity of N-acetyl-lactosamine synthesis, using the β-galactosidase from *Bacillus circulans* in hydro-organic media. *Biotechnology Progress, 26*(5), 1278–1289.

Cameron, N. R., Spain, S. G., Kingham, J. A., Weck, S., Albertin, L., Barker, C. A., et al. (2008). Synthesis of well-defined glycopolymers and some studies of their aqueous solution behaviour. *Faraday Discussions, 139*, 359–368.

Chiffoleau-Giraud, V., Spangenberg, P., & Rabiller, C. (1997). β-galactosidase transferase activity in ice and use of vinyl-β-D-galactoside as donor. *Tetrahedron Asymmetry, 8*(12), 2017–2023.

Clark, D. P., & Pazdernik, N. J. (2016). *Proteomics. In Biotechnology* (2nd ed., pp. 295–333). London: Academic Cell. https://doi.org/10.1016/B978-0-12-385015-7.00009-0.

de Oliveira, H. F. N., & Felisberti, M. I. (2013). Amphiphilic copolymers of sucrose methacrylate and acrylic monomers: Bio-based materials from renewable resource. *Carbohydrate Polymers*, *94*(1), 317–322.

Gandini, A., & Lacerda, T. M. (2015). From monomers to polymers from renewable resources: Recent advances. *Progress in Polymer Science*, *48*, 1–39.

Giacomini, C., Irazoqui, G., Gonzalez, P., Batista-viera, F., & Brena, B. M. (2002). Enzymatic synthesis of galactosyl-xylose by *Aspergillus oryzae* beta-galactosidase. *Journal of Molecular Catalysis B: Enzymatic*, *19–20*, 159–165.

Gill, I., & Valivety, R. (2000). Enzymatic glycosylation in plasticized glass phases: A novel and efficient route to O-glycosides. *Angewandte Chemie, International Edition*, *39*(21), 3804–3808.

Gloster, T. M. (2014). Advances in understanding glycosyltransferases from a structural perspective. *Current Opinion in Structural Biology*, *28*, 131–141.

Granville, A. M., Quémener, D., Davis, T. P., Barner-Kowollik, C., & Stenzel, M. H. (2007). Chemo-enzymatic synthesis and RAFT polymerization of 6-O-methacryloyl mannose: A suitable glycopolymer for binding to the tetrameric lectin concanavalin A? *Macromolecular Symposia*, *255*, 81–89.

Gross, R. A., Kumar, A., & Kalra, B. (2001). Polymer synthesis by in vitro enzyme catalysis. *Chemical Reviews*, *101*(7), 2097–2124.

Hořejší, V., Chaloupecká, O., & Kocourek, J. (1978). Studies on lectins: XLIII. Isolation and characterization of the lectin from restharrow boots (*Ononis hircina* Jacq.). *Biochimica et Biophysica Acta (BBA)—General Subjects*, *539*(3), 287–293.

Hushegyi, A., Klukova, L., Bertok, T., & Tkac, J. (2015). Carbohydrate nanotechnology and its application to biosensor development. In K. J. Stine (Ed.), *Carbohydrate nanotechnology* (1st ed., pp. 387–421). Hoboken, NJ: John Wiley & Sons, Inc.

Jiang, Y., & Loos, K. (2016). Enzymatic synthesis of biobased polyesters and polyamides. *Polymers*, *8*(7), 243.

Kadereit, D., & Waldmann, H. (2001). Enzymatic protecting group techniques. *Chemical Reviews*, *101*(11), 3367–3396.

Kaftzik, N., Wasserscheid, P., & Kragl, U. (2002). Use of ionic liquids to increase the yield and enzyme stability in the β-galactosidase catalysed synthesis of N-acetyllactosamine. *Organic Process Research & Development*, *6*(4), 553–557.

Kiessling, L. L., & Grim, J. C. (2013). Glycopolymer probes of signal transduction. *Chemical Society Reviews*, *42*(10), 4476–4491.

Kim, J., Haam, S., Park, D.-W., Ahn, I.-S., Lee, T. G., Kim, H.-S., et al. (2004). Biocatalytic esterification of β-methylglucoside for synthesis of biocompatible sugar-containing vinyl esters. *Chemical Engineering Journal*, *99*(1), 15–22.

Kitagawa, M., & Tokiwa, Y. (1998). Synthesis of polymerizable sugar ester possessing long spacer catalyzed by lipase from Alcaligenes sp. and its chemical polymerization. *Biotechnology Letters*, *20*(7), 627–630.

Kloosterman, W. M. J., Brouwer, S. G. M., & Loos, K. (2014). Enzyme-catalyzed synthesis of saccharide acrylate monomers from nonedible biomass. *Chemistry—An Asian Journal*, *9*(8), 2156–2161.

Kloosterman, W. M. J., Jovanovic, D., Brouwer, S. G. M., & Loos, K. (2014). Amylase catalyzed synthesis of glycosyl acrylates and their polymerization. *Green Chemistry*, *16*(1), 203–210.

Kloosterman, W. M. J., Roest, S., Priatna, S. R., Stavila, E., & Loos, K. (2014). Chemo-enzymatic synthesis route to poly(glucosyl-acrylates) using glucosidase from almonds. *Green Chemistry*, *16*(4), 1837–1846.

Kloosterman, W. M. J., Van Dijk, G. S., & Loos, K. (2014). Biocatalytic synthesis of maltodextrin-based acrylates from starch and α-cyclodextrina. *Macromolecular Bioscience*, *14*(9), 1268–1279.

Kumar, P., Barrett, D. M., Delwiche, M. J., & Stroeve, P. (2009). Methods for pretreatment of lignocellulosic biomass for efficient hydrolysis and biofuel production. *Industrial & Engineering Chemistry Research*, *48*(8), 3713–3729.

Kwan, D. H. (2017). Structure-guided directed evolution of glycosidases: A case study in engineering a blood group antigen-cleaving enzyme. In B. Imperiali (Ed.), *Vol. 597, Methods in enzymology* (1st ed., pp. 25–53). London: Academic Press.

La Ferla, B. (2002). Lipases as useful tools for the stereo- and regioselective protection and deprotection of carbohydrates. *Monatshefte für Chemie*, *133*(4), 351–368.

Lairson, L. L., Henrissat, B., Davies, G. J., & Withers, S. G. (2008). Glycosyltransferases: Structures, functions, and mechanisms. *Annual Review of Biochemistry*, 77(1), 521–555.

Lang, M., Kamrat, T., & Nidetzky, B. (2006). Influence of ionic liquid cosolvent on transgalactosylation reactions catalyzed by thermostable β-glycosylhydrolase CelB from *Pyrococcus furiosus*. *Biotechnology and Bioengineering*, *95*(6), 1093–1100.

Li, Y., & Rethwisch, D. G. (2002). Scale-up of pseudo solid-phase enzymatic synthesis of α-methyl glucoside acrylate. *Biotechnology and Bioengineering*, *79*(1), 15–22.

Lichtenthaler, F. W., & Peters, S. (2004). Carbohydrates as green raw materials for the chemical industry. *Comptes Rendus Chimie*, 7(2), 65–90.

Loos, K. (Ed.), (2010). *Biocatalysis in polymer chemistry* (1st ed.). Weinheim: Wiley-VCH Verlag GmbH & Co. KGaA.

Mcnaught, A. D. (1996). Nomenclature of carbohydrates. *Pure and Applied Chemistry*, *68*(10), 1919–2008.

Millqvist-Fureby, A., MacManus, D. A., Davies, S., & Vulfson, E. N. (1998). Enzymatic transformations in supersaturated substrate solutions: II. Synthesis of disaccharides via transglycosylation. *Biotechnology and Bioengineering*, *60*(2), 197–203.

Miura, Y., Hoshino, Y., & Seto, H. (2016). Glycopolymer nanobiotechnology. *Chemical Reviews*, *116*(4), 1673–1692.

Miura, Y., Ikeda, T., & Kobayashi, K. (2003). Chemoenzymatically synthesized glycoconjugate polymers. *Biomacromolecules*, *4*(2), 410–415.

Miura, Y., Ikeda, T., Wada, N., Sato, H., & Kobayashi, K. (2003). Chemoenzymatic synthesis of glycoconjugate polymers: Greening the synthesis of biomaterials. *Green Chemistry*, *5*(5), 610–614.

Miura, Y., Wada, N., Nishida, Y., Mori, H., & Kobayashi, K. (2004). Chemoenzymatic synthesis of glycoconjugate polymers starting from nonreducing disaccharides. *Journal of Polymer Science, Part A: Polymer Chemistry*, *42*(18), 4598–4606.

Moss, G. P., Smith, P. A. S., & Tavernier, D. (1995). Glossary of class names of organic compounds and reactivity intermediates based on structure. *Pure and Applied Chemistry*, *67*, 1307–1375.

Nakajima, H., Dijkstra, P., & Loos, K. (2017). The recent developments in biobased polymers toward general and engineering applications: Polymers that are upgraded from biodegradable polymers, analogous to petroleum-derived polymers, and newly developed. *Polymers*, *9*(10), 523.

Olofsson, K., Bertilsson, M., & Lidén, G. (2008). A short review on SSF—An interesting process option for ethanol production from lignocellulosic feedstocks. *Biotechnology for Biofuels*, *1*(1), 1–14.

Park, D. W., Haam, S., Lee, T. G., Kim, H.-S., & Kim, W.-S. (2004). Chemoenzymatic synthesis of sugar-containing biocompatible hydrogels: Crosslinked poly(β-methylglucoside acrylate) and poly(β-methylglucoside methacrylate). *Journal of Biomedical Materials Research. Part A*, *71A*(3), 497–507.

Park, O. J., Kim, D. Y., & Dordick, J. S. (2000). Enzyme-catalyzed synthesis of sugar-containing monomers and linear polymers. *Biotechnology and Bioengineering*, *70*(2), 208–216.

Pellis, A., Comerford, J. W., Maneffa, A. J., Sipponen, M. H., Clark, J. H., & Farmer, T. J. (2018). Elucidating enzymatic polymerisations: Chain-length selectivity of *Candida antarctica* lipase B towards various aliphatic diols and dicarboxylic acid diesters. *European Polymer Journal, 106*, 79–84.

Pérez, J., Muñoz-Dorado, J., De La Rubia, T., & Martínez, J. (2002). Biodegradation and biological treatments of cellulose, hemicellulose and lignin: An overview. *International Microbiology, 5*(2), 53–63.

Pérez-Victoria, I., & Morales, J. C. (2007). One- and two-step enzymatic synthesis of polymerizable vinyladipoyl mono- and diesters of non-reducing trisaccharides. *Journal of Molecular Catalysis B: Enzymatic, 48*(1–2), 8–15.

Perin, G. B., & Felisberti, M. I. (2018). Enzymatic synthesis and structural characterization of methacryloyl-D-fructose- and methacryloyl-D-glucose-based monomers and poly (methacryloyl-D-fructose)-based hydrogels. *Journal of Chemical Technology and Biotechnology, 93*(6), 1694–1704.

Petrović, D. M., Kok, I., Woortman, A. J. J., Ćirić, J., & Loos, K. (2015). Characterization of oligocellulose synthesized by reverse phosphorolysis using different cellodextrin phosphorylases. *Analytical Chemistry, 87*(19), 9639–9646.

Potier, P., Bouchu, A., Descotes, G., & Queneau, Y. (2000). Proteinase N-catalysed transesterifications in DMSO–water and DMF–water: Preparation of sucrose monomethacrylate. *Tetrahedron Letters, 41*(19), 3597–3600.

Potier, P., Bouchu, A., Gagnaire, J., & Queneau, Y. (2001). Proteinase N-catalysed regioselective esterification of sucrose and other mono- and disaccharides. *Tetrahedron: Asymmetry, 12*(17), 2409–2419.

Qi, W., Deshui, L., Zhichun, C., Guohua, Y., & Xianfu, L. (2003). Enzyme-catalyzed regioselective synthesis of novel raffinose vinyl esters. *Chemical Journal of Chinese Universities, 24*(10), 1806–1810.

Raku, T., & Tokiwa, Y. (2003). Chemoenzymatic synthesis of fucose- or rhamnose-branched polymer. *Macromolecular Bioscience, 3*(3–4), 151–156.

Rijcken, C. J. F., Veldhuis, T. F. J., Ramzi, A., Meeldijk, J. D., van Nostrum, C. F., & Hennink, W. E. (2005). Novel fast degradable thermosensitive polymeric micelles based on PEG-block-poly(N-(2-hydroxyethyl)methacrylamide-oligolactates). *Biomacromolecules, 6*(4), 2343–2351.

Rosales, E., Pazos, M., & Ángeles Sanromán, M. (2018). Solid-state fermentation for food applications. In A. Pandey, C. Larroche, & C. R. Soccol (Eds.), *Current developments in biotechnology and bioengineering* (1st ed., pp. 319–355). Amsterdam: Elsevier.

Santin, M., Rosso, F., Sada, A., & Peluso, G. (1996). Enzymatic synthesis of 2-p-D-galactopyranosyloxy ethyl met hacrylate (GalEMA) by the thermophilic archeon *Sulfolobus solfataricus*. *Biotechnology and Bioengineering, 49*, 217–222.

Satyanarayana, K. G., Arizaga, G. G. C., & Wypych, F. (2009). Biodegradable composites based on lignocellulosic fibers—An overview. *Progress in Polymer Science, 34*(9), 982–1021.

Schnaar, R. L., & Lee, Y. C. (1975). Polyacrylamide gels copolymerized with active esters. A new medium for affinity systems. *Biochemistry, 14*(7), 1535–1541.

Schuster, M. C., Mortell, K. H., Hegeman, A. D., & Kiessling, L. L. (1997). Neoglycopolymers produced by aqueous ring-opening metathesis polymerization: Decreasing saccharide density increases activity. *Journal of Molecular Catalysis A: Chemical, 116*(1–2), 209–216.

Shoda, S., Uyama, H., Kadokawa, J., Kimura, S., & Kobayashi, S. (2016). Enzymes as green catalysts for precision macromolecular synthesis. *Chemical Reviews, 116*(4), 2307–2413.

Sunasee, R., Adokoh, C. K., Darkwa, J., & Narain, R. (2014). Therapeutic potential of carbohydrate-based polymeric and nanoparticle systems. *Expert Opinion on Drug Delivery, 11*(6), 867–884.

Sym, E. A. (1933). Über die Esterasewirkung. III. *Biochemische Zeitschrift, 258*, 304–324.

Sym, E. A. (1936). A method for enzymic ester synthesis. *Enzymologia, 1*, 156–160.

Therisod, M., & Klibanov, A. M. (1986). Facile enzymatic preparation of monoacylated sugars in pyridine. *Journal of the American Chemical Society, 108*(18), 5638–5640.

Therisod, M., & Klibanov, A. M. (1987). Regioselective acylation of secondary hydroxyl groups in sugars catalyzed by lipases in organic solvents. *Journal of the American Chemical Society, 109*(13), 3977–3981.

Tokiwa, Y., Kitagawa, M., Fan, H., Raku, T., Hiraguri, Y., Shibatani, S., et al. (1999). Synthesis of vinyl arabinose ester catalyzed by protease from Streptomyces sp. *Biotechnology Techniques, 13*(3), 173–176.

Utama, R. H., Jiang, Y., Zetterlund, P. B., & Stenzel, M. H. (2015). Biocompatible glycopolymer nanocapsules via inverse miniemulsion periphery RAFT polymerization for the delivery of gemcitabine. *Biomacromolecules, 16*(7), 2144–2156.

Van Rantwijk, F., Woudenberg-Van Oosterom, M., & Sheldon, R. A. (1999). Glycosidase-catalysed synthesis of alkyl glycosides. *Journal of Molecular Catalysis B: Enzymatic, 6*(6), 511–532.

Woudenberg- van Oosterom, M., Van Belle, H. J. A., Van Rantwijk, F., & Sheldon, R. A. (1998). Immobilised β-galactosidases and their use in galactoside synthesis. *Journal of Molecular Catalysis A: Chemical, 134*(1–3), 267–274.

Wang, J., Niu, J., Sawada, T., Shao, Z., & Serizawa, T. (2017). A bottom-up synthesis of vinyl-cellulose nanosheets and their nanocomposite hydrogels with enhanced strength. *Biomacromolecules, 18*(12), 4196–4205.

Wang, X., Wu, Q., Wang, N., & Lin, X.-F. (2005). Chemo-enzymatic synthesis of disaccharide-branched copolymers with high molecular weight. *Carbohydrate Polymers, 60*(3), 357–362.

Wu, Q., Chen, Z. C., Lu, D. S., & Lin, X.-F. (2006). Chemo-enzymatic synthesis of raffinose-branched polyelectrolytes and self-assembly application in microcapsules. *Macromolecular Bioscience, 6*(1), 78–83.

Wu, Q., Lu, D., Cai, Y., Xue, X., Chen, Z., & Lin, X.-F. (2001). Regio- and stereo-selective synthesis of vinyl glucose ester catalyzed by an alkaline protease of *Bacillus subtilis*. *Biotechnology Letters, 23*(24), 1981–1985.

Wu, Q., Wang, N., Xiao, Y.-M., Lu, D.-S., & Lin, X.-F. (2004). Regiospecific alkaline protease-catalyzed divinyl acyl transesterifications of primary hydroxyl groups of mono- and di-saccharides in pyridine. *Carbohydrate Research, 339*(12), 2059–2067.

Wulff, G., Schmid, J., & Venhoff, T. (1996). The synthesis of polymerizable vinyl sugars. *Macromolecular Chemistry and Physics, 197*(1), 259–274.

Xiao, Y., Wu, Q., Wang, N., & Lin, X.-F. (2004). Regioselective monoacylation of cyclomaltoheptaose at the C-2 secondary hydroxyl groups by the alkaline protease from *Bacillus subtilis* in nonaqueous media. *Carbohydrate Research, 339*(7), 1279–1283.

Yilmaz, G., & Becer, C. R. (2015). Glyconanoparticles and their interactions with lectins. *Polymer Chemistry, 6*(31), 5503–5514.

Zaks, A., & Klibanov, A. M. (1984). Enzymatic catalysis in organic media at 100°C. *Science, 224*(4654), 1249–1251.

Zaks, A., & Klibanov, A. M. (1985). Enzyme-catalyzed processes in organic solvents. *Proceedings of the National Academy of Sciences of the United States of America, 82*(10), 3192–3196.

Zeng, X., Qu, K., & Rehman, A. (2016). Glycosylated conductive polymer: A multimodal biointerface for studying carbohydrate-protein interactions. *Accounts of Chemical Research, 49*(9), 1624–1633.

Zeuner, B., Riisager, A., Mikkelsen, J. D., & Meyer, A. S. (2014). Improvement of trans-sialylation versus hydrolysis activity of an engineered sialidase from *Trypanosoma rangeli* by use of co-solvents. *Biotechnology Letters, 36*(6), 1315–1320.

Zhu, Y., Romain, C., & Williams, C. K. (2016). Sustainable polymers from renewable resources. *Nature, 540*(7633), 354–362.

CHAPTER NINE

Enzyme-initiated free radical polymerizations of vinyl monomers using horseradish peroxidase

Kyle J. Rodriguez[a], Michela M. Pellizzoni[a], Robert J. Chadwick[b], Chao Guo[b], Nico Bruns[a,b,*]
[a]Adolphe Merkle Institute, University of Fribourg, Fribourg, Switzerland
[b]Department of Pure and Applied Chemistry, University of Strathclyde, Glasgow, United Kingdom
*Corresponding author: e-mail address: nico.bruns@strath.ac.uk

Contents

Abstract

In this chapter, we highlight the use of horseradish peroxidase (HRP) as a catalyst to initiate free radical polymerizations of vinyl monomers under benign reaction conditions. A variety of vinyl monomers, including 4-acryloylmorpholine (AM), 2-hydroxyethyl methacrylate (HEMA), and poly(ethylene glycol) methyl ether acrylate (PEGA) were polymerized. The enzyme converts exogenous hydrogen peroxide into a usable radical source, which when coupled with a β-diketone, yields a radical that initiates chain growth in the presence of monomers. The resulting polymers were characterized using nuclear magnetic resonance (NMR) spectroscopy and gel permeation chromatography (GPC). By using enzymatic free radical polymerizations, polymers can be generated in a sustainable, environmentally-friendly, and scalable fashion.

Methods in Enzymology, Volume 627
ISSN 0076-6879
https://doi.org/10.1016/bs.mie.2019.08.013

1. Introduction

Free radical polymerizations (FRPs) are used every year worldwide to manufacture almost 45% of synthetic polymers such as rubber and plastics using peroxy compounds, redox systems or azo initiators (Binder, 2019; Braun, 2009). From a historical perspective, FRP was first reported by Simon (1839) and later by Blyth and Hofmann (1845) in 1839 and 1845, respectively. They observed the formation of glass like solids starting from a styrene containing natural resin isolated from the wounded bark of *Liquidambar orientalis* in the presence of light. The concept of polymers and radicals were introduce in the same century (by Berzelius, Berthelot and Wurtz) even if the meaning and nature of these entities remained ambiguous until the reports of Gomberg (1897) and Staudinger (1920) at the beginning of last century. The mechanism of FRP was postulated by Flory (1936) when several synthetic polymers obtained by condensation of formaldehyde with phenol, urea and proteins (e.g., Bakelite, Pollopas and Galalit), as well as cellulose derivatives were already on the market (Mark, 1964).

Nowadays, due to their simple reaction conditions, FRPs have been implemented for generating materials such as hydrogels, drug vehicles, and packaging. These applications have resulted in FRPs being implemented as a common synthetic tool for generating highly valuable materials. Additionally, in the last decades, new methodologies have been developed to control radical polymerizations (reversible-deactivation radical polymerization techniques such as nitroxide mediated polymerizations (NMP), atom transfer radical polymerizations (ATRP), and reversible-addition-fragmentation chain transfer (RAFT) polymerizations) but their application on industrial scale is still limited and FRPs remain the most widely used techniques due to the well establish and simple reaction conditions (Binder, 2019).

Incorporating biocatalysis into traditional chemical processes allows for a more sustainable approach for generating materials necessary for the food and pharmaceutical industries where biocompatibility is necessary. Harnessing the power of biology can be a very powerful tool, e.g., to repurpose enzymes in perform non-native transformations, which result in efficient catalysis through biological mechanisms.

After the first report in 1951 by Parravano, who reported the polymerization of methyl methacrylate in the presence of Xantine oxidase and formaldehyde (Parravano, 1951) many enzymes like oxidases and peroxidases were found to be able to catalyze radical polymerizations of a variety of monomers ranging from phenolic compounds to those containing vinyl groups (Derango, Chiang, Dowbenko, & Lasch, 1992; Hollmann, 2010; Hollmann & Arends, 2012; Singh, Ma, & Kaplan, 2000; Zavada, Battsengel, & Scott, 2016).

The use of biocatalysts to initiate and/or mediate polymerizations not only pursues a greener catalytic approach but also allows to target bionano science. Thanks to the multiple mechanisms of molecular amplification (e.g., enzyme turnover, radical chain propagations) and the capability to localize the polymerization at microscale level, new applications for biosensing, high-throughput screening, and chemical imaging emerged (Malinowska & Nash, 2016). As an example, HRP-mediated polymerization was exploited to prepare molecularly imprinted polymers (MIP) for the recognition of 2,4-dichlorophenoxyacetic acid (an herbicide) and salicylic acid, starting from the immobilized HRP catalyst in combination with H_2O_2, methacrylate, vinyl monomers, cross-linkers and acetylacetone as the mediator (Daoud Attieh, Zhao, Elkak, Falcimaigne-Cordin, & Haupt, 2017). The capability of HRP to promote the formation of a fluorescent alginate hydrogel in the presence of hydrogen peroxide, produced in situ by glucose oxidase (GO_x), was recently used by Nash and coworkers to developed a platform for the directed evolution of GO_x (Vanella, Ta, & Nash, 2019).

The mechanisms by which these reactions occur are driven by the heme prosthetic group found in the enzymes' active site (Fig. 1). Exogenous hydrogen peroxide (H_2O_2) is converted into a usable radical source especially when the reaction is coupled with β-diketone mediators. These polymerizations proceed through the mechanism by which nature processes peroxides, the oxidoreductive pathway. Utilizing hydrogen peroxide, a hydroxyl radical is generated which reacts with the mediator acetylacetone to a more stable acetylacetone-based radical. The β-diketone mediator plays two roles in the reaction: (1) It is a reducing substrate for the enzyme, and (2) it forms a radical capable of initiating the polymerization of vinyl monomers (Durand, Lalot, Brigodiot, & Maréchal, 2000; Emery, Lalot, Brigodiot, & Maréchal, 1997). This radical species then initiates the chain growth process. HRP undergoes its biological role when exposed to peroxides, forming compound I and compound II before it returns to its original Fe^{III} species which is required for peroxidase activity.

Furthermore, compared to AIBN or photo initiated FRPs, HRP-initiated polymerizations offer access to the same materials but under benign reaction conditions without the need for elevated temperatures or photo irradiation (Hollmann, 2010; Hollmann & Arends, 2012; Kobayashi, Uyama, & Kimura, 2001). This can be highly advantageous for the production of temperature and light sensitive materials.

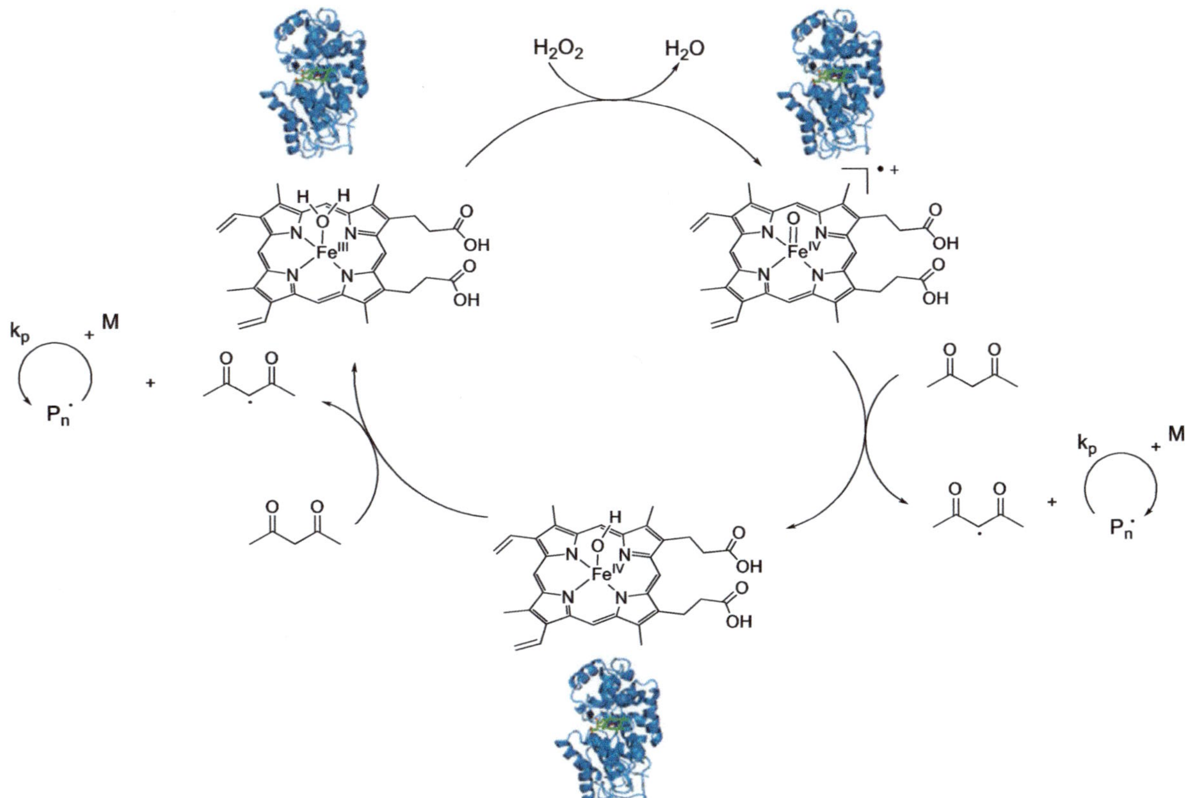

Fig. 1 Catalytic cycle for the free radical polymerization of vinyl monomers using HRP as a catalyst. (M, monomer; P_n, propagating polymer radical; k_p, rate constant of propagation (i.e., polymer chain growth)).

Herein, we describe a comprehensive synthetic strategy for enzyme-initiated free radical polymerizations of vinyl monomers that is based on work by Maréchal (Durand et al., 2000; Emery et al., 1997; Lalot, Brigodiot, & Maréchal, 1999), Gross (Kalra & Gross, 2000) and Kobayashi (Kobayashi et al., 2001). The aim is to provide the reader with know-how on what is required for generating polymeric materials using this method.

2. Free radical polymerizations of vinyl monomers using horseradish peroxidase

2.1 Polymerizations

Equipment
- Analytical balance
- Schlenk line outfitted with five ports equipped with inert gas (Ar or N_2)
- Schlenk flask (10 mL)
- Stirring plate
- Magnetic stirring bar
- Syringe (1 and 2 mL)
- Micropipette (20, 100, 1000 μL)
- Needles (ø 0.80 × 50 mm and ø 0.80 × 120 mm)
- Glass Pasteur pipette (230 mm)
- Cotton or glass wool
- Scintillation vial (6 mL)
- 3 × rubber septum

Chemicals
- Horseradish peroxidase, (HRP; highly stabilized, essential salt-free, lyophilized powder, 200–300 units/mg solid, P2088) (Sigma)
- 4-Acryloylmorpholine (AM) (Sigma)
- 2-Hydroxyethyl methacrylate (HEMA) (Sigma)
- Poly(ethylene glycol) methyl ether acrylate (PEGA; $M_n = 480\,g\,mol^{-1}$) (Sigma)
- Acetylacetone (ACAC) (Sigma)
- Hydrogen peroxide (H_2O_2; 30 wt% in H_2O) (Sigma)
- Aluminum oxide activated, basic, Brockmann I (Sigma)
- Aluminum oxide activated, neutral, Brockmann I (Sigma)
- Milli-Q water
- Dimethyl sulfoxide (DMSO) (Sigma)
- Diethyl ether (Sigma)

Procedure

The free radical polymerizations of vinyl monomers using horseradish peroxidase as a catalyst (Scheme 1) were performed as follows:

1. A Schlenk flask, equipped with a magnetic stirring bar and rubber septum, was attached to a Schlenk line in a fume hood. The flask was then evacuated by vacuum and back filled with argon three times.
2. The monomers were purified of inhibitors by passing them through a small plug of basic aluminum oxide via a positive flow of compressed air into 6 mL scintillation vials. The plugs were prepared by inserting a small piece of cotton into a Pasteur pipette and filling it half way with basic aluminum oxide.
3. To the open Schlenk flask under a slight positive flow of argon, HRP (6.1 mg, 77.6 μmol) was added. 10 mmol monomer (PEGA 5 mmol) was added using a 1000 μL micropipette. Once both were added, 700 μL of Milli-Q water and 300 μL of DMSO were added to the flask to dissolve the HRP and monomer. The flask was then sealed with a rubber septum.
4. Using a ø 0.80 × 120 mm needle and closing the stopcock of the Schlenk flask, argon was passed through the solution to degas the mixture for 15 min with an exhaust needle (ø 0.80 × 50 mm) through the septum.
5. In a scintillation vial (6 mL), ACAC (25.0 mg, 0.25 mmol) and H_2O_2 (8.5 mg, 0.25 mmol) were added using a 100 μL micropipette and 20 μL micropipette, respectively, and dissolved in 700 μL of Milli-Q water and 300 μL of DMSO. The vial was then sealed using a rubber septum and was degassed by inserting a needle (ø 0.80 × 120 mm) into the solution and passing argon through the liquid for 15 min with an exhaust needle (ø 0.80 × 50 mm) through the septum.
6. To start the reaction, the contents of the scintillation vial were transferred to the Schlenk flask containing the enzyme and monomer using an argon purged syringe (2 mL).
7. Once all of the reactants were added, a 0.3 mL aliquot of the reaction mixture was taken using an argon purged syringe (1 mL). The aliquot was then passed over a plug of neutral aluminum oxide (prepared the same way as step 2) to quench the reaction and remove the HRP. The plug was rinsed with 1 mL of D_2O (for p(HEMA) DMSO-d_6) into a scintillation vial (6 mL) for NMR and GPC analysis (Section 3).
8. The reaction mixture was stirred at room temperature for a total of 21 h under argon.
9. After the first hour and the final time point, a 0.3 mL aliquot was taken using an argon purged syringe (1 mL) equipped with a needle (ø 0.80 × 120 mm) from the reaction mixture for NMR and GPC analysis (Section 3). The aliquot was passed through a plug of neutral aluminum oxide as mentioned in Step 7. After the final time point, the reaction was quenched

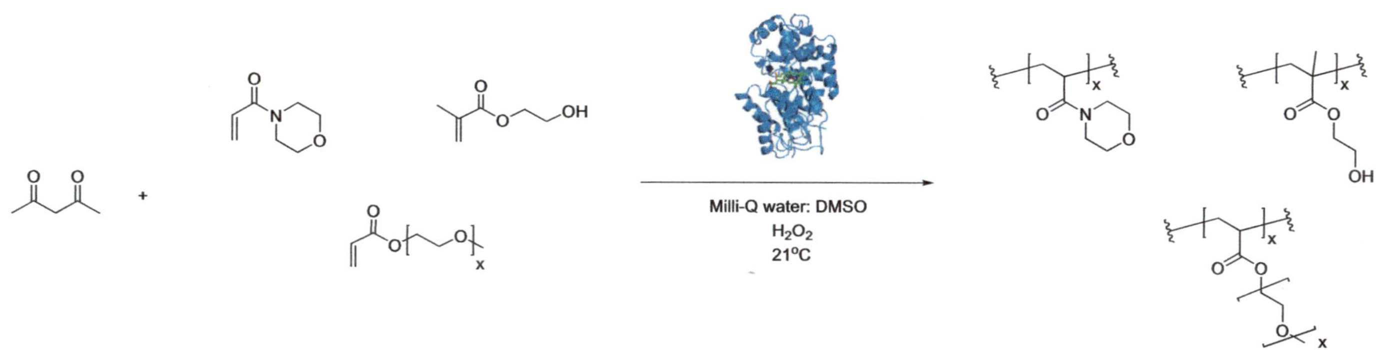

Scheme 1 Synthesis of p(AM), p(HEMA), and p(PEGA) under free radical polymerization conditions using HRP as a catalyst.

by opening the reaction flask to air and turning off the flow of argon. The remaining reaction mixture was redissolved in a minimal amount of DMSO and precipitated into cold diethyl ether. The resulting product was filtered and dried under vacuum.

3. Polymer characterization

Equipment

- NMR spectrometer, e.g., Bruker Ascend 400 with a 5 mm sample head operating at 400.13 MHz
- NMR tubes, e.g., Wilmad 535-PP-7 and tube cap
- Processing software, e.g., Mestrelab Research: MestReNova (Mnova)
- GPC system, e.g., PSS SECurity system equipped with a refractive index (RI) detector
- Scintillation vial (6 mL)
- PTFE syringe filter 0.2 μm pore size
- Syringe (2 mL)
- GPC vials

Chemicals

- Deuterium oxide (D_2O) (Cambridge Isotope Labs)
- DMSO-d_6 (Cambridge Isotope Labs)
- Magnesium sulfate (Sigma)
- Lithium bromide (Sigma)
- Dimethylformamide (DMF) (Fisher Scientific)

3.1 Nuclear magnetic resonance (NMR) spectroscopy

NMR measurements are used to characterize a variety of parameters in polymer chemistry. Here we highlight the use of ^{1}H NMR spectroscopy to determine the conversion of the polymerization. The ^{1}H NMR spectra were recorded in D_2O (for p(HEMA) DMSO-d_6). The content of the scintillation vials of the aliquots diluted with D_2O (for p(HEMA) DMSO-d_6) (c.f. Section 2.1, Steps 7 and 8) were transferred into NMR tubes. The measurements were collected on a Bruker Ascend 400 spectrometer operating at 400.13 MHz. The spectra were recorded using the following parameters: relaxation delay: 1 s and transients: 32. The resulting spectra were processed using Mestrelab Research: MestReNova v. 11.0.1–17801. All spectra were baseline and phase corrected. Fig. 2 shows the ^{1}H NMR spectrum of the final time point of p(AM) to illustrate the integrated peak regions to determine the conversion of the polymerization.

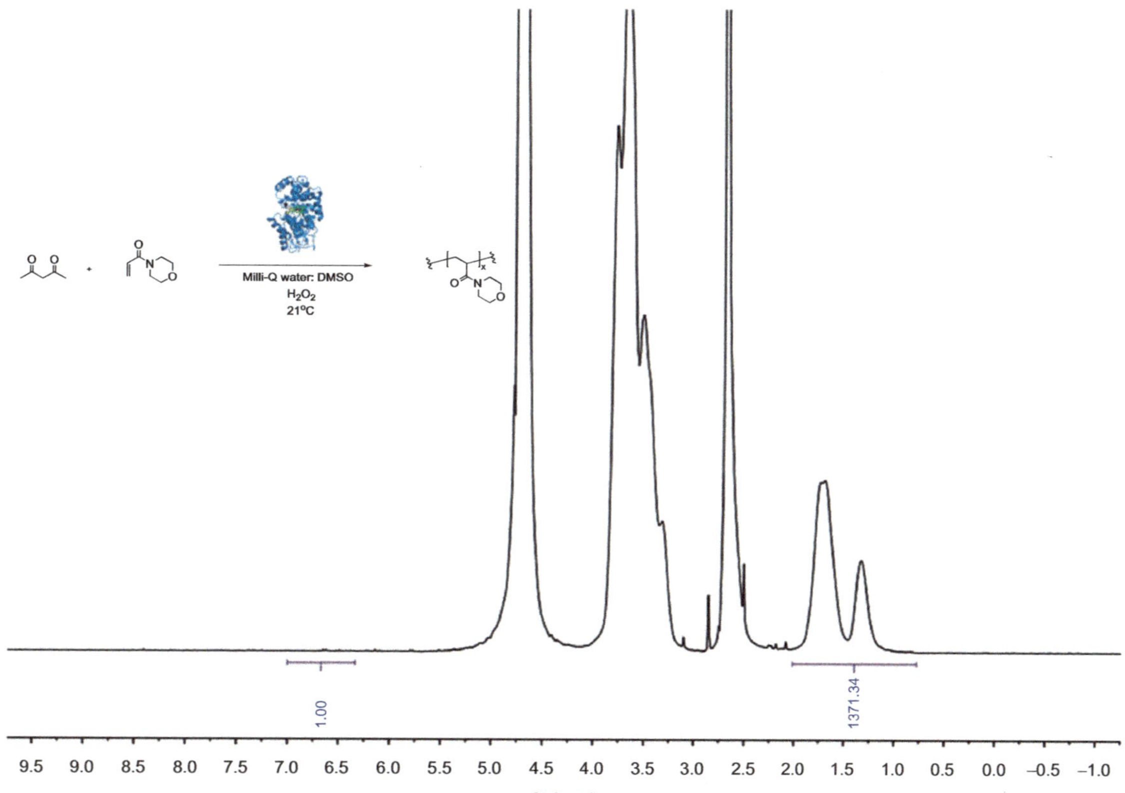

Fig. 2 ^{1}H NMR spectra of the final time point for p(AM); integral regions are used to calculate the extent of monomer conversion.

The NMR data collected for the polymers provides insight into the extent of monomer conversion of this polymerization method. Traditionally, free radical polymerizations result in high conversions and large molecular weight polymers, and even polymeric gels. It is imperative to understand the spectral regions for analyzing these data. In terms of conversion, it is common in the polymerizations of vinyl monomers to compare a vinyl proton of the monomer to a well-known peak in the polymer to determine conversion. Fig. 2 illustrates this analytical practice using p(AM) as an example.

The spectral region 7.00–6.40 ppm corresponds to one vinyl proton of the monomer, and the region 1.90–0.75 ppm corresponds to the two backbone protons adjacent to the morpholino-pendent group. By integrating these two regions, conversion can be calculated using Eq. (1).

$$\%Conversion = \left(\frac{\frac{\int poly}{\#Hpoly}}{\frac{\int poly}{\#Hpoly} + \frac{\int mono}{\#Hmono}} \right) \times 100\% \quad (1)$$

HRP-assisted free radical polymerizations suffer from slow initiation due to the self-quenching mechanism afforded by the biological pathway of peroxidases, where the resulting hydroxyl radical reacts with itself to quench further radical formation (Durand et al., 2000); therefore, to ensure high conversion, the reactions were allowed to stir for 21 h. Final time points, represented here in Fig. 3, were taken after the allotted reaction time. All three spectra show almost full monomer consumption (Table 1) and broadened peaks corresponding to the polymer.

3.2 Gel permeation chromatography (GPC)

GPC measurements are performed to determine the number average molecular weight (M_n) and dispersity (Đ). These data help elucidate the molecular weights achieved during a polymerization. GPC was performed on a PSS SECurity GPC system equipped with an RI detector using DMF (0.05 M LiBr) as an eluent. Measurements were performed at 60 °C with a flow rate of 1.0 mL/min. The system was outfitted with two Agilent PolarGel M columns (ID = 7.5 mm, L = 300 mm, particle size = 8 μm) and an Agilent

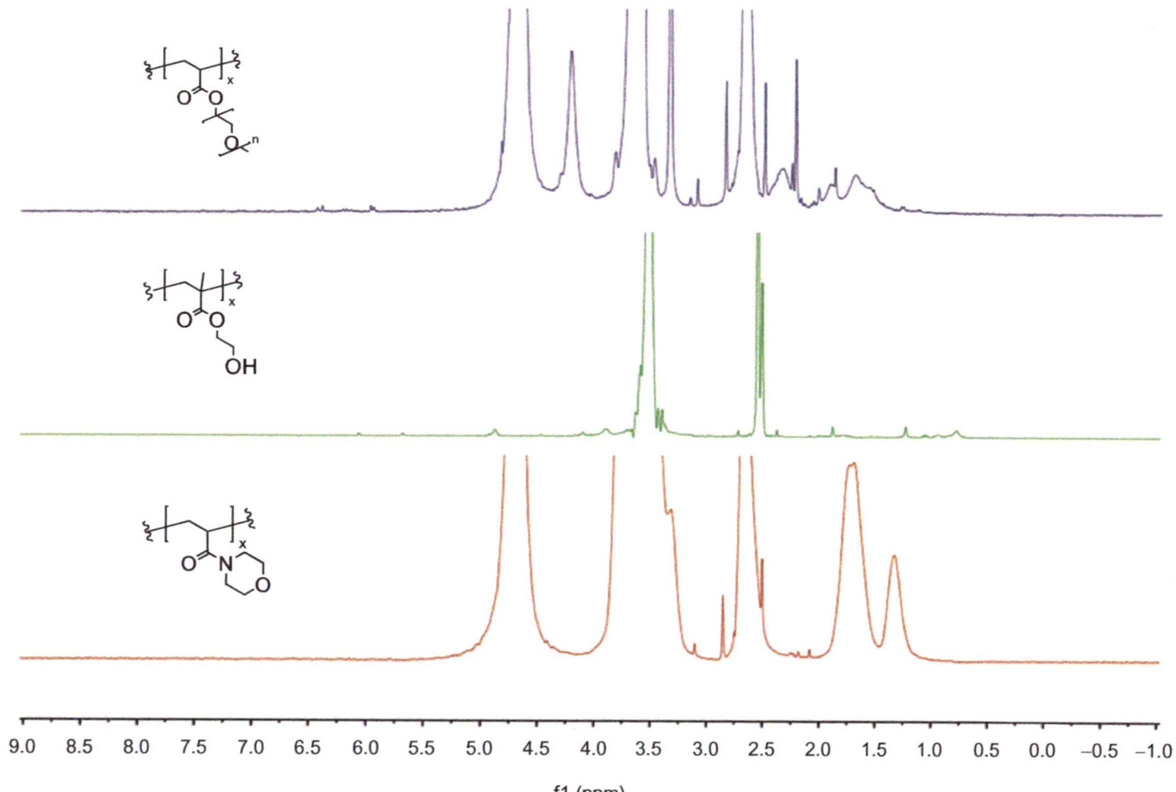

Fig. 3 ^{1}H NMR spectra of final time points of HRP-initiated p(AM) (red), p(HEMA) (green), and p(PEGA) (blue) polymerizations.

Table 1 Final polymer characterization determined by GPC.

Polymer	**Conversion (%)**[a]	M_n **(kDa)**	**Dispersity (Đ)**
p(AM)	99.9	75.7	2.93
p(HEMA)	96.8	128.2	2.96
p(PEGA)	97.6	324.3	1.91

[a]Determined by ^{1}H NMR spectroscopy.

PolarGel M guard column (particle size = 8 μm). The calibration was performed using poly(ethylene oxide) (PEO) standards.

Procedure

GPC samples were prepared as follows:

1. Following NMR analysis, the NMR tubes were collected and their content was transferred to individual scintillation vials (6 mL).
2. 2 mL of GPC elution solvent (DMF + 0.05 M LiBr) was added to each vial.
3. To remove residual H_2O/D_2O, magnesium sulfate was added to each sample and stirred until dry.
4. The slurry-like mixture was then aspirated using a syringe (2 mL) which was then equipped with a PTFE syringe filter (0.2 μm pore size). The solution was injected into a GPC vial.
5. The samples were then placed in the autosampler, and for each measurement 100 μL of the sample was injected onto the column.

The GPC measurements of the final polymers illustrate the power of this synthetic technique for generating high molecular weight polymers with typical dispersities seen for FRPs (Table 1). These data represent the size and uniformity of the polymer chains generated, which can dictate the utility of these materials.

4. Conclusions

The use of horseradish peroxidase in free radical polymerization reactions represents a bio-initiation for the polymerization of vinyl monomers. This technique harnesses the power of naturally occurring biological processes for synthetic polymer chemistry. The enzyme facilitates an efficient catalysis, which leads to high conversion of monomer. Moreover, it represents an environmentally friendly catalyst. As reported in this chapter, this technique is robust to a variety of functional groups and provides benign reaction conditions for generating various polymers. The simplicity of the reaction set up makes this technique an extremely valuable tool for generating high value, functional materials. While this technique has a variety of advantages compared to conventional FRP methods, it still suffers some limitations. Due to the need to extract the enzymes from their natural source, the scalability of these reactions has so far been limited. Efforts continue to be made to circumvent this method's shortcomings. Additionally, this method has influenced others to utilize enzymes for more advanced radical polymerization techniques such as ATRP and RAFT by harnessing the promiscuity of metalloproteins and other biomolecules (see chapters "Biocatalytic ATRP in solution and on surfaces" by Rodriguez et al. and "Enzyme-initiated reversible addition−fragmentation chain transfer (RAFT) polymerization: Precision polymer synthesis via enzymatic catalysis" by Wang and An). These findings have resulted in the exploration of a variety of enzymes' ability to catalyze or initiate non-native radical polymerizations that have improved reaction times, efficiencies, and materials that can be generated (Rodriguez et al., 2018).

Acknowledgments

This work was supported by the Swiss National Science Foundation through grants Nos. PP00P2_144697, PP00P2_172927 and Ambizione program grant 179865, as well as through the Department of Pure and Applied Chemistry of the University of Strathclyde.

References

Binder, W. H. (2019). The past 40 years of macromolecular sciences: Reflections on challenges in synthetic polymer and material science. *Macromolecular Rapid Communications*, *40*(1), 1800610. https://doi.org/10.1002/marc.201800610.

Blyth, J., & Hofmann, A. W. (1845). Ueber das Styrol und einige seiner Zersetzungsprodukte. *Justus Liebigs Annalen der Chemie*, *53*(3), 289–329. https://doi.org/10.1002/jlac.18450530302.

Braun, D. (2009). Origins and development of initiation of free radical polymerization processes. *International Journal of Polymer Science, 2009*, 893234. https://doi.org/10.1155/2009/893234.

Daoud Attieh, M., Zhao, Y., Elkak, A., Falcimaigne-Cordin, A., & Haupt, K. (2017). Enzyme-initiated free-radical polymerization of molecularly imprinted polymer nanogels on a solid phase with an immobilized radical source. *Angewandte Chemie*, *56*(12), 3339–3343. https://doi.org/10.1002/anie.201612667.

Derango, R. A., Chiang, L.-C., Dowbenko, R., & Lasch, J. G. (1992). Enzyme-mediated polymerization of acrylic monomers. *Biotechnology Techniques*, *6*(6), 523–526. https://doi.org/10.1007/BF02447825.

Durand, A., Lalot, T., Brigodiot, M., & Maréchal, E. (2000). Enzyme-mediated initiation of acrylamide polymerization: Reaction mechanism. *Polymer*, *41*(23), 8183–8192. https://doi.org/10.1016/S0032-3861(00)00204-4.

Emery, O., Lalot, T., Brigodiot, M., & Maréchal, E. (1997). Free-radical polymerization of acrylamide by horseradish peroxidase-mediated initiation. *Journal of Polymer Science Part A: Polymer Chemistry*, *35*(15), 3331–3333. https://doi.org/10.1002/(SICI)1099-0518(19971115)35:15<3331::AID-POLA27>3.0.CO;2-B.

Flory, P. J. (1936). Molecular size distribution in linear condensation polymers. *Journal of the American Chemical Society*, *58*(10), 1877–1885. https://doi.org/10.1021/ja01301a016.

Gomberg, M. (1897). Tetraphenylmethan. *Berichte der Deutschen Chemischen Gesellschaft*, *30*(2), 2043–2047. https://doi.org/10.1002/cber.189703002177.

Hollmann, F. (2010). Enzymatic polymerization of vinylpolymers. In K. Loos (Ed.), *Biocatalysis in polymer chemistry* (pp. 143–505). Weinheim: Wiley-VCH.

Hollmann, F., & Arends, I. W. C. E. (2012). Enzyme initiated radical polymerizations. *Polymers*, *4*(1), 759–793. https://doi.org/10.3390/polym4010759.

Kalra, B., & Gross, R. A. (2000). Horseradish peroxidase mediated free radical polymerization of methyl methacrylate. *Biomacromolecules*, *1*(3), 501–505. https://doi.org/10.1021/bm005576v.

Kobayashi, S., Uyama, H., & Kimura, S. (2001). Enzymatic polymerization. *Chemical Reviews*, *101*(12), 3793–3818. https://doi.org/10.1021/cr990121l.

Lalot, T., Brigodiot, M., & Maréchal, E. (1999). A kinetic approach to acrylamide radical polymerization by horse radish peroxidase-mediated initiation. *Polymer International*, *48*(4), 288–292. https://doi.org/10.1002/(sici)1097-0126(199904)48:4<288::Aid-pi145>3.0.Co;2-i.

Malinowska, K. H., & Nash, M. A. (2016). Enzyme- and affinity biomolecule-mediated polymerization systems for biological signal amplification and cell screening. *Current Opinion in Biotechnology*, *39*, 68–75. https://doi.org/10.1016/j.copbio.2016.01.007.

Mark, H. (1964). Macromolecules: Origin, development, significance. *Biophysical Journal*, *4*(1 Pt. 2), 5–7. https://doi.org/10.1016/S0006-3495(64)86923-X.

Parravano, G. (1951). Chain reactions induced by enzymic systems. *Journal of the American Chemical Society*, *73*, 183–184. https://doi.org/10.1021/ja01145a063.

Rodriguez, K. J., Gajewska, B., Pollard, J., Pellizzoni, M. M., Fodor, C., & Bruns, N. (2018). Repurposing biocatalysts to control radical polymerizations. *ACS Macro Letters*, 7(9), 1111–1119. https://doi.org/10.1021/acsmacrolett.8b00561.

Simon, E. (1839). Ueber den flüssigen Storax (Styrax liquidus). *Annalen der Pharmacie*, *31*(3), 265–277. https://doi.org/10.1002/jlac.18390310306.

Singh, A., Ma, D., & Kaplan, D. L. (2000). Enzyme-mediated free radical polymerization of styrene. *Biomacromolecules*, *1*(4), 592–596. https://doi.org/10.1021/bm005537j.

Staudinger, H. (1920). Über polymerisation. *Berichte der deutschen chemischen Gesellschaft (A and B Series)*, *53*(6), 1073–1085. https://doi.org/10.1002/cber.19200530627.

Vanella, R., Ta, D. T., & Nash, M. A. (2019). Enzyme-mediated hydrogel encapsulation of single cells for high-throughput screening and directed evolution of oxidoreductases. .*Biotechnology and Bioengineering*, *116*(8), 1878–1886. https://doi.org/10.1002/bit.27002.

Zavada, S. R., Battsengel, T., & Scott, T. F. (2016). Radical-mediated enzymatic polymerizations. *International Journal of Molecular Sciences*, *17*(2), 195. https://doi.org/10.3390/ijms17020195.

CHAPTER TEN

Biocatalytic ATRP in solution and on surfaces

Kyle J. Rodriguez[a], Michela M. Pellizzoni[a], Mohammad Divandari[b], Edmondo M. Benetti[b,*], Nico Bruns[c,*]

[a]Adolphe Merkle Institute, University of Fribourg, Fribourg, Switzerland
[b]Laboratory for Surface Science and Technology, Department of Materials, ETH Zürich, Zürich, Switzerland
[c]Department of Pure and Applied Chemistry, University of Strathclyde, Glasgow, United Kingdom
*Corresponding authors: e-mail address: edmondo.benetti@mat.ethz.ch; nico.bruns@strath.ac.uk

Contents

Abstract

The promiscuity of enzymes allows for their implementation as catalysts for non-native chemical transformations. Utilizing the redox activity of metalloenzymes under activator regenerated by electron transfer (ARGET) ATRP conditions, well-controlled and defined polymers can be generated. In this chapter, we review bioATRP in solution and on surfaces and provide experimental protocols for hemoglobin-catalyzed ATRP and for surface-initiated biocatalytic ATRP. This chapter highlights the polymerization of acrylate and acrylamide monomers and provides detailed experimental protocols for the characterization of the polymers and of the polymer brushes.

Methods in Enzymology, Volume 627
ISSN 0076-6879
https://doi.org/10.1016/bs.mie.2019.08.014

1. Introduction

The introduction of atom transfer radical polymerizations (ATRP) (Kato, Kamigaito, Sawamoto, & Higashimura, 1995; Wang & Matyjaszewski, 1995) and reversible addition fragmentation chain transfer (RAFT) polymerizations (Chiefari et al., 1998) in the 1990s transformed the field of polymer synthesis, because these reversible-deactivation radical polymerizations (RDRPs) allow to synthesize styrenic, acrylate, methaclyate and acrylamide polymers of predefined molecular weight and narrow molecular weight distribution. Importantly, they are facile synthetic routes to block copolymers, bioconjugates, polymer brushes on surfaces, and other advanced polymer architectures (Keddie, 2014; Matyjaszewski, 2012; Matyjaszewski & Tsarevsky, 2014; Moad, 2019; Zoppe et al., 2017). RDRPs are also often referred to as controlled radical polymerizations (CRPs), even though IUPAC recommends to use the former term (Jenkins, Jones, & Moad, 2009). In RDRPs, there are numerous ways in which the concentration of radicals during a reaction is kept low, thereby avoiding chain termination reactions. In ATRP, this is achieved by halogen atom transfer between a catalyst (usually a transition metal complex based on copper, iron, or ruthenium) and the radical-bearing growing polymer chain to yield a halogen-terminated dormant chain, followed by the reactivation of the chain by halogen transfer back to the catalyst. In RAFT polymerizations, a stoichiometric amount of a thiocarbonylthio compound leads to reversible chain transfer. The field of RDRPs has matured considerably since its beginnings and has resulted in highly active catalytic systems for ATRP and extremely efficient methods for RAFT polymerizations. Both methods tolerate the presence of proteins, can be used to synthesize protein-polymer conjugates, and have even been conducted in the vicinity of living cells (Keddie, 2014; Matyjaszewski, 2012; Matyjaszewski & Tsarevsky, 2014; Moad, 2019). However, reversible-deactivation radical polymerizations that are catalyzed or mediated by enzymes were unknown until Bruns and coworkers, concurrently with di Lena and coworkers, found that heme proteins such as horseradish peroxidase (HRP), hemoglobin (Hb) and catalase, as well as the copper-containing laccases are able to catalyze ATRP (Ng, di Lena, & Chai, 2011a, 2011b; Sigg et al., 2011; Silva et al., 2013). Radical polymerizations are known to occur in nature, such as the synthesis of lignin (d'Ischia et al., 2013; Hatfield & Vermerris, 2001), plant cell wall formation and cuticle sclerotization in insects (Riva, 2006). Peroxidases and laccases have also long been known to initiate free radical polymerizations of vinyl-type monomers by oxidation reactions fueled by peroxides or oxygen, respectively (Aktaş, Kibarer, & Tanyolaç, 2000; Emery, Lalot, Brigodiot, & Maréchal, 1997; Hollmann & Arends, 2012; Kalra & Gross, 2000; Lalot, Brigodiot, & Maréchal, 1999; Shoda, Uyama, Kadokawa, Kimura, & Kobayashi, 2016; Wang et al., 2018) (see chapter "Enyzme-initiated free radical polymerizations of vinyl monomers using horseradish peroxidase" by Rodriguez et al. for the corresponding methods). The difference between enzymatic free radical polymerization and biocatalytic ATRP (bioATRP) is that the enzyme not only catalytically initiates the polymerization, but also catalytically engages in reversible deactivation and activation reactions of the polymer chain. The enzyme initiates the polymerization by abstracting a halogen atom from a typical ATRP initiator such as 2-hydroxyethyl α-bromoisobutyrate (HEBIB) (Fig. 1). This oxidizes the metal center of the enzyme and creates a carbon-centered radical on the initiator. Radical chain growth reaction occurs until the enzymes transfers a halogen atom back onto the radical at the chain end of the growing polymer chain.

As a result, a halogen-terminated polymer chain (i.e., a dormant chain) and the enzyme in its reduced state is obtained. Eventually, the chain is reactivated by the enzyme, continues to grow, and the process repeats. An excess of a reducing agent such as sodium ascorbate is essential for the reactions in order to bring the enzyme into its reduced form at the beginning of the reaction and to regenerate this activating species during the polymerization. Otherwise, the oxidized enzyme accumulates during the reaction due to unavoidable termination side reactions. Thus, bioATRP is conducted

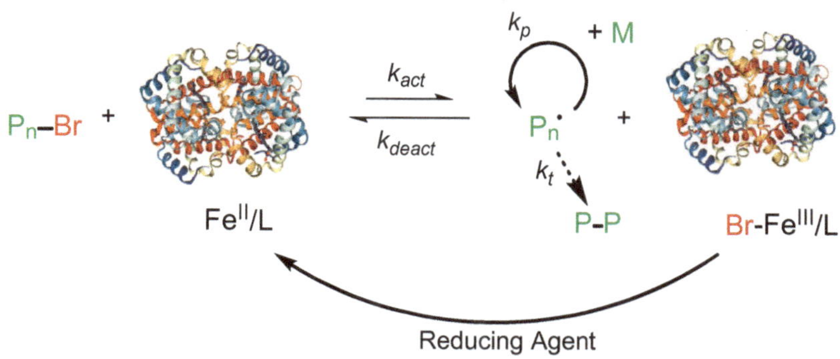

Fig. 1 Proposed mechanism of bioATRP using hemoglobin as the *ATRPase* catalyst under ARGET ATRP conditions.

under activator regenerated by electron transfer (ARGET) ATRP conditions (Matyjaszewski, 2012) and involves dehalogenation and halogenation reactions via radical mechanisms. It is therefore a new-to-nature enzymatic activity of these oxidoreductases, and an example for the promiscuity of peroxidases, heme proteins, and laccases. The discovery of enzymes that catalyze ATRP activity, which we termed "ATRPases" (Sigg et al., 2011), inspired ourselves and others to investigate this novel enzymatic activity, to explore enzyme-mimetic catalysts for ATRP, and to take advantage of the unique properties of enzymes in polymerization reactions (Rodriguez et al., 2018). An obvious advantage of biocatalytic ATRP is that the involved catalysts are proteins or biologically derived compounds that are derived from renewable resources and that are non-toxic. Moreover, a variety of methods exist to quantitatively remove proteins from a reaction mixture, including salt- or solvent-induced precipitation, so that the biocatalyst can be separated from the polymer product (Fodor et al., 2016). Thus, bioATRP could be used as an environmentally friendly and biocompatible alternative to conventional ATRP catalysis, leading to metal-free materials. They are essential for biomedical-, energy- and food-grade applications of ATRP-derived polymers (Matyjaszewski, 2012). As enzymes are large molecules that display a rich variety of functional groups on their surface, they can be physically adsorbed to surfaces (Divandari, Pollard, Dehghani, Bruns, & Benetti, 2017), enclosed into polymeric vesicles (Dinu et al., 2015), or chemically conjugated into protein cages (Renggli et al., 2017), thereby confining polymerizations into nanoscale reaction volumes, e.g., to create cell-mimetic nanoreactors (Dinu et al., 2015). Moreover, bioATRP has been used to create molecularly imprinted hemoglobin biosensors (Sun et al., 2016; Sun, Zhang, Li, Zhao, & Liu, 2017) and to develop an ultrasensitive turbidity assay for hemozoin (Rifaie-Graham et al., 2019), which is a heme biocrystal that is found in malaria parasites. These examples show the potential of bioATRP in applications ranging from polymer synthesis, surface modifications, to biosensing and diagnostics. Here, we review bioATRP in solution and on surfaces and provide the reader with detailed methods for hemoglobin-catalyzed bioATRP in solution and on surfaces.

2. BioATRP in solution

In our first report of bioATRP, we used HRP to catalyze ARGET ATRP of *N*-isopropyl acrylamide (NIPAM) using the initiator HEBIB (Sigg et al., 2011). The presence of bromine-terminated polymer chains and dispersities below 1.5 at the beginning of the reaction show that the reaction involved reversible transfer of bromine from the initiator to the enzymes and back to the polymers. However, the molecular weight of poly(*N*-isopropyl acrylamide) (PNIPAM) did not increase with conversion and the dispersity (Đ) increased during the reaction, indicating that the degree of control of the polymerization was poor. Better results were obtained with hemoglobin as the catalyst and the monomers poly(ethylene glycol) methyl ether acrylate (PEGA) and poly(ethylene glycol) methyl ether methacrylate (PEGMA) (Silva et al., 2013). The reactions were carried out at room temperature. Linear first order kinetics, low dispersities, and a linear increase of the molecular weight of the polymers with conversion indicated that the Hb controlled the reaction. In the case of NIPAM the molecular weights did not increase with conversion. However, polymers were obtained that were partially bromine-terminated and that could be reinitiated to synthesize a second block, so that also here reversible bromine transfer occurred. NIPAM is known to be a monomer that is difficult to control in aqueous ATRP (Ye & Narain, 2009). Therefore, it is not surprising to see a marked difference in the ability of Hb to control the ATRP of PEGA and PEGMA vs NIPAM. Interestingly, when Hb-catalyzed bio-ATRP is not carried out in solution but on a surface, the degree of control is excellent (Divandari et al., 2017; Gao, Karaaslan, Kadla, & Ko, 2014) and no signs of chain termination were observed in these surface-initiated polymerizations (vide infra).

As hemoglobin has cysteines exposed on its surface that can act as irreversible chain transfer agents, the protein had to be reacted with a maleimide reagent in order to block the cysteines (Silva et al., 2013). Otherwise, chains start to grow from the thiols on the protein, resulting in polymer-protein conjugates as a side reaction. Cys-blocked Hb worked best in acidic conditions that arose from the addition of ascorbic acid to water, resulting in a pH of just below 4. In a later study the reason for this was investigated (Divandari et al., 2017). At pH 4, hemoglobin transitions from its native conformation to a more loose acidic conformation. This transition is reversible if the protein is brought back to neutral pH, and does not involve the release of heme into the surrounding medium. This relaxed acidic protein conformation of Hb is beneficial for bioATRP, most likely because it widens the access to the heme group and therefore allows initiator and polymer chain end to interact more efficiently with the metal center of the active site.

Di Lena and coworkers first reported that radical polymerizations can be initiated from ATRP initiators by laccase catalysis (Ng et al., 2011a).

Control over the reactions was achieved by adding a RAFT chain transfer agent to the polymerizations, making this the first example of enzyme-initiated RAFT polymerizations. Others, including An, Konkolewicz, and Qiao have followed up with substantial work on bioRAFT polymerizations (Danielson et al., 2018, 2016; Liu, Lv, & An, 2017; Liu, Lv, Zhu, & An, 2018; Reyhani et al., 2018; Zhang et al., 2015), which is reviewed and discussed in chapter "Enzyme-initiated reversible addition–fragmentation chain transfer (RAFT) polymerization: Precision polymer synthesis via enzymatic catalysis" by Wang and An of this Methods in Enzymology volume. Di Lena continued his work to demonstrate the controlled polymerization of PEGA under ARGET ATRP conditions, as indicated by a linear increase of molecular weight with conversion (Ng et al., 2011b). Dispersities were low at the beginning and remained below 1.7 at the end of the reaction. Most of the experiments were carried out with catalase and the highly active initiator 2-bromoproprionitrile (BPN) at 60 °C, but laccase and HRP were also used. Matyjaszewski and coworkers confirmed that enzymes (using catalase as an example) can produce narrowly dispersed polymers (polyPEGA with Đ = 1.19 at a conversion of 49%) (Simakova, Mackenzie, Averick, Park, & Matyjaszewski, 2013). They elaborated in detail that also the heme group itself, as well as PEGylated heme, are efficient ATRP catalysts (Fu, Simakova, Fantin, Wang, & Matyjaszewski, 2018; Simakova et al., 2013; Smolne et al., 2016).

An opportunity for bioATRP lies in the polymerization of monomers that cannot be polymerized by conventional ATRP. An example is *N*-vinylimidazole (NVIm), a monomer that is produced on an industrial scale. Poly(*N*-vinylimidazole) and its copolymers find applications as dye transfer inhibitors in washing formulations, in cosmetics, for filtration membranes to remove copper ions from wine and beer during winemaking and beer-brewing, or as precursor for polymerized ionic liquids. NVIm is polymerized by free radical polymerization, while conventional ATRP fails to control NVIm polymerizations. Therefore, complex polymer architectures with PVIm blocks cannot be synthesized. One reason is that the polymer complexes metal ions and thus deactivates the ATRP catalysts. In contrast, the metal co-factors of enzymes are embedded in the protein. Laccase from *Trametes versicolor* catalyzed the controlled polymerization of NVIm under ARGET ATRP conditions (Fodor et al., 2016). The molecular weight increased with conversion, and the dispersities were between 1.27 at the beginning of the reaction and 1.57 at its end. Purified PVIm had a dispersity of only 1.07 according to mass spectrometry.

Thus, laccase-catalyzed bioATRP of NVIm allows to synthesize narrowly dispersed PVIm with predefined molecular weight.

When planning to conduct bioATRP, Cys-blocked hemoglobin from bovine blood is a good choice as a catalyst because the heme protein is very cheap and readily available at high purities. Bovine blood is a by-product of meat production and therefore an abundantly available raw material. Several methods can be used to isolate Hb (Riggs, 1981). The simplest one is the following (Merck Index, 1996). Erythrocytes are collected from blood by centrifugation, and lysed by the addition of ether. Centrifugation sediments the cell debris, and the hemoglobin remain in solution. The hemoglobin is lyophilized and sold. Bovine hemoglobin is a tetramer composed of two different subunits ($\alpha_2\beta_2$ dimer) with a total molecular weight of 64,500 $g\,mol^{-1}$ (Merck Index, 1996). Hemoglobin, even though its main function is to bind and transport oxygen, is also a peroxidase (Brunori, 1999; Reeder, 2010). As we demonstrate here, it is also a suitable catalyst for bioATRP. One thing to take note when conducting experiments with Hb is that lyophilized hemoglobin bought from commercial suppliers is mostly methemoglobin, i.e., with iron in its +III oxidation state, because dry hemoglobin oxidizes in air. The oxidation state and the coordination around the iron in the active site can be easily investigated by UV–vis spectroscopy (Zijlstra & Buursma, 1997). Furthermore, in order for Hb to be used as an ATRPase, the surface cysteine residues must be blocked to avoid chain transfer to the thiols and therefore incorporation of the catalyst into polymer chains. Below, detailed experimental procedures for the blocking of the cysteine residues and for bioATRP in solution are described.

2.1 Procedure to block cysteine residues on the surface of hemoglobin

Equipment

- Analytical balance
- Round bottom flask (500 mL)
- Incubator
- Amicon Ultra-15 spin centrifugal filter units (molecular weight cutoff (MWCO) 10,000 kDa) (Millipore)
- Centrifuge
- UV–Vis spectrophotometer
- Scintillation vial (6 mL)
- Quartz cuvette (1 cm path length)

Chemicals

- Hemoglobin from bovine blood (Hb; lyophilized powder) (Sigma)
- *N*-(2-hydroxyethyl)maleimide (HEMI) (Sigma)
- 5,5′-Dithiobis-(2-nitrobenzoic) acid (DTNB) (Sigma)
- 100 mM sodium phosphate buffer (pH 6.5, 150 mM NaCl)
- Tris (Sigma)
- Acetyl cysteine (Sigma)
- Milli-Q water

Procedure

The blocking of the surface cysteine residues of hemoglobin (Cys-blocked Hb) was performed as follows:

1. Hemoglobin (551 mg, 8.54 μmol) and *N*-(2-hydroxyethyl)maleimide (32.8 mg, 0.225 mmol) were added to a 500 mL round bottom flask and dissolved in 300 mL of 100 mM sodium phosphate buffer (pH 6.5, 150 mM NaCl).
2. The reaction flask was then placed in an incubator set at 37 °C for 2 h.
3. The reaction was quenched by desalting the Cys-blocked Hb by spin diafiltration using an Amicon Ultra-15 spin centrifugal filter unit (molecular weight cutoff (MWCO) 10,000 $g\,mol^{-1}$, Millipore) firstly against 100 mM sodium phosphate buffer (pH 6.0, 150 mM NaCl) two times; followed by centrifugation against Milli-Q water three times.
4. The final concentration of the Cys-blocked Hb stock solution following the diafiltration was 11 $mg\,mL^{-1}$.
5. The extent of functionalization was determined using UV–vis spectroscopy through Ellman's test, measuring the absorption at 412 nm for both Hb and Cys-blocked Hb.
6. Ellman's test (Bulaj, Kortemme, & Goldenberg, 1998; Ellman, 1959) was performed by preparing a DTNB stock solution with 50 mM sodium acetate and 2 mM DTNB final concentrations in Milli-Q water, and a Tris stock solution with a final concentration of 1 M Tris and pH 8.0.
7. A calibration curve was generated by using acetyl cysteine as a standard with an initial concentration of 10 μM.
8. The DTNB working solution was prepared by adding 50 μL of the DTNB stock solution, 100 μL of the Tris stock solution, and 840 μL of Milli-Q water into a quartz cuvette and placed into the spectrophotometer.

9. 10 μL of sample (Hb and Cys-blocked Hb) was mixed with the 990 μL of the DTNB working solution. The sample was mixed at room temperature and allowed to equilibrate for 5 min.
10. After the 5 min, a triplicate series of spectra were recorded at 412 nm. The molarity of the thiol groups were calculated by dividing the absorbance average of the triplicate measurement by 13,600 M^{-1} cm^{-1}.

2.2 Polymerization of poly(PEGA) under ARGET ATRP conditions using Cys-blocked Hb as a catalyst in solution

Equipment
- Analytical balance
- Schlenk line outfitted with 5 ports equipped with inert gas (Ar or N_2)
- 2 × Schlenk flask (25 mL)
- Stirring plate
- 2 × Magnetic stirring bar
- Syringe (1 and 2 mL)
- Micropipette (20, 100, 1000 μL)
- Needles (ø 0.80 × 50 mm and ø 0.80 × 120 mm)
- Glass Pasteur pipette (230 mm)
- Cotton or glass wool
- Scintillation vial (6 mL)
- 3 × rubber septum

Chemicals
- Cys-blocked hemoglobin (Section 2.1)
- Poly(ethylene glycol) methyl ether acrylate (PEGA; $M_n = 480\,g\,mol^{-1}$) (Sigma)
- 2-Hydroxyethyl-2-bromoisobutyrate (HEBIB) (Sigma)
- L-Ascorbic acid (99%) (Sigma)
- Aluminum oxide activated, basic, Brockmann I (Sigma)
- Aluminum oxide activated, neutral, Brockmann I (Sigma)
- Milli-Q water

Procedure

The polymerization of poly(PEGA) under ARGET ATRP using Cys-blocked hemoglobin as a catalyst was performed as follows:

1. Two Schlenk flasks, equipped with magnetic stirring bars and rubber septa, were attached to a Schlenk line in a fume hood. The flasks were then evacuated by vacuum and back filled with argon three times.

2. The monomer, PEGA, was purified of inhibitors by passing it through a small plug of basic aluminum oxide via a positive flow of compressed air into a 6 mL scintillation vial. The plug was prepared by inserting a small piece of cotton into a Pasteur pipette and filling it half way with basic aluminum oxide.
3. To one of the Schlenk flasks under a positive flow of argon, HEBIB (11.1 mg, 0.05 mmol) was injected using a 20 μL micropipette, and PEGA (1.8 g, 3.75 mmol) was added using a 1000 μL micropipette. Once both were added, 6 mL of Milli-Q water was added to the flask. The flask was then sealed with a rubber septum.
4. Using a ø 0.80 × 120 mm needle and closing the stopcock of the Schlenk flask, argon was passed through the solution to degas the mixture for 30 min with an exhaust needle (ø 0.80 × 50 mm) through the septum.
5. To the other Schlenk flask, 2 mL of the Cys-blocked Hb in Milli-Q water (11 mg mL^{-1}) was added using a 1000 μL micropipette and sealed with a rubber septum. The solution was degassed by closing the stopcock and inserting a needle (ø 0.80 × 120 mm) into the solution and passing argon through the liquid for 30 min with an exhaust needle (ø 0.80 × 50 mm) through the septum.
6. Additionally, in a separate scintillation vial (6 mL), a solution of L-ascorbic acid was prepared by weighing out 30 mg and dissolving it in 2 mL of Milli-Q water (15 mg mL^{-1}). The vial was then sealed off using a rubber septum and was degassed by inserting a needle (ø 0.80 × 120 mm) into the solution and passing argon through the liquid for 30 min with an exhaust needle (ø 0.80 × 50 mm) through the septum.
7. To start the reaction, 2 mL of the Cys-blocked Hb was removed from the catalyst Schlenk flask using an argon purged syringe (2 mL) and inserted into the reaction flask containing the initiator and monomer. Following this, 1 mL of the sodium L-ascorbate solution was transferred to the reaction flask using an argon purged syringe (1 mL).
8. Once all of the reactants were added, the reaction was stirred at room temperature under an argon atmosphere for the duration of the polymerization. A 0.3 mL aliquot of the reaction mixture was taken using the same syringe used for adding the reducing agent. The aliquot was then passed over a plug of neutral aluminum oxide (prepared the same way as step 2) to quench the reaction and remove the Cys-blocked Hb and rinsed with 1 mL of D_2O into a scintillation vial (6 mL) for NMR and GPC analysis (Section 2.3).

9. Every hour, one 0.3 mL aliquot was taken using an argon purged syringe (1 mL) equipped with a needle (ø 0.80 × 120 mm) from the reaction mixture for NMR and GPC analysis (Section 2.3). The aliquot was passed through a plug of neutral aluminum oxide as mentioned in Step 8. Eventually, the reaction was quenched by opening the reaction flask to air and turning off the flow of argon. The remaining reaction mixture can be purified of Cys-blocked Hb by passing it through a plug of neutral aluminum oxide, and stored for further analysis as an end point measurement.

2.3 Polymer characterization

The characterization of the polymerization of poly(PEGA) using NMR and GPC was performed as follows:

Equipment

- NMR spectrometer, e.g., Bruker Ascend 400 with a 5 mm sample head operating at 400.13 MHz
- NMR tubes, e.g., Wilmad 535-PP-7 and tube cap
- Processing software, e.g., Mestrelab Research: MestReNova (Mnova)
- GPC system, e.g., PSS SECurity system equipped with a refractive index (RI) detector
- Scintillation vial (6 mL)
- PTFE syringe filter 0.2 µm pore size
- Syringe (2 mL)
- GPC vials

Chemicals

- Deuterium oxide (D_2O) (Cambridge Isotope Labs)
- Magnesium sulfate (Sigma)
- Tetrahydrofuran (THF) (Scharlau)

2.3.1 Proton nuclear magnetic resonance (^1H NMR) spectroscopy

NMR measurements are used to characterize a variety of parameters in polymer chemistry. Here we highlight the use of ^1H NMR and its application in determining conversion and kinetic data. The ^1H NMR spectra were recorded in D_2O, which was highlighted in Section 2.2, Steps 8 and 9. The scintillation vials of the aliquots diluted with D_2O were transferred into NMR tubes. The measurements were collected on a Bruker Ascend 400 spectrometer operating at 400.13 MHz. The spectra were recorded using the following parameters: relaxation delay: 1 s and transients: 32. The resulting spectra were processed using Mestrelab Research: MestReNova

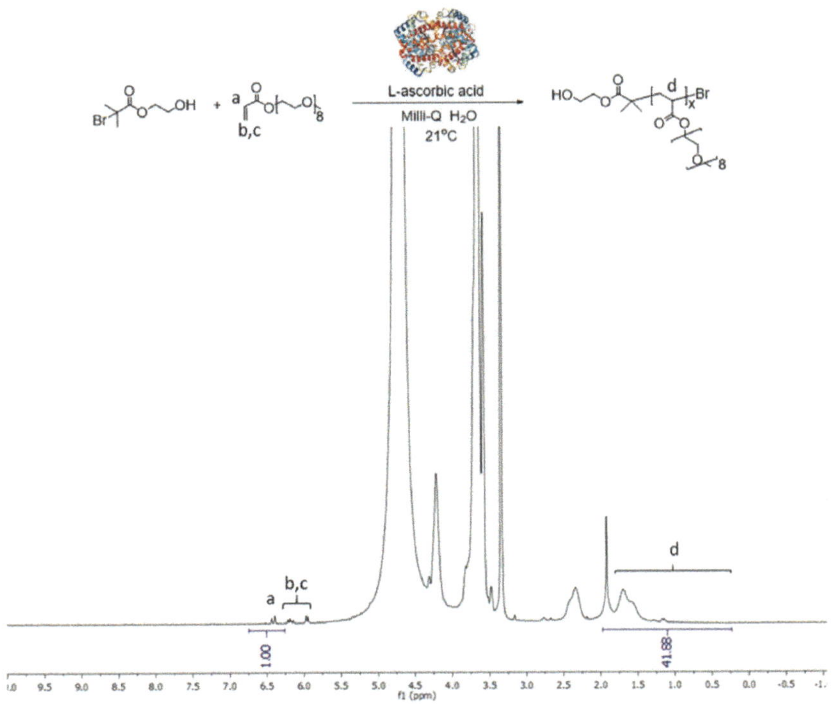

Fig. 2 ^{1}H NMR spectrum of the reaction mixture of polyPEGA after 6 h. By comparing the integral of the proton corresponding to a vinyl proton of the monomer and that of the polymer backbone allows for the calculation of conversion and consequently of the reaction kinetics.

v. 11.0.1–17801. All spectra were baseline and phase corrected. Fig. 2 shows the ^{1}H NMR spectrum of the final time point to illustrate the integrated peak regions to determine the conversion and kinetics of the polymerization.

The NMR data collected for poly(PEGA) provides important information regarding the extent of monomer conversion and the reaction kinetics. Controlled radical polymerizations yield high monomer conversion over time due to a propagating radical that is mediated by an activation and deactivation. This results in a controlled growth of uniformed chains, from which the reaction proceeds through first order kinetics. It is important to integrate the proper spectral regions for analyzing this data. In terms of conversion, it is common for polymerizations of vinyl monomers to compare a vinyl proton of the monomer to a well-known peak in the polymer. Fig. 2 illustrates this process for the final time point of the polyPEGA polymerization (6 h).

The spectral region 6.75–6.25 ppm corresponds to one vinyl proton of the monomer (*mono*), and the region 1.00–2.00 ppm corresponds to two backbone protons adjacent to the PEG-pendent group (*poly*). By integrating these two regions, conversion can be calculated using Eq. (1), where both integral regions (*poly* and *mono*) are divided by the number of protons these peaks correspond to. In the instance of poly(PEGA), $\#H_{poly}$ and $\#H_{mono}$ equate to 2 and 1, respectively.

$$\%Conversion = \left(\frac{\frac{\int poly}{\#Hpoly}}{\frac{\int poly}{\#Hpoly} + \frac{\int mono}{\#Hmono}} \right) \times 100\% \tag{1}$$

After determining the % conversion for each time point, the kinetics of the reaction can be evaluated by applying Eq. (2). With this data for each time point, one can determine whether or not the reaction proceeds through first order kinetics, a requirement for controlled radical polymerizations.

$$ln[M]_0/[M] = \left(\frac{1}{\left(1 - \left(\frac{\%Conversion}{100}\right)\right)} \right) \tag{2}$$

Plotting $\ln[M]_0/[M]$ vs time results in a graphical representation of the reaction kinetics. For the polymerization of PEGA, the reaction follows first order kinetics, as can be seen in Fig. 3.

2.3.2 Gel permeation chromatography (GPC)

GPC measurements were performed to determine the molecular weight (M_n) and dispersity (Đ). These data help elucidate the level of control and the molecular weights achieved during a polymerization. GPC was performed on a PSS SECurity GPC system equipped with an RI detector in THF at 30 °C with a flow rate of 1.0 mL/min. Two columns and a guard column were used and the calibration was performed using polystyrene standards.

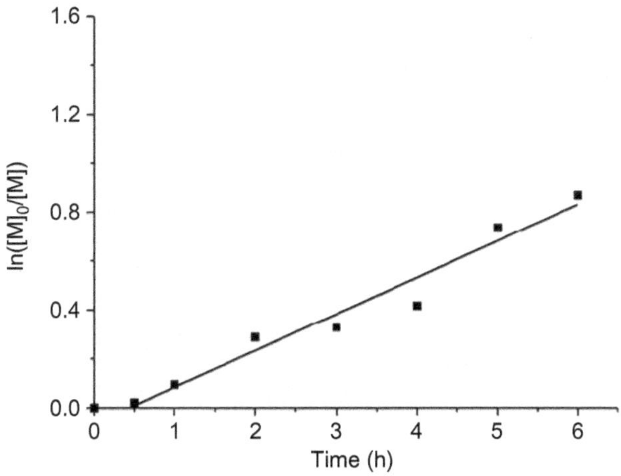

Fig. 3 Reaction kinetics of the polymerization of PEGA catalyzed by Cys-blocked Hb under ARGET ATRP conditions. *Adapted with permission from Silva, T. B., Spulber, M., Kocik, M. K., Seidi, F., Charan, H., Rother, M., et al. (2013). Hemoglobin and red blood cells catalyze atom transfer radical polymerization.* Biomacromolecules, 14 *(8), 2703–2712. doi: 10.1021/bm400556x). Copyright 2013 American Chemical Society.*

Procedure

GPC samples were prepared as follows:

1. Following NMR analysis, the NMR tubes were collected and their content was transferred to individual scintillation vials (6 mL).
2. 2 mL of GPC elution solvent (THF) was added to each vial.
3. To remove residual H_2O/D_2O, magnesium sulfate was added to each sample and stirred until dry.
4. The slurry-like mixture was then aspirated using a syringe (2 mL) and which was then equipped with a PTFE syringe filter (0.2 μm pore size). The solution was injected into a GPC vial.
5. The samples were then placed in the autosampler, and for each measurement 100 μL of the sample was injected onto the column.

The GPC measurements for polyPEGA show the molecular weight distributions of the polymers. By utilizing the % conversion information obtained through ^{1}H NMR, it is possible to relate the % conversion for each kinetic time point to the molecular weight (M_n) obtained through GPC (Fig. 4A).

Furthermore, GPC traces provide insight on the dispersity of these polymerizations and by overlapping the traces it is possible to monitor the shift in molecular weights over time (Fig. 4B). The GPC traces show that the distributions have a tail toward lower molecular weight, indicating that the

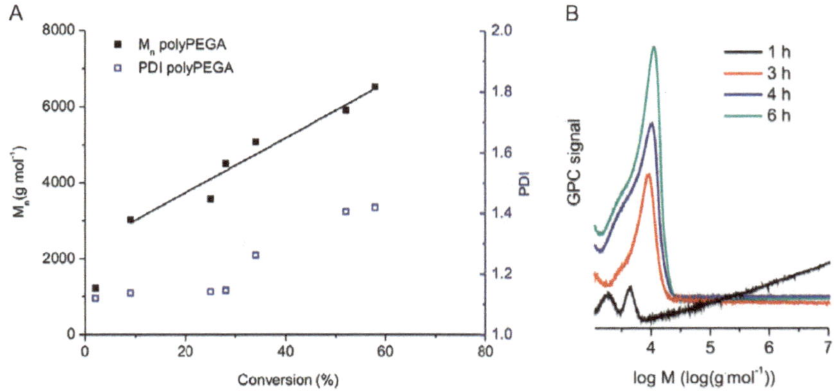

Fig. 4 GPC results of polyPEGA: (A) molecular weight (black) and PDI (blue) vs conversion (%). (B) GPC traces of various time points. *Adapted with permission from Silva, T. B., Spulber, M., Kocik, M. K., Seidi, F., Charan, H., Rother, M., et al. (2013). Hemoglobin and red blood cells catalyze atom transfer radical polymerization.* Biomacromolecules, 14 *(8), 2703–2712. doi:10.1021/bm400556x). Copyright 2013 American Chemical Society.*

degree of control exerted by the protein is modest. Most likely, slow initiation occurred. However, molecular weights increased with conversion. Taken collectively, GPC is a powerful tool for characterizing controlled radical polymerizations by providing information on the size and dispersity of the polymers.

3. BioATRP on surfaces

The application of biomolecular catalysts to trigger the grafting of polymers from solid supports previously functionalized with initiator functions represents an extremely convenient strategy to generate polymer interfaces with controlled physicochemical properties. In particular, bioATRP enables the grafting of polymers under conditions that are compatible with physiological environments, overcoming the standing drawbacks characterizing ATRP processes catalyzed by toxic transition metal complexes (Zoppe et al., 2017).

The first example of surface-initiated bioATRP (SI-bioATRP) was reported by di Lena and coworkers, who polymerized PEGMA from ATRP initiators that were immobilized on crosslinked polystyrene microparticles (Ng et al., 2011a), although the degree of control in the grafting of polymers was not assessed. In a subsequent report, Ko et al. exploited SI-bioATRP catalyzed by HRP, laccase, and catalase in order to graft PNIPAM chains from electrospun lignin nanofibers, finally obtaining polymer brush-based coatings of more than 100 nm thickness (Gao et al., 2014). After cleavage of the polymers synthesized by laccase from the surface, their analysis revealed that the polymers had a molecular weight of more than 1,000,000 g mol^{-1}, monomodal molecular weight distribution, and dispersity as low as 1.3. Taken together, these results demonstrate that surface-initiated bioATRP of NIPAM progressed to very high molecular weights with high degree of control.

Xing and coworkers used surface-initiated bioATRP of 2-hydroxypropyl methacrylate to cover silk fibers with a thick layer of polymer that increased the mass of the fabric by up to 39 wt% (Yang, Lu, Xing, & Chen, 2018). The polymer-grafts on the silk enhanced the wrinkle recovery properties of the fabric, while whiteness, breaking strength, and moisture regain decreased.

To prepare biosensors for hemoglobin, surface-initiated bioATRP was conducted on electrodes by Sun et al. (2016, 2017). Hemoglobin acted as catalyst for the polymerization and simultaneously as template for the resulting molecularly imprinted polymer networks on the electrodes. Acrylamide was copolymerized with the crosslinker *N,N′*-methylene bis-acrylamide. Reactions were carried out on planar electrodes (Sun et al., 2016) and on three-dimensional gold nanodendites (Sun et al., 2017). The resulting biosensors were highly sensitive to detect hemoglobin (lower limit of detection: $3.2 \times 10^{-14}\ \text{mg mL}^{-1}$) and had a linear range that spans Hb concentrations of 12 orders of magnitude.

In our study, we investigated how the interactions between Hb catalyst and surfaces can be tuned in order to modulate the growth of polymer brushes by SI-bioATRP (Divandari et al., 2017). In particular, the affinity of PNIPAM brushes toward Hb could be efficiently switched by shifting the temperature of the polymerization medium across the lower critical solution temperature (LCST) of the polymer (Jones, Smith, Huck, & Alexander, 2002). Via this strategy, the interaction between Hb and the propagating PNIPAM grafts is reversibly triggered to enable the regeneration of catalysts at the surface of the polymer brushes, and to facilitate the access of monomers to the propagating chain ends. High-molar mass PNIPAM brushes were synthesized through multiple cycles of reinitiation (at $T >$ LCST) and rinsing (at $T <$ LCST). The polymer layer reached ~30 nm of dry thickness through linear brush-thickening rates. No signs of radical chain termination were observed, as the surface-initiated bioATRP could be reinitiated multiple times. The results demonstrate that enzymatic polymerizations at surfaces can be modulated and controlled by tuning the interfacial physicochemical properties of the polymer, i.e., the affinity of the surface toward proteins.

3.1 Fabrication of initiator-bearing surfaces

Equipment

- Analytical balance
- Quartz crystal microbalance with dissipation (QCM-D) sensors (QSX 310 Ti, Q-Sense, Gothenburg, Sweden).
- $1 \times 1\,cm^2$ silicon oxide substrates (P/B ⟨100⟩, Si-Mat Silicon Wafers, Germany)
- Beaker
- Pipette

Chemicals

- Nitrodopamine (ND)
- Ultra-pure water
- Ethanol
- Dry dichloromethane (DCM)
- 2-Bromoisobutyryl bromide (BIBB)
- Triethylamine (TEA)

Procedure

In order to form chemically robust ATRP initiator layers on inorganic substrates, we used substituted catechol-based anchors and exploited their strong affinity toward metal oxide surfaces (Amstad, Gillich, Bilecka, Textor, & Reimhult, 2009; Malisova, Tosatti, Textor, Gademann, & Zürcher, 2010; Morgese et al., 2015). Self assembled monolayers (SAMs) exposing primary amino groups were obtained as follow:

1. $1 \times 1\,cm^2$ silicon oxide substrates were coated by reactive magnetron sputtering (at Paul Scherrer Institute PSI, Villigen, Switzerland) with 20 nm-thick TiO_2 layers.
2. TiO_2-coated substrates were incubated overnight in a $0.1\,mg\,mL^{-1}$ solution of ND in ultra-pure water, followed by extensive rinsing with ultra-pure water and ethanol, in order to remove any unreacted, physisorbed adsorbate, and final drying under a stream of N_2.
3. ND-functionalized surface was incubated for 3 h in a 20 mL solution of DCM containing 0.2 mL of BIBB and 0.2 mL of TEA, yielding ATRP initiator-bearing SAMs.

A similar fabrication protocol was applied on TiO_2-coated QCM-D sensors.

3.2 Surface characterization methods

The dry thicknesses of ATRP initiator-functionalized substrates, PNIPAM and PPEGA brushes were analyzed in air by variable angle spectroscopic ellipsometry (VASE), using a M-2000F Woollam variable angle spectroscopic ellipsometer (J.A. Woollam Co. U.S.). The values of the optical parameters Ψ (amplitude) and Δ (phase) were acquired as a function of wavelength (which was included between 350 and 800 nm) using the WVASE32 software package (LOT Oriel GmbH). The optical properties of the organic/polymer layers were modeled by applying a Cauchy model, $n = A + B \cdot \lambda^{-2}$; where n is the refractive index, λ is the wavelength, A and B are two constants that for transparent organic films corresponded to 1.45 and 0.01, respectively (Hilfiker, Syoowicki, Bungay, & Carpio, 1998). The underlying substrate was considered as made of bulk silicon, supporting a 2 nm-thick SiO_2 layer and a 20 nm-thick TiO_2 coating. Bulk dielectric constants were employed for these materials.

Fourier-transform infrared spectroscopy (F-TIR) was used to analyze the chemical composition of PNIPAM and PPEGA brushes synthesized by SI-bioATRP using a IFS 66V spectrometer (Bruker Optik GmbH, Germany) equipped with a liquid-nitrogen-cooled MCT detector.

Static water contact angles (CA) were measured by employing a Model 100, Ramé-Hart goniometer (Ramé-Hart Instrument Co., USA).

The nanomorphology of PNIPAM and PPEGA brushes were analyzed by atomic force microscopy (AFM) using a Bruker Icon Dimension (Bruker, USA) in tapping mode, with OMCL-AC160TS-C3 (Olympus, Japan) AFM cantilevers presenting a normal spring constant of $26\,N\,m^{-1}$.

Quartz crystal microbalance with dissipation monitoring (QCM-D) was applied to monitor in situ the growth of PNIPAM and PPEGA brushes and their interaction with Hb, by employing a E4 instrument (Q-Sense AB, Västra Frölunda, Sweden). Data analysis was performed with a Q-Soft 301 software (Q-Sense AB). An extended viscoelastic model (Reimhult, Höök, & Kasemo, 2003) was used to fit the frequency and dissipation shifts (Δf and ΔD, respectively) using three overtones, to evaluate the hydrated protein mass adsorbed on the brush-bearing sensors.

3.3 Surface-initiated bioATRP of NIPAM monitored by ex situ VASE

Equipment

- Conical flasks
- pH meter
- Schlenk line outfitted with five ports equipped with inert gas (Ar or N_2)

- Schlenk flask (25 mL)
- Stirring plate
- Magnetic stirring bar
- Syringe (1 and 2 mL)

Chemicals

- ATRP initiator-functionalized TiO_2 substrates (see Section 3.1 for fabrication)
- NIPAM
- Sodium ascorbate
- Cys-blocked Hb
- Acetate buffer (pH 4, 0.1 M)
- Ultra-pure water

Procedure

1. In a Schlenk flask, the polymerization mixture comprising NIPAM (206 mg, 1.817 mmol, sodium ascorbate (NaAsc) (5.40 mg, 27.2 μmol), Cys-blocked Hb (5.9 mg, 91 nmol) and 3.2 mL of 0.1 M acetate buffer (pH 4) was preliminarily deoxygenated by N_2 bubbling for 30 min.
2. Later on, the polymerization mixture was poured into conical flasks kept under inert atmosphere and containing ATRP initiator-functionalized TiO_2 substrates via a degassed syringe.
3. After determined reaction times the substrates were removed from the flasks, copiously rinsed with ultra-pure water, and analyzed by VASE in order to determine PNIPAM brush thickness.

The growth of polymer brushes by SI-bioATRP requires a close contact between the Hb-based catalyst and the initiator/dormant alkyl halide-bearing functions at the substrate surface. For this reason, the chemical composition of the growing polymer chains and the interfacial properties of the subsequently generated brushes must be carefully considered to dissect the mechanism of the grafting process. Protein resistant polymer brushes prevent surface interactions with adsorbing biomolecules. Hence, their synthesis by SI-bioATRP results a difficult task, and films presenting just limited thicknesses can be efficiently grown (Divandari et al., 2017). In contrast, the grafting of more bio-adhesive brushes would be hindered by a significant surface contamination by proteins adsorbing from the polymerization medium, leading to a slowing down of brush growth due to the limited diffusion of reactants toward the propagating chain ends. PNIPAM brushes emerge as the most suitable assemblies to test SI-bioATRP, since their affinity toward the biomolecular catalyst can be reversibly modulated by switching their swelling properties across the polymer's LCST (Burkert et al., 2010; Yin, Zhang, Jiang, & Zhu, 2009).

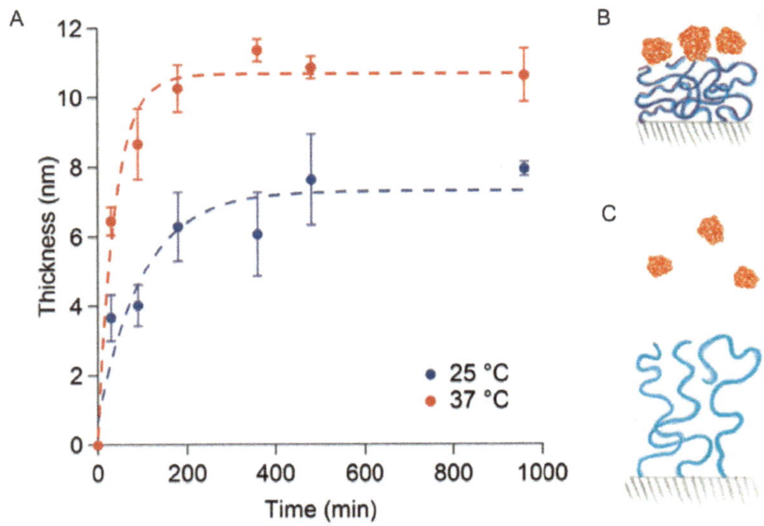

Fig. 5 (A) The growth of PNIPAM brushes by SI-bioATRP was recorded ex situ measuring brush dry thickness by VASE. (B) Schematic depiction of collapsed and protein adherent PNIPAM polymer brushes at 37 °C. (C) Schematic depiction of hydrated and protein repellent PNIPAM brushes at 25 °C. *Adapted with permission from Divandari, M., Pollard, J., Dehghani, E., Bruns, N., & Benetti, E. M. (2017). Controlling enzymatic polymerization from surfaces with switchable bioaffinity.* Biomacromolecules, 18 *(12), 4261–4270. doi: 10.1021/acs.biomac.7b01313). Copyright 2017 American Chemical Society.*

Following the simplest experimental and analytical procedures, the synthesis of PNIPAM brushes by SI-bioATRP can be monitored ex situ by employing VASE. When the polymerization was performed at 25 °C, that is below the LCST of PNIPAM, polymer brushes rapidly grew during the first 100 min of reaction time, reaching ~7 nm of dry thickness, after which the brush-thickening rate attained a plateau (Fig. 5A). The apparent termination of the grafting process was due to the biopassive character of PNIPAM brushes below their LCST, which substantially decelerates brush growth as soon as the polymer-layer thickness reached values that are sufficient to efficiently prevent the adsorption of Hb at the brush-medium interface (Fig. 5B) (Halperin, 1999).

The growth of PNIPAM brushes at 37 °C, which is above their LCST, was significantly faster during the first stages of reaction, the grafted films reaching ~11 nm of dry thickness after 100 min of polymerization (Fig. 5A). However, also in this case, termination of brush thickening was recorded, presumably due to the formation of a dense protein layer on the collapsed PNIPAM brushes, which hindered further diffusion of reactants, and prevented the contact between the catalyst and the dehydrated dormant/propagating grafts (Fig. 5C).

In order to accomplish a progressive growth of relatively thick PNIPAM brushes by SI-bioATRP, a multistep polymerization process was devised. Namely, SI-bioATRP was performed for 1 h at 37 °C, after which the substrates were removed from the reagent solution. They were rinsed with water at room temperature in order to induce PNIPAM swelling and the simultaneous detachment of the physisorbed Hb layer. After drying under a stream of N_2, brush thickness was measured by VASE and the samples were subsequently re-incubated in a freshly prepared polymerization mixture previously set to 37 °C, and left reacting for another hour. This cycle was repeated several times, as schematically depicted in Fig. 6A and B.

The step-by-step synthesis of PNIPAM brushes by SI-bioATRP monitored by VASE showed a linear growth in brush thickness ($R^2 = 0.99$, from linear regression to the VASE data), which increased by ~4 nm following each rinsing/polymerization step (Fig. 6C). The progressive thickening of PNIPAM brushes did not show any sign of irreversible termination reactions, and suggested a well-controlled radical polymerization process (Husseman et al., 1999).

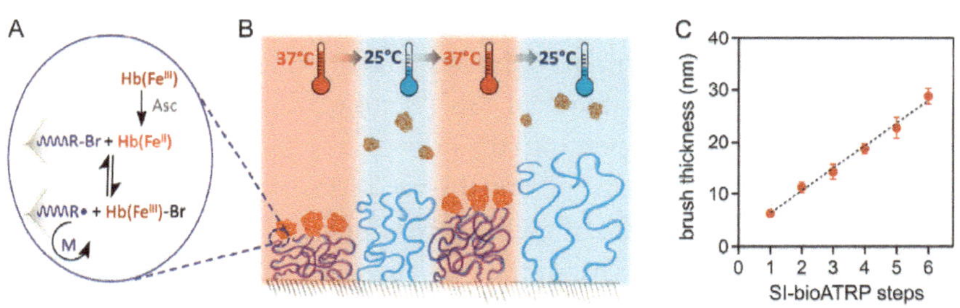

Fig. 6 (A and B) Scheme depicting multistep SI-bioATRP of NIPAM. Growth of PNIPAM brushes is carried out at 37 °C. A collapsed brush conformation enables a close interaction between Hb and the polymer interface. Reducing the temperature to 25 °C leads to PNIPAM swelling and the consequent desorption of Hb from the polymer surface. The Hb layer on PNIPAM brushes can be refreshed by incubating the surfaces in a new polymerization mixture at 37 °C. (C) Through multiple polymerization/rinsing steps the controlled growth of PNIPAM brushes can be successfully accomplished. *Adapted with permission from Divandari, M., Pollard, J., Dehghani, E., Bruns, N., & Benetti, E. M. (2017). Controlling enzymatic polymerization from surfaces with switchable bioaffinity.* Biomacromolecules, 18 *(12), 4261–4270. doi: 10.1021/acs.biomac.7b01313). Copyright 2017 American Chemical Society.*

3.4 Surface-initiated bioATRP monitored in situ by QCM-D

Equipment

- Temperature controller
- pH meter
- Schlenk line outfitted with five ports equipped with inert gas (Ar or N_2)
- Schlenk flask (25 mL)
- Stirring plate
- Magnetic stirring bar
- Syringe (1 and 2 mL)

Chemicals

- ATRP initiator-functionalized TiO_2-coated QCM-D sensors (see Section 3.1 for fabrication)
- NIPAM
- Sodium ascorbate
- Cys-blocked Hb
- Acetate buffer (pH 4, 0.1 M)
- Ultra-pure water

Procedure

1. ATRP initiator-functionalized TiO_2-coated QCM-D sensors were inserted inside liquid flow cells (volume = 140 μL), which were equipped with a temperature controller. First, deoxygenated ultra-pure water kept at 37 °C was injected into the cells, and the system was left equilibrating until stable baselines of Δf and ΔD were recorded.
2. A previously deoxygenated mixture containing NIPAM (1.8 mmol), NaAsc (27 μmol), Cys-blocked Hb (90 nmol) in 0.1 M acetate buffer (3.2 ml, pH 4) was injected into the cells via a syringe, while the temperature was maintained to 37 °C and the variations of Δf and ΔD associated to the growth of PNIPAM grafts were simultaneously recorded.
3. Following 1 h of polymerization, the cells were flushed with cold ultra-pure water, in order to swell the PNIPAM brushes and remove physisorbed Hb. After this step, the temperature was increased to 37 °C and a fresh polymerization mixture was injected into the cells.

3.4.1 Results

The multistep growth of PNIPAM brushes by SI-bioATRP could be conveniently monitored in situ using QCM-D. ATRP initiator-functionalized TiO_2-coated QCM-D sensors were placed into the liquid flow cells of the QCM-D. After equilibration, deoxygenated reaction mixture was injected into the cells at 37 °C. The polymerizations were carried out for 1 h, after

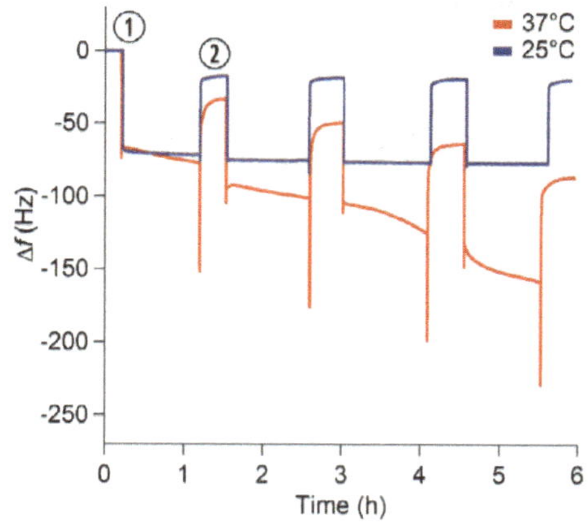

Fig. 7 The synthesis of PNIPAM brushes by SI-bioATRP following multiple polymerization/rinsing steps could be followed in situ by QCM-D. The decrease in Δf recorded after each grafting step was correlated to the increment in the hydrated mass of PNIPAM brushes. *Adapted with permission from Divandari, M., Pollard, J., Dehghani, E., Bruns, N., & Benetti, E. M. (2017). Controlling enzymatic polymerization from surfaces with switchable bioaffinity.* Biomacromolecules, 18 *(12), 4261–4270. doi: 10.1021/acs.biomac.7b01313). Copyright 2017 American Chemical Society.*

which the cells were flushed with cold ultra-pure water. The temperature was set to 37 °C again, and a fresh polymerization mixture was injected into the cells. As displayed in Fig. 7, during each polymerization step at 37 °C a progressive decrease in Δf indicated a concomitant increment in the hydrated mass of PNIPAM brushes growing from the QCM-D sensors. In contrast, when the multiple polymerization steps were conducted at 25 °C, a much modest reduction in Δf suggested how the exclusion of Hb from the interface of hydrated PNIPAM brushes substantially slowed down their growth.

3.5 Discussion

The SI-bioATRP of NIPAM highlights how the fine control of interfacial physicochemical properties of molecularly designed surfaces can be exploited to govern the biocatalytic synthesis of grafted macromolecules.

The adhesion of Hb, which is capable of catalyzing ATRP, on initiator-bearing surfaces can be exploited to synthesize PNIPAM brushes in a highly controlled fashion. A close contact between biocatalyst and the dormant chain ends on propagating brushes is required to achieve an efficient polymer grafting. The affinity of PNIPAM brushes toward Hb can be reversibly switched by shifting the medium temperature across the polymer's LCST. Hence, following a multistep polymerization process, polymer growth is achieved at 37 °C when Hb adsorbs on the collapsed PNIPAM. At 25 °C the polymer brushes swell. This detaches the physisorbed proteins and allows for a new layer of catalyst to form on the polymer surface when the temperature is increased above LCST again. The multistep growth of PNIPAM brushes can be efficiently monitored ex situ, by employing VASE, as well as in situ, within a QCM-D cell equipped with ATRP initiator-functionalized sensors. In both these cases, PNIPAM brush growth was progressive, indicating a well-controlled grafting process in fully aqueous and biocompatible conditions, where transition metal-catalyzed ATRP suffers from an intrinsic lack of control. Apart from giving fundamental insights into surface-confined enzymatic polymerizations, these findings also show a new way to control macromolecular functionalization of materials through tailored bioadhesion.

4. Conclusions

BioATRP introduced reversible-deactivation radical polymerizations to the field of biocatalytic polymerizations, thereby greatly increasing the scope and possibilities of enzyme-catalyzed radical polymerizations. Under the right reaction conditions, heme enzymes such as HRP, Hb, and catalase, as well as laccases can catalyze a reversible atom transfer between halogenated organic compounds, halogen-terminated polymer chains and the enzyme. This establishes an ATRP equilibrium. In solution, the polymerizations proceed with first order kinetics. Depending on the monomer and the reaction conditions, the molecular weight of the polymers increases throughout the reaction and dispersities are low, but not excellent. In contrast, surface-initiated bioATRP proceeds with high chain end fidelity and therefore an excellent degree of control.

All reported bioATRP reactions use commercial enzymes. Obviously, neither HRP, hemoglobin, catalase, nor laccases have evolved to catalyze ATRP reactions. Thus, it is not surprising that in many cases the degree of control that the enzymes display over the reaction is only moderate, especially when the reactions are carried out in solution. While for many applications other benefits of enzymes, such as their benign character, might be the reason to choose bioATRP over conventional ATRP, it is clear that future work has to increase the performance of enzymes in ATRP. Luckily, a wide variety of tools exist to achieve this, including the engineering of enzymes by rational design or directed evolution, and the use of artificial metalloenzymes.

Acknowledgments

This work was supported by the Swiss National Science Foundation through grants Nos. PP00P2_144697, PP00P2_172927 and Ambizione program grant 179865, as well as through the Department of Pure and Applied Chemistry of the University of Strathclyde.

References

Aktaş, N., Kibarer, G., & Tanyolaç, A. (2000). Effects of reaction conditions on laccase-catalyzed α-naphthol polymerization. *Journal of Chemical Technology & Biotechnology, 75*(9), 840–846. https://doi.org/10.1002/1097-4660(200009)75:9<840::Aid-jctb292->3.0.Co;2-9.

Amstad, E., Gillich, T., Bilecka, I., Textor, M., & Reimhult, E. (2009). Ultrastable iron oxide nanoparticle colloidal suspensions using dispersants with catechol-derived anchor groups. *Nano Letters, 9*(12), 4042–4048. https://doi.org/10.1021/nl902212q.

Brunori, M. (1999). Hemoglobin is an honorary enzyme. *Trends in Biochemical Sciences, 24*(4), 158–161. https://doi.org/10.1016/S0968-0004(99)01380-8.

Bulaj, G., Kortemme, T., & Goldenberg, D. P. (1998). Ionization-reactivity relationships for cysteine thiols in polypeptides. *Biochemistry, 37*(25), 8965–8972. https://doi.org/10.1021/bi973101r.

Burkert, S., Bittrich, E., Kuntzsch, M., Müller, M., Eichhorn, K.-J., Bellmann, C., et al. (2010). Protein resistance of PNIPAAm brushes: Application to switchable protein adsorption. *Langmuir, 26*(3), 1786–1795. https://doi.org/10.1021/la902505q.

Chiefari, J., Chong, Y. K., Ercole, F., Krstina, J., Jeffery, J., Le, T. P. T., et al. (1998). Living free-radical polymerization by reversible addition–fragmentation chain transfer: The RAFT process. *Macromolecules, 31*(16), 5559–5562. https://doi.org/10.1021/ma9804951.

Danielson, A. P., Van Kuren, D. B., Lucius, M. E., Makaroff, K., Williams, C., Page, R. C., et al. (2016). Well-defined macromolecules using horseradish peroxidase as a RAFT initiase. *Macromolecular Rapid Communications, 37*(4), 362–367. https://doi.org/10.1002/marc.201500633.

Danielson, A., Van-Kuren, D., Bornstein, J., Kozuszek, C., Berberich, J., Page, R., et al. (2018). Investigating the mechanism of horseradish peroxidase as a RAFT-initiase. *Polymers, 10*(7), 741. https://doi.org/10.3390/polym10070741.

Dinu, M. V., Spulber, M., Renggli, K., Wu, D., Monnier, C. A., Petri-Fink, A., et al. (2015). Filling polymersomes with polymers by peroxidase-catalyzed atom transfer radical polymerization. *Macromolecular Rapid Communications, 36*(6), 507–514. https://doi.org/10.1002/marc.201400642.

d'Ischia, M., Wakamatsu, K., Napolitano, A., Briganti, S., Garcia-Borron, J. C., Kovacs, D., et al. (2013). Melanins and melanogenesis: Methods, standards, protocols. *Pigment Cell & Melanoma Research, 26*(5), 616–633. https://doi.org/10.1111/pcmr.12121.

Divandari, M., Pollard, J., Dehghani, E., Bruns, N., & Benetti, E. M. (2017). Controlling enzymatic polymerization from surfaces with switchable bioaffinity. *Biomacromolecules, 18*(12), 4261–4270. https://doi.org/10.1021/acs.biomac.7b01313.

Ellman, G. L. (1959). Tissue sulfhydryl groups. *Archives of Biochemistry and Biophysics, 82*(1), 70–77. https://doi.org/10.1016/0003-9861(59)90090-6.

Emery, O., Lalot, T., Brigodiot, M., & Maréchal, E. (1997). Free-radical polymerization of acrylamide by horseradish peroxidase-mediated initiation. *Journal of Polymer Science Part A: Polymer Chemistry, 35*(15), 3331–3333. https://doi.org/10.1002/(SICI)1099-0518(19971115)35:15<3331::AID-POLA27>3.0.CO;2-B.

Fodor, C., Gajewska, B., Rifaie-Graham, O., Apebende, E. A., Pollard, J., & Bruns, N. (2016). Laccase-catalyzed controlled radical polymerization of N-vinylimidazole. *Polymer Chemistry*, 7(43), 6617–6625. https://doi.org/10.1039/C6PY01261B.

Fu, L., Simakova, A., Fantin, M., Wang, Y., & Matyjaszewski, K. (2018). Direct ATRP of methacrylic acid with iron-porphyrin based catalysts. *ACS Macro Letters*, 7(1), 26–30. https://doi.org/10.1021/acsmacrolett.7b00909.

Gao, G., Karaaslan, M. A., Kadla, J. F., & Ko, F. (2014). Enzymatic synthesis of ionic responsive lignin nanofibres through surface poly(N-isopropylacrylamide) immobilization. *Green Chemistry*, *16*(8), 3890–3898. https://doi.org/10.1039/C4GC00757C.

Halperin, A. (1999). Polymer brushes that resist adsorption of model proteins: Design parameters. *Langmuir*, *15*(7), 2525–2533. https://doi.org/10.1021/la981356f.

Hatfield, R., & Vermerris, W. (2001). Lignin formation in plants. The dilemma of linkage specificity. *Plant Physiology*, *126*(4), 1351–1357. https://doi.org/10.1104/pp.126.4.1351.

Hilfiker, J. N., Syoowicki, R. A., Bungay, C. L., & Carpio, R. (1998). Spectroscopic ellipsometry for polymer thin films. *Solid State Technology*, *41*(10), 101.

Hollmann, F., & Arends, I. W. C. E. (2012). Enzyme initiated radical polymerizations. *Polymers*, *4*(1), 759–793. https://doi.org/10.3390/polym4010759.

Husseman, M., Malmström, E. E., McNamara, M., Mate, M., Mecerreyes, D., Benoit, D. G., et al. (1999). Controlled synthesis of polymer brushes by "living" free radical polymerization techniques. *Macromolecules*, *32*(5), 1424–1431. https://doi.org/10.1021/ma981290v.

Jenkins, A. D., Jones, R. G., & Moad, G. (2009). Terminology for reversible-deactivation radical polymerization previously called "controlled" radical or "living" radical polymerization (IUPAC recommendations 2010). *Pure and Applied Chemistry*, *82*(2), 483–491. https://doi.org/10.1351/pac-rep-08-04-03.

Jones, D. M., Smith, J. R., Huck, W. T. S., & Alexander, C. (2002). Variable adhesion of micropatterned thermoresponsive polymer brushes: AFM investigations of poly(N-isopropylacrylamide) brushes prepared by surface-initiated polymerizations. *Advanced Materials*, *14*(16), 1130–1134. https://doi.org/10.1002/1521-4095(20020816)14:16<1130::Aid-adma1130>3.0.Co;2-7.

Kalra, B., & Gross, R. A. (2000). Horseradish peroxidase mediated free radical polymerization of methyl methacrylate. *Biomacromolecules*, *1*(3), 501–505. https://doi.org/10.1021/bm005576v.

Kato, M., Kamigaito, M., Sawamoto, M., & Higashimura, T. (1995). Polymerization of methyl methacrylate with the carbon tetrachloride/dichlorotris- (triphenylphosphine)-ruthenium(II)/methylaluminum Bis(2,6-di-tert-butylphenoxide) initiating system: Possibility of living radical polymerization. *Macromolecules*, *28*(5), 1721–1723. https://doi.org/10.1021/ma00109a056.

Keddie, D. J. (2014). A guide to the synthesis of block copolymers using reversible-addition fragmentation chain transfer (RAFT) polymerization. *Chemical Society Reviews*, *43*(2), 496–505. https://doi.org/10.1039/C3CS60290G.

Lalot, T., Brigodiot, M., & Maréchal, E. (1999). A kinetic approach to acrylamide radical polymerization by horse radish peroxidase-mediated initiation. *Polymer International*, *48*(4), 288–292. https://doi.org/10.1002/(sici)1097-0126(199904)48:4<288::Aid-pi145>3.0.Co;2-i.

Liu, Z., Lv, Y., & An, Z. (2017). Enzymatic cascade catalysis for the synthesis of multiblock and ultrahigh-molecular-weight polymers with oxygen tolerance. *Angewandte Chemie International Edition*, *56*(44), 13852–13856. https://doi.org/10.1002/anie.201707993.

Liu, Z., Lv, Y., Zhu, A., & An, Z. (2018). One-enzyme triple catalysis: Employing the promiscuity of horseradish peroxidase for synthesis and functionalization of well-defined polymers. *ACS Macro Letters*, 7(1), 1–6. https://doi.org/10.1021/acsmacrolett.7b00950.

Malisova, B., Tosatti, S., Textor, M., Gademann, K., & Zürcher, S. (2010). Poly(ethylene glycol) adlayers immobilized to metal oxide substrates through catechol derivatives: Influence of assembly conditions on formation and stability. *Langmuir, 26*(6), 4018–4026. https://doi.org/10.1021/la903486z.

Matyjaszewski, K. (2012). Atom transfer radical polymerization (ATRP): Current status and future perspectives. *Macromolecules, 45*(10), 4015–4039. https://doi.org/10.1021/ma3001719.

Matyjaszewski, K., & Tsarevsky, N. V. (2014). Macromolecular engineering by atom transfer radical polymerization. *Journal of the American Chemical Society, 136*(18), 6513–6533. https://doi.org/10.1021/ja408069v.

Merck Index. (1996). In S. Budavari (Ed.), *Royal Society of Chemistry* (p. 794). Merck Index.

Moad, G. (2019). A critical survey of dithiocarbamate reversible addition-fragmentation chain transfer (RAFT) agents in radical polymerization. *Journal of Polymer Science Part A: Polymer Chemistry, 57*(3), 216–227. https://doi.org/10.1002/pola.29199.

Morgese, G., Causin, V., Maggini, M., Corrà, S., Gross, S., & Benetti, E. M. (2015). Ultrastable suspensions of polyoxazoline-functionalized ZnO single nanocrystals. *Chemistry of Materials, 27*(8), 2957–2964. https://doi.org/10.1021/acs.chemmater.5b00252.

Ng, Y.-H., di Lena, F., & Chai, C. L. L. (2011a). Metalloenzymatic radical polymerization using alkyl halides as initiators. *Polymer Chemistry, 2*(3), 589–594. https://doi.org/10.1039/C0PY00139B.

Ng, Y.-H., di Lena, F., & Chai, C. L. L. (2011b). PolyPEGA with predetermined molecular weights from enzyme-mediated radical polymerization in water. *Chemical Communications (Cambridge, United Kingdom), 47*(22), 6464–6466. https://doi.org/10.1039/C1CC10989H.

Reeder, B. J. (2010). The redox activity of hemoglobins: From physiologic functions to pathologic mechanisms. *Antioxidants & Redox Signaling, 13*(7), 1087–1123. https://doi.org/10.1089/ars.2009.2974.

Reimhult, E., Höök, F., & Kasemo, B. (2003). Intact vesicle adsorption and supported biomembrane formation from vesicles in solution: Influence of surface chemistry, vesicle size, temperature, and osmotic pressure. *Langmuir, 19*(5), 1681–1691. https://doi.org/10.1021/la0263920.

Renggli, K., Sauter, N., Rother, M., Nussbaumer, M. G., Urbani, R., Pfohl, T., et al. (2017). Biocatalytic atom transfer radical polymerization in a protein cage nanoreactor. *Polymer Chemistry, 8*(14), 2133–2136. https://doi.org/10.1039/C6PY02155G.

Reyhani, A., Nothling, M. D., Ranji-Burachaloo, H., McKenzie, T. G., Fu, Q., Tan, S., et al. (2018). Blood-catalyzed RAFT polymerization. *Angewandte Chemie, International Edition, 57*, 10288–10292. https://doi.org/10.1002/anie.201802544.

Rifaie-Graham, O., Pollard, J., Raccio, S., Balog, S., Rusch, S., Hernández-Castañeda, M. A., et al. (2019). Hemozoin-catalyzed precipitation polymerization as an assay for malaria diagnosis. *Nature Communications, 10*(1), 1369. https://doi.org/10.1038/s41467-019-09122-z.

Riggs, A. (1981). Preparation of blood hemoglobins of vertebrates. In *Vol. 76. Methods in Enzymology* (pp. 5–29). Academic Press.

Riva, S. (2006). Laccases: Blue enzymes for green chemistry. *Trends in Biotechnology, 24*(5), 219–226. https://doi.org/10.1016/j.tibtech.2006.03.006.

Rodriguez, K. J., Gajewska, B., Pollard, J., Pellizzoni, M. M., Fodor, C., & Bruns, N. (2018). Repurposing biocatalysts to control radical polymerizations. *ACS Macro Letters, 7*(9), 1111–1119. https://doi.org/10.1021/acsmacrolett.8b00561.

Shoda, S.-i., Uyama, H., Kadokawa, J.-i., Kimura, S., & Kobayashi, S. (2016). Enzymes as green catalysts for precision macromolecular synthesis. *Chemical Reviews (Washington, DC, United States), 116*(4), 2307–2413. https://doi.org/10.1021/acs.chemrev.5b00472.

Sigg, S. J., Seidi, F., Renggli, K., Silva, T. B., Kali, G., & Bruns, N. (2011). Horseradish peroxidase as a catalyst for atom transfer radical polymerization. *Macromolecular Rapid Communications*, *32*(21), 1710–1715. https://doi.org/10.1002/marc.201100349.

Silva, T. B., Spulber, M., Kocik, M. K., Seidi, F., Charan, H., Rother, M., et al. (2013). Hemoglobin and red blood cells catalyze atom transfer radical polymerization. *Biomacromolecules*, *14*(8), 2703–2712. https://doi.org/10.1021/bm400556x.

Simakova, A., Mackenzie, M., Averick, S. E., Park, S., & Matyjaszewski, K. (2013). Bioinspired iron-based catalyst for atom transfer radical polymerization. *Angewandte Chemie, International Edition*, *52*(46), 12148–12151. https://doi.org/10.1002/anie.201306337.

Smolne, S., Buback, M., Demeshko, S., Matyjaszewski, K., Meyer, F., Schroeder, H., et al. (2016). Kinetics of Fe–mesohemin–(MPEG500)2-mediated RDRP in aqueous solution. *Macromolecules*, *49*(21), 8088–8097. https://doi.org/10.1021/acs.macromol.6b01774.

Sun, Y., Du, H., Lan, Y., Wang, W., Liang, Y., Feng, C., et al. (2016). Preparation of hemoglobin (Hb) imprinted polymer by Hb catalyzed eATRP and its application in biosensor. *Biosensors & Bioelectronics*, 77, 894–900. https://doi.org/10.1016/j.bios.2015.10.067.

Sun, Y., Zhang, J., Li, J., Zhao, M., & Liu, Y. (2017). Preparation of protein imprinted polymers via protein-catalyzed eATRP on 3D gold nanodendrites and their application in biosensors. *RSC Advances*, 7(45), 28461–28468. https://doi.org/10.1039/C7RA03772D.

Wang, X., Chen, S., Wu, D., Wu, Q., Wei, Q., He, B., et al. (2018). Oxidoreductase-initiated radical polymerizations to design hydrogels and micro/nanogels: Mechanism, molding, and applications. *Advanced Materials (Weinheim, Germany)*, *30*(17), 1705668. https://doi.org/10.1002/adma.201705668.

Wang, J.-S., & Matyjaszewski, K. (1995). Controlled/"living" radical polymerization. Atom transfer radical polymerization in the presence of transition-metal complexes. *Journal of the American Chemical Society*, *117*(20), 5614–5615. https://doi.org/10.1021/ja00125a035.

Yang, J., Lu, S., Xing, T., & Chen, G. (2018). Preparation, structure, and properties of silk fabric grafted with 2-hydroxypropyl methacrylate using the HRP biocatalyzed ATRP method. *Polymers*, *10*(5), 557. https://doi.org/10.3390/polym10050557.

Ye, J., & Narain, R. (2009). Water-assisted atom transfer radical polymerization of N-isopropylacrylamide: Nature of solvent and temperature. *The Journal of Physical Chemistry B*, *113*(3), 676–681. https://doi.org/10.1021/jp808905s.

Yin, Z., Zhang, J., Jiang, L.-P., & Zhu, J.-J. (2009). Thermosensitive behavior of poly (N-isopropylacrylamide) and release of incorporated hemoglobin. *The Journal of Physical Chemistry C*, *113*(36), 16104–16109. https://doi.org/10.1021/jp903589a.

Zhang, B., Wang, X., Zhu, A., Ma, K., Lv, Y., Wang, X., et al. (2015). Enzyme-initiated reversible addition–fragmentation chain transfer polymerization. *Macromolecules*, *48*(21), 7792–7802. https://doi.org/10.1021/acs.macromol.5b01893.

Zijlstra, W. G., & Buursma, A. (1997). Spectrophotometry of hemoglobin: Absorption spectra of bovine oxyhemoglobin, deoxyhemoglobin, carboxyhemoglobin, and methemoglobin. *Comparative Biochemistry and Physiology, Part B: Biochemistry & Molecular Biology*, *118*(4), 743–749. https://doi.org/10.1016/s0305-0491(97)00230-7.

Zoppe, J. O., Ataman, N. C., Mocny, P., Wang, J., Moraes, J., & Klok, H.-A. (2017). Surface-initiated controlled radical polymerization: State-of-the-art, opportunities, and challenges in surface and interface engineering with polymer brushes. *Chemical Reviews (Washington, DC, United States)*, *117*(3), 1105–1318. https://doi.org/10.1021/acs.chemrev.6b00314.

CHAPTER ELEVEN

Enzyme-initiated reversible addition – fragmentation chain transfer (RAFT) polymerization: Precision polymer synthesis via enzymatic catalysis

Xiao Wang[a], Zesheng An[b,*]

[a]Institute of Nanochemistry and Nanobiology, College of Environmental and Chemical Engineering, Shanghai University, Shanghai, China

[b]State Key Laboratory of Supramolecular Structure and Materials, College of Chemistry, Jilin University, Changchun, China

*Corresponding author: e-mail address: anzesheng@jlu.edu.cn

Contents

Methods in Enzymology, Volume 627
ISSN 0076-6879
https://doi.org/10.1016/bs.mie.2019.05.021

Abstract

Enzyme-initiated reversible addition–fragmentation chain transfer (RAFT) polymerization provides a sustainable strategy for efficient production of well-defined polymers under mild conditions. Horseradish peroxidase (HRP), a heme-containing metalloenzyme, catalyzes oxidation of acetylacetone (ACAC) by hydrogen peroxide (H_2O_2) to generate ACAC radicals, initiating polymerization of vinyl monomers. This HRP/H_2O_2/ACAC ternary initiating system is applied to RAFT polymerization of different types of vinyl monomers. Furthermore, to overcome the inherent limitation of necessity for oxygen-free conditions, another enzyme, glucose oxidase (GOx) or pyranose 2-oxidase (P2Ox), with excellent deoxygenation capability, is introduced to consume oxygen by catalyzing oxidation of glucose to generate H_2O_2. The generated H_2O_2 is directly supplied to HRP catalysis for radical generation. Both GOx-HRP and P2Ox-HRP cascade catalysis afford RAFT polymerization with oxygen tolerance. In this chapter, we mainly focus on detailed synthetic protocols of RAFT polymerizations initiated by HRP/H_2O_2/ACAC ternary initiating system and P2Ox-HRP cascade catalysis. The general characterization and analytical methods used in these enzyme-initiated RAFT polymerizations are also included.

1. Introduction

Reversible deactivation radical polymerization (RDRP) techniques provide general access to well-defined polymers with predetermined molecular weight (MW) and architecture (Leibfarth, Mattson, Fors, Collins, & Hawker, 2013; Moad, Rizzardo, & Thang, 2008; Rosen & Percec, 2009). Although thermal activation or initiation of RDRP has been much successful, there is an urgency to minimize energy consumption and to tolerate thermally sensitive species. Thus polymer synthesis with high efficiency at ambient conditions is highly desired. Moreover, oxygen tolerance is also appealing for radical polymerization synthesis exposed to air, which would eliminate the need for specific equipment and save time and effort needed for strict prior deoxygenation. Methods that can initiate RDRP at ambient conditions, particularly those can control RDRP even in the presence of oxygen, have attracted significant attention. Recent developments along this theme include photoinitiation, photoredox catalysis, and enzymatic catalysis (Guo & An, 2018; Rodriguez et al., 2018; Wang, Fu, & Matyjaszewski, 2018; Wang, Shen, & An, 2018; Yeow & Boyer, 2017).

Visible light mediated RDRP is highly attractive for its high efficiency, low energy consumption, and working at ambient conditions. Temporal and spatial control over polymerization can be achieved by manipulation of various parameters of incident light such as intensity, wavelength, and light switching on/off (Chen, Zhong, & Johnson, 2016; Dadashi-Silab,

Doran, & Yagci, 2016; Yeow & Boyer, 2017). Remarkably, sunlight has also been demonstrated for photo-induced RDRP processes, suggesting the great potential for sustainable production of well-defined polymers by harnessing the abundant daylight (Anastasaki et al., 2014; Konkolewicz, Schröder, Buback, Bernhard, & Matyjaszewski, 2012; Wang et al., 2016). Significantly, some visible light mediated RDRP processes have been demonstrated to be oxygen tolerant, realized by photoreduction of oxygen and/or the use of a sacrificial reducing agent (Xu, Jung, Atme, Shanmugam, & Boyer, 2014; Xu, Jung, & Boyer, 2014; Xu, Shanmugam, Duong, & Boyer, 2015; Yeow & Boyer, 2017).

Employing enzymatic catalysis for radical generation and deoxygenation in RDRP is another promising option for polymer synthesis (Guo & An, 2018; Rodriguez et al., 2018). Enzymatic catalysis in vivo is the most widespread process in nature to maintain diverse life activities in a mild, efficient, and selective manner (Torrelo, Hanefeld, & Hollmann, 2015). Biocatalytic processes employing single-enzyme or multi-enzyme cascade systems in vitro are also widely applied in organic synthesis (Reetz, 2013; Straathof, 2014; Torrelo et al., 2015). RDRP mediated by enzymatic catalysis for the preparation of precision polymers has emerged just over the past decade. In early studies, metalloproteins including HRP, hemoglobin, catalase, as well as laccase with active transition metal complexes were explored as atom transfer radical polymerization (ATRP) catalysts (Fodor et al., 2016; Ng, di Lena, & Chai, 2011a, 2011b; Sigg et al., 2011; Silva et al., 2013). Later, enzymes, especially HRP, catalyzing radical generation, were used to initiate RAFT polymerization (Danielson et al., 2016; Reyhani et al., 2018; Zhang et al., 2015) or activate ATRP catalyst (Enciso, Fu, Lathwal, et al., 2018; Enciso, Fu, Russell, & Matyjaszewski, 2018).

Moreover, oxidases that catalyze oxidation of substrates by consuming oxygen, including glucose oxidase (GOx) and pyranose 2-oxidase (P2Ox), have been employed to deoxygenate polymerization solution, thus endowing RDRP with oxygen tolerance (Chapman, Gormley, Herpoldt, & Stevens, 2014; Enciso, Fu, Lathwal, et al., 2018; Enciso, Fu, Russell, & Matyjaszewski, 2018; Liu, Lv, & An, 2017; Lv, Liu, Zhu, & An, 2017; Schneiderman et al., 2018; Wang, Fu, & Matyjaszewski, 2018; Wang, Shen, & An, 2018; Zhang et al., 2015). Cascade systems composed of enzymatic and/or organic catalysis were also explored to realize simultaneous oxygen removal and initiation for RDRP (Enciso, Fu, Lathwal, et al., 2018; Enciso, Fu, Russell, & Matyjaszewski, 2018; Liu et al., 2017; Lv et al., 2017; Reyhani et al., 2018; Zhang et al., 2015).

RAFT polymerization represents a robust and versatile synthetic pathway for its applicability to a wide range of monomers and excellent compatibility with various polymerization conditions (Chiefari et al., 1998). Compared to traditional free radical polymerization, a suitable chain transfer agent (CTA), such as dithiobenzoate, trithiocarbonate or xanthate, is simply added to lower the concentration of active radicals to attain a controlled radical polymerization (Warren & Armes, 2014). Initially, di Lena and coworkers reported a laccase-catalyzed radical polymerization using laccase/ascorbic acid/alkyl halide, but the control of the polymerization was poor and a reasonable degree of control was realized only when a CTA (2-cyano-2-propyl dithiobenzoate) was added (Ng et al., 2011a, 2011b). Although it seemed that RAFT played a major role in the polymerization control, the detailed mechanism was unclear since alkyl halide itself is a typical ATRP controlling agent. Enzyme-initiated RAFT polymerization has been recently developed by incorporating an HRP initiating system to the RAFT mechanism (Danielson et al., 2016; Zhang et al., 2015). The single-enzyme HRP/H_2O_2/ACAC ternary initiating system is well compatible with RAFT polymerization in aqueous buffer solution at pH close to neutral conditions. Bienzymatic GOx-HRP and P2Ox-HRP cascade catalysis was subsequently developed to simultaneously achieve high initiating efficiency and oxygen tolerance (Liu et al., 2017; Zhang et al., 2015). More recently, enzyme-mediated RAFT was also applied to heterogeneous conditions for polymerization-induced self-assembly (PISA) to provide block copolymer nanoparticles with various morphologies (Tan et al., 2017, 2018; Zhang et al., 2015). These and other cases will be discussed in more detail in this chapter to provide a primary profile of enzyme-catalyzed RAFT polymerization with an aim of widening its applicability and inspiring further development.

1.1 HRP-initiated RAFT polymerization

Peroxidase and oxidase are known to have remarkable utility in producing radical species for polymer synthesis (Fodor et al., 2016; Kobayashi & Makino, 2009). Ternary systems composed of an enzyme, an oxidant and a reducing substrate can provide a radical flux and initiate radical polymerization of vinyl monomers (Veitch, 2004). The HRP/H_2O_2/ACAC ternary initiating system is one of such systems that have been often used for the synthesis of radical polymers (Emery, Lalot, Brigodiot, & Maréchal, 1997; Kobayashi & Makino, 2009). On the basis of the enzyme

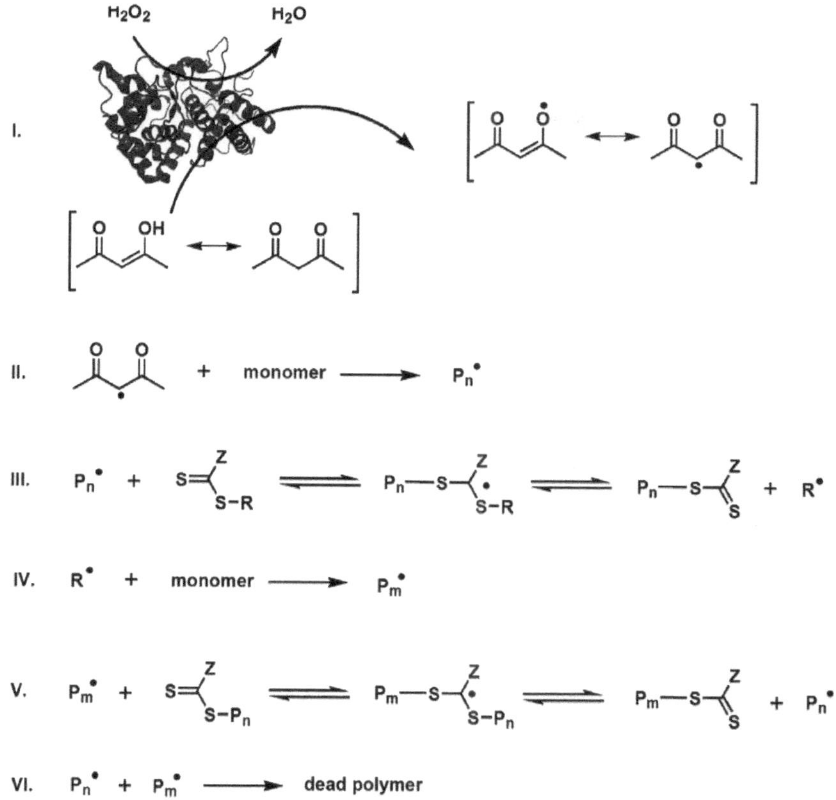

Fig. 1 Schematic representation of initiation and polymerization mechanism of HRP-initiated RAFT polymerization. *Reprinted with permission from Zhang, B., Wang, X., Zhu, A., Ma, K., Lv, Y., Wang, et al. (2015). Enzyme-initiated reversible addition–fragmentation chain transfer polymerization.* Macromolecules, *48(21), 7792–7802. Copyright (2015) American Chemical Society.*

initiating systems used in traditional radical polymerization, it was natural to apply them to initiate RAFT polymerization. The combination of enzyme catalysis and RAFT polymerization provides new a route for the synthesis of well-defined polymers under environmentally benign conditions. The mechanism of HRP-initiated RAFT polymerization is depicted in Fig. 1 (Zhang et al., 2015). In the presence of H_2O_2, HRP catalyzes oxidation of ACAC to generate two ACAC radicals (step I). Then the generated ACAC radicals initiate radical polymerization of vinyl monomers (step II). In the presence of a suitable CTA, the polymerization further proceeds according to the well-established RAFT mechanism (steps III – VI).

No detectable conversion was observed in control experiments in the absence of each one of the components, confirming that each component is indispensable for this ternary system (Zhang et al., 2015). The enzymatic process is believed to be confined to the first step but not involved in subsequent RAFT steps, making this enzymatic RAFT polymerization facile and well-controlled for various more and less activated monomers under both homogeneous and heterogeneous dispersion conditions. Besides buffer solution, biological milieu can also be used for this enzyme-initiated RAFT polymerization without any interfering effect.

Interestingly, HRP shows a remarkable versatility regarding substrate type and can efficiently catalyze different reactions under mild conditions (Kobayashi & Makino, 2009; Zavada, McHardy, & Scott, 2014; Zhang et al., 2016). More recently, HRP-mediate triple catalysis for initiating RAFT polymerization and subsequent postpolymerization modifications by thiol-ene and Diels-Alder reactions was demonstrated (Liu, Lv, Zhu, & An, 2018). Employing the promiscuity of an enzyme for precision polymer synthesis and functionalization via mild, efficient "bioclick" reactions is a new concept in polymer chemistry and is expected to be useful for design and synthesis of biomaterials.

1.2 Cascade catalysis for oxygen tolerant RAFT polymerization

Many in vivo biological processes require cascade catalysis of two or more enzymes. The product(s) produced by former enzyme catalysis can be directly used for latter enzyme catalysis as the substrate(s). Given the high efficiency and selectivity of enzymatic catalysis, complex transformations can be readily realized via a series of enzymatic cascade catalysis without intermediate interruption or purification (Lin, Wu, Huang, Ren, & Qu, 2015; van Oers, Rutjes, & van Hest, 2014; Vriezema et al., 2007). The concept of enzymatic catalysis has been adapted to polymer synthesis and has garnered intense interest.

Typically, RDRP synthesis requires careful deoxygenation prior to and during polymerization, since oxygen can either quench radicals or deteriorate polymerization control. This deoxygenation procedure severely limits the potential for large-scale polymer production via RDRP. Several strategies have been explored to overcome this oxygen-free constraint, including the use of sacrificial reductants or photocatalysts to eliminate oxygen dissolved in polymerization solution (Matyjaszewski, Coca, Gaynor, Wei, & Woodworth, 1998; Rosen & Percec, 2009; Shanmugam, Xu, & Boyer, 2015; Shen, Lu,

Zhu, Lv, & An, 2017; Wang et al., 2016; Xu et al., 2015; Xu, Jung, Atme, et al., 2014; Xu, Jung, & Boyer, 2014). Alternatively, enzyme-catalyzed deoxygenation, such as glucose oxidase (GOx) catalyzing oxidation of glucose by oxygen, has also been utilized to perform RDRP process without special degassing equipment (vacuum pump, inert gas etc.) (Chapman et al., 2014; Chapman, Gormley, Stenzel, & Stevens, 2016; Wang, Fu, & Matyjaszewski, 2018; Wang, Shen, & An, 2018; Xu et al., 2015). Notably, in the presence of HRP and ACAC, the H_2O_2 generated in GOx deoxygenation can be immediately utilized for following initiation according to the HRP initiating mechanism. In this regard, enzymatic cascade catalysis of GOx-HRP or P2Ox-HRP has been elegantly exploited to conduct RAFT polymerization without any external deoxygenation operation (Liu et al., 2017; Zhang et al., 2015).

1.2.1 GOx-HRP cascade catalysis

GOx has been widely explored for deoxygenation of radical polymerization due to its high efficiency. Low concentrations of GOx (~μM level) were reported to be able to sufficiently deoxygenate the system, allowing for RAFT polymerization to be conducted in open reaction vessels (Chapman et al., 2014). Moreover, the produced H_2O_2 can be used for subsequent reactions, allowing reaction cascade to be established for in situ radical generation. For example, hydroxyl radicals can be efficiently generated via the Fenton reaction (reaction between H_2O_2 and Fe^{3+}) or via a redox reaction between H_2O_2 and ascorbic acid (Vc). These two processes have been demonstrated to initiate RAFT polymerization in open reaction vessels (Lv et al., 2017; Reyhani et al., 2018; Shenoy, Tibbitt, Anseth, & Bowman, 2013). Enzymatic catalysis has been constructed by combining GOx deoxygenation and HRP initiation. In this cascade scheme, utilizing the H_2O_2 generated from GOx deoxygenation, HRP catalyzes radical generation according to the HRP/H_2O_2/ACAC initiating mechanism, which resulted in a high polymerization rate, high monomer conversion, and low MW dispersity for the polymerization of *N,N*-dimethylacrylamide (DMA) in open vessels (Zhang et al., 2015). This GOx-HRP enzymatic cascade has also been exploited to activate ATRP process with polymerization directly exposed to air (Enciso, Fu, Lathwal, et al., 2018; Enciso, Fu, Russell, & Matyjaszewski, 2018).

Additionally, GOx was also reported to be impressively robust in a wide range of water/organic solvent mixtures (Chapman et al., 2014; Schneiderman et al., 2018). Relatively high enzyme activity as well as high deoxygenation capability was retained even in up to 80% tert-butanol, acetonitrile, or in more complex aqueous solutions. Given the appealing characteristics including low cost, excellent compatibility with organic solvents, and good thermal stability (up to ~50 °C) (Lv et al., 2017; O'malley & Ulmer, 1973), GOx catalysis may be potentially applied to other synthetic systems with complex conditions and hopefully for large-scale polymer production.

1.2.2 P2Ox-HRP cascade catalysis

Pyranose 2-oxidase (P2Ox) oxidizes a number of monosaccharides, including glucose, by oxygen to yield H_2O_2 (Giffhorn, 2000). Similar to GOx-HRP, P2Ox-HRP cascade catalysis was also developed to conduct RAFT polymerization with oxygen tolerance (Liu et al., 2017). Different from GOx, P2Ox consumes oxygen by catalyzing oxidation of D-glucose on carbon-2 to generate 2-keto-D-glucose and H_2O_2 (Giffhorn, 2000). Interestingly, root fungi make use of a P2Ox-peroxidase cascade to decompose lignin to supply nutrient for their growth. Inspired by this natural enzymatic cascade, a P2Ox-HRP cascade was devised to conduct oxygen-tolerant RAFT polymerization, in which the H_2O_2 generated by P2Ox deoxygenation is supplied to HRP-catalyzed reaction to produce ACAC radicals, and thus initiating RAFT polymerization (Fig. 2).

Remarkably, well-defined multiblock copolymers (up to 10 blocks) with MWs of 68–97 kg/mol in open vessels directly exposed to air were successfully prepared via iterative polymerizations with near-quantitative

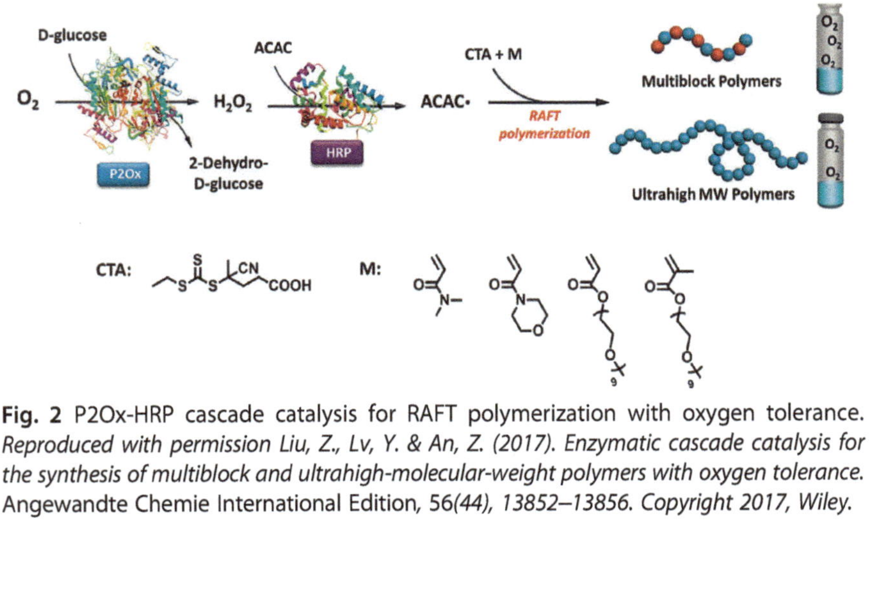

Fig. 2 P2Ox-HRP cascade catalysis for RAFT polymerization with oxygen tolerance. *Reproduced with permission Liu, Z., Lv, Y. & An, Z. (2017). Enzymatic cascade catalysis for the synthesis of multiblock and ultrahigh-molecular-weight polymers with oxygen tolerance.* Angewandte Chemie International Edition, *56(44), 13852–13856. Copyright 2017, Wiley.*

conversion for each step. Thanks to the high polymerization rate and the one-pot synthetic strategy, such high-order block copolymers could be obtained within as short as 10.5 h. Furthermore, ultrahigh-molecular-weight (UHMW) polymers (up to $2.3 \times 10^6 \, g \, mol^{-1}$) in closed vessels without extra degassing were prepared via RAFT synthesis mediated by P2Ox-HRP cascade catalysis. The successful preparation of these difficult-to-access polymers highlights the power and efficiency of the P2Ox-HRP enzymatic cascade, which also proves that the merger of biochemistry and polymer chemistry indeed is a very fruitful direction that fully deserves more research efforts from both areas.

1.3 Enzymatic catalysis for RAFT-PISA

Polymerization-induced self-assembly (PISA) via RDRP has been recognized as a powerful and versatile approach for efficient synthesis of nano-objects with controlled morphologies (Wang, Fu, & Matyjaszewski, 2018; Wang, Shen, & An, 2018; Warren & Armes, 2014). Currently, most PISA syntheses are conducted via RAFT dispersion polymerization for its excellent accessibility to higher order morphologies compared to emulsion formulations (Canning, Smith, & Armes, 2016; Jennifer et al., 2016; Wang & An, 2019; Warren & Armes, 2014; Zhang, D'Agosto, Boyron, Rieger, & Charleux, 2012). The first reported case of enzyme-initiated RAFT-PISA formulation employed HRP/H_2O_2/ACAC initiating system using poly(*N,N*-dimethylacrylamide) (PDMA) and poly(poly(ethylene glycol) methyl ether methacrylate) (PPEGMA) as stabilizers for dispersion polymerization of 2-methoxyethylacrylate (MEA) at 30 °C (Zhang et al., 2015). Nanospheres of PDMA-*b*-PMEA and PPEGMA-*b*-PMEA block copolymers with predetermined MWs and low dispersities were obtained. Fast polymerization was achieved at a low HRP concentration (~0.4 μM), suggesting the high efficiency of HRP initiation under dispersion polymerization conditions.

Later, Tan and coworkers combined GOx deoxygenation and photo-initiation and conducted oxygen-tolerant PISA to prepare 2-hydroxypropyl methacrylate (HPMA) based nano-objects with various morphologies at room temperature (Tan et al., 2017). Moreover, the oxygen-tolerance nature allowed for high-throughput low-volume synthesis using microtiter plates, benefiting effective construction of phase diagrams for PISA formulations. More recently, further advances were made on PISA syntheses via enzymatic cascade catalysis for simultaneous deoxygenation and initiation. GOx-HRP cascade catalysis was also adapted to RAFT dispersion PISA to prepare polymorphic block copolymer nanoparticles (Tan et al., 2018). In situ encapsulation of inorganic nanoparticles and even active proteins were demonstrated via enzyme mediated PISA under mild conditions. Given the compatibility with biological media, enzyme catalyzed PISA may offer a mild and efficient platform for the synthesis of nanoparticles with potential value in bio-related applications.

Herein, we focus on polymer synthesis via RAFT polymerization using HRP/H_2O_2/ACAC initiating system and P2Ox-HRP cascade catalysis. pH and temperature are important synthetic parameters for these enzyme-mediated RAFT formulations. General characterization and analysis methods involved are also included.

2. Protocols for RAFT polymerization using HRP/H_2O_2/ACAC initiating system

In this section, detailed protocols for synthesis of homopolymers and multiblock copolymers employing HRP/H_2O_2/ACAC ternary initiating system are described. Other synthesis cases similarly applying this HRP-initiating system to various monomers, in heterogeneous dispersion polymerization and in fetal bovine serum (FBS) are also covered briefly.

2.1 Molar ratio of initiating system

HRP acts on both H_2O_2 and ACAC as its substrates. In order to achieve a good RAFT control, H_2O_2 is used as the limiting reagent to maintain a low concentration of radicals with a molar ratio of $[H_2O_2]$:[CTA] <1. ACAC is typically used in excess to accelerate the rate of enzymatic catalysis. A molar ratio of [CTA]:[HRP]:$[H_2O_2]$:[ACAC] being 1:0.00054:0.2:4 was reported for the RAFT syntheses to afford well-defined polymers with a high polymerization rate and a low MW dispersity (Zhang et al., 2015). Relatively low HRP concentrations (even down to 10^{-8} M) are sufficient for efficient initiation. The concentration of the whole ternary initiating system as well as the molar ratio of the components can be manipulated to tune the efficiency of radical generation and thus adjust the polymerization rate.

2.2 Solution pH

Suitable pH of the solution is also an important concern for this HRP initiating system. As with most enzymes, the stability and activity of HRP are closely related to solution pH. Moreover, the enol tautomer of ACAC is proposed to be the actual substrate for radical generation (step I in Fig. 1), and the tautomerization toward enol against keto is promoted at higher pH (Teixeira, Lalot, Brigodiot, & Maréchal, 1999). Therefore, the initiation efficiency is also highly dependent on pH in this ternary system. And for a RAFT process, basic solution should also be avoided, because RAFT polymer chain-ends are susceptible to hydrolysis in aqueous media especially above pH 7 (McCormick & Lowe, 2004; Warren & Armes, 2014). Investigations on HRP-initiated RAFT polymerizations of DMA in buffer solutions at pH 3.5–7 indicated that both high conversions (>90% within 2 h) and narrow MW distributions ($Đ < 1.15$) were observed at pH 6–7 (Zhang et al., 2015). This suggests high initiation efficiency as well as good RAFT control are attainable under these conditions.

2.3 Homopolymer synthesis

The synthesis of poly(*N,N*-dimethylacrylamide) (PDMA) with a target degree of polymerization (DP) of 200 in PBS (pH 7) is taken as an example to illustrate the general procedure of HRP-initiated RAFT homopolymerization.

2.3.1 Materials

Phosphate buffer solution (PBS, pH 7, 20 mM).
Nitrogen.
Horseradish peroxidase (HRP, ≥100 U/mg).
Hydrogen peroxide (H_2O_2, 30 wt% aqueous solution).
Acetylacetone (ACAC).
N,N-dimethylacrylamide (DMA, passed through a Al_2O_3 column to remove the inhibitor prior to polymerization).
4-Cyano-4-(ethylsulfanylthiocarbonyl)sulfanylpentanoic acid (CTA1 in Fig. 7, synthesized as previously reported (Thang, Chong, Mayadunne, Moad, & Rizzardo, 1999)).
2,6-Di-tert-butyl-4-methylphenol (BHT), used as a polymerization quenching agent.

2.3.2 Polymerization procedure

The target $PDMA_{200}$ was synthesized at a molar ratio of [DMA]:[CTA1]:[HRP]:[H_2O_2]:[ACAC] = 200:1:0.00054:0.2:4.

(1) HRP, ACAC and H_2O_2 stock solutions were prepared in PBS (pH 7) prior to polymerization. The H_2O_2 stock solution was degassed with nitrogen for 30 min.

(2) CTA1 (0.0039 g, 0.015 mmol), DMA (0.2961 g, 2.99 mmol), an HRP stock solution (158 μL, 0.051 mM), and an ACAC stock solution (141 μL, 0.43 M) were added into a flask.
(3) PBS (pH 7, 2.61 mL) was added to make the final monomer concentration [M] 10% w/v.
(4) The flask was sealed, followed by degassing with nitrogen for 30 min, and then immersed into a preheated oil bath at 30 °C.
(5) After the temperature was stabilized, a degassed H_2O_2 (50 μL, 0.06 M) solution was injected by a microsyringe into the solution to initiate the polymerization.
(6) Finally, the sealing rubber plug was removed and the polymerization was quenched by addition of a small amount of BHT.

2.3.3 UHMW polymer synthesis

Synthesis of polymers with an ultrahigh MW was also conducted employing this synthesis protocol. When targeting a higher DP, the polymerization time was prolonged (to 4–30 h) to reach a high monomer conversion. Alternatively, the concentration of H_2O_2 could also be increased reasonably to accelerate the polymerization rate.

2.3.4 Block copolymer synthesis

Block copolymers were also accessible via sequential HRP-initiated RAFT polymerization without any intermediate purification procedures. This also validated the high end-group fidelity and "living" characteristics of this HRP-initiated RAFT process. In specific, the molar ratio of [CTA]:[HRP]:[H_2O_2]:[ACAC] was maintained as 200:1:0.00054:0.2:4, while the concentration of total monomers was maintained at 10% w/v. Therefore, after addition of monomers for subsequent blocks, a same amount of HRP, ACAC and H_2O_2 as that for the first block were supplied.

2.3.5 Compatibility with biological media

This HRP/H_2O_2/ACAC ternary initiating system was also applicable in biological media, such as fetal bovine serum (FBS). For similar polymerizations conducted in FBS, the phosphate buffer should be replaced with FBS.

2.4 Polymerization kinetic study

Kinetic studies of HRP-initiated RAFT polymerization were carried out to characterize the level of RAFT control and the molecular parameters of the obtained polymers including homopolymers, block copolymers and UHMW polymers. For example, during HRP-initiated RAFT synthesis targeting $PDMA_{200}$ mentioned in Section 2.3.2, samples were carefully withdrawn at pre-determined time intervals and used for proton nuclear magnetic resonance (^{1}H NMR) and gel permeation chromatography (GPC) tests.

2.4.1 ^{1}H NMR spectroscopy

Samples (~50 μL) were quenched by adding a small amount of BHT and dissolved in D_2O (0.5 mL). ^{1}H NMR spectra were performed on a JEOL resonance ECZ400S 400 MHz spectrometer. ^{1}H NMR analysis indicated that near-quantitative conversion ($\alpha > 95\%$) was achieved within 1 h.

The $\ln([M]_0/[M])$ values were calculated according to:

$$\ln([M]_0/[M]) = \ln\left(\frac{[M]_0}{[M]_0 - \Delta[M]}\right) = \ln\left(\frac{[M]_0/[M]_0}{[M]_0/[M]_0 - \Delta[M]/[M]_0}\right) = \ln\left(\frac{1}{1-\alpha}\right)$$

As shown in Fig. 3A, a kinetic plot of $\ln([M]_0/[M])$ vs polymerization time indicates excellent linear polymerization kinetics without an obvious induction period. This is consistent with a pseudo-living polymerization, indicating that the polymerization proceeded with good RAFT control.

2.4.2 GPC measurement

DMF (HPLC, containing LiBr 1.75 mg/mL) was used as the eluent (0.7 mL/min) for GPC measurements. Samples (~50 μL) quenched by a small amount of BHT were directly dried at 60 °C for 12 h and then dissolved in abovementioned DMF (1 mL). GPC tests were conducted on a Waters Alliance e2695 instrument with a refractive index (RI) detector (e2414, Waters Alliance) and a light scattering (LS) detector (90° and 7° angles) (TDA 305 Malvern Instruments Ltd., UK). Two columns (D2500 and D5000, Malvern Instruments) were used for separation. PMMA standards were used for the molecular weight calibration. MW and dispersity ($Đ = M_w/M_n$) of obtained polymers were reported by an OMNISEC V10 software.

GPC analysis of samples with different conversions provides molecular parameters of the synthesized polymers, including MW and dispersity. Symmetric and monomodal GPC traces with negligible tailing (Fig. 3C) suggest no obvious premature termination by a quencher, such as oxygen. And the linear evolution of MW and gradually narrowed dispersity ($Đ < 1.2$) (Fig. 3B) indicates all the characteristics of a well-controlled RAFT process

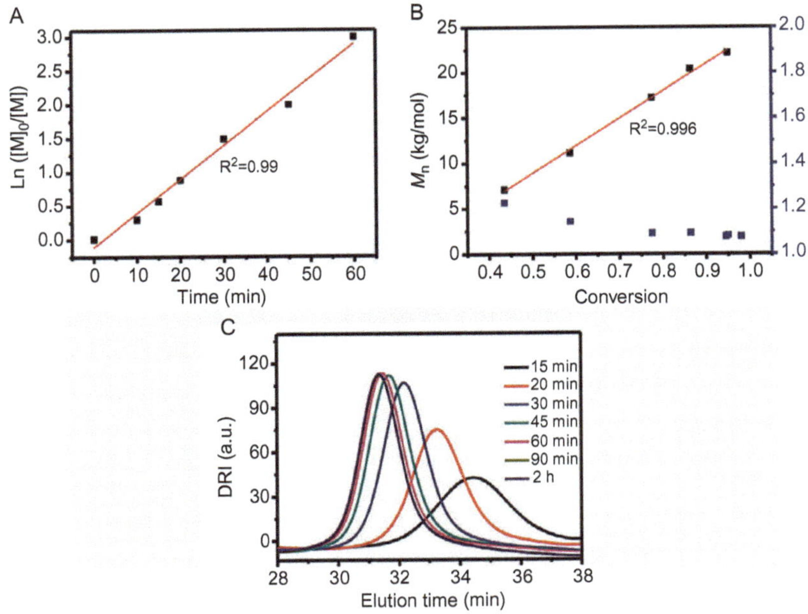

Fig. 3 Polymerization kinetics of HRP-initiated RAFT polymerization of DMA. ([DMA]:[CTA1]:[HRP]:[H_2O_2]:[ACAC] = 200:1:0.00054:0.2:4, [DMA] = 10% w/v, buffer solution (pH 7), 30 °C) (A) Pseudo-first-order kinetic plot of ln([M]$_0$/[M]) vs. time. (B) Evolution of MW and dispersity with conversion. (C) GPC traces (RI-GPC, DMF). *Reprinted with permission from Zhang, B., Wang, X., Zhu, A., Ma, K., Lv, Y., Wang, et al. (2015). Enzyme-initiated reversible addition–fragmentation chain transfer polymerization.* Macromolecules, *48(21), 7792–7802. Copyright (2015) American Chemical Society.*

were retained for HRP-initiated polymerization under these conditions. Therefore, the livingness of this HRP-initiated RAFT polymerization was confirmed by pseudo-first-order kinetics, predictable MW and low MW dispersity.

2.5 End-group fidelity

Structural analyses via ^{1}H NMR, MALDI-ToF, and UV–vis absorption were performed to confirm the end-group fidelity of polymers synthesized by HRP-initiated RAFT polymerization. A $PDMA_{32}$ was synthesized by HRP-initiated RAFT polymerization targeting a DP of 50, which was stopped at a conversion of ~63%. The theoretical DP was given as follows:

$$DP_{th} = \text{target DP} \times \text{conversion} = 50 \times 63\% \approx 32$$

The polymer was dialyzed against deionized water and then dried via freeze-drying for further structural analysis.

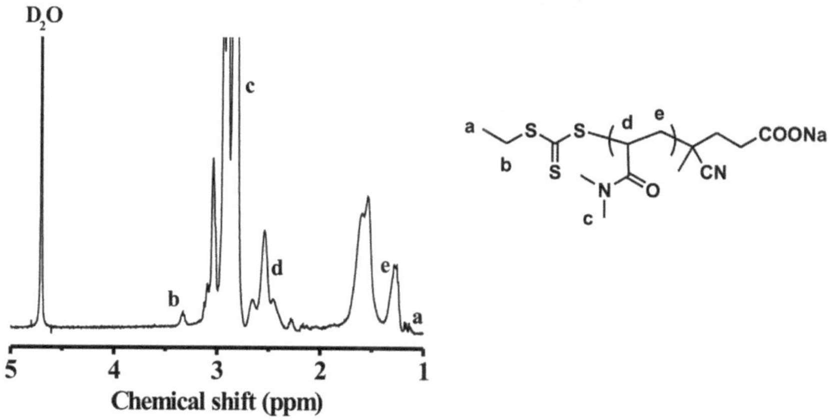

Fig. 4 ^{1}H NMR spectrum ($CDCl_3$) of $PDMA_{32}$ synthesized by HRP-initiated RAFT polymerization. *Reprinted with permission from Zhang, B., Wang, X., Zhu, A., Ma, K., Lv, Y., Wang, et al. (2015). Enzyme-initiated reversible addition–fragmentation chain transfer polymerization.* Macromolecules, *48(21), 7792–7802. Copyright (2015) American Chemical Society.*

2.5.1 ^{1}H NMR spectrum

As shown in Fig. 4, signals for CH_3– (a) and –CH_2– (b) of the *Z*-group of CTA1 are observed. This indicates that the CTA end-group was retained for the synthesized PDMA. And a DP_{NMR} can be calculated by analyzing the integral signals according to:

$$DP_{NMR} = \frac{(c+d)\times 2}{7\times b} \approx 33$$

2.5.2 MALDI-ToF spectrum

Matrix-assisted laser desorption/ionization-time of flight (MALDI-ToF) mass spectrum of purified $PDMA_{32}$ was recorded using an AB SCIEX 5800. α-Cyano-4-hydroxycinnamic acid was used as the matrix. As shown in Fig. 5, the major mass peaks are separated by one DMA unit ($m/z = 99$). And the observed mass agrees well with the calculated value of $PDMA_{32}$. This confirms that the exact structure of $PDMA_{32}$ backbone with the expected end-group were present.

2.5.3 UV–vis absorption spectrum

UV–vis spectroscopy of purified $PDMA_{32}$ in deionized water was performed using a Hitachi U-3010 UV–vis spectrometer. A strong absorption centered at 307 nm corresponds to the trithiocarbonate of CTA end-group (Fig. 6).

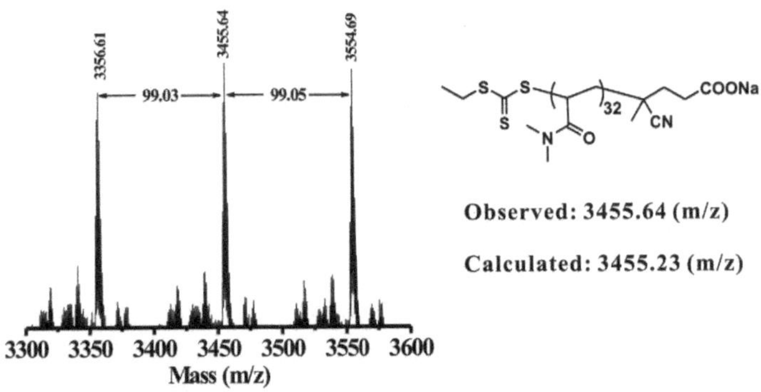

Fig. 5 MALDI-ToF mass spectrum of $PDMA_{32}$ synthesized by HRP-initiated RAFT polymerization. *Adapted with permission from Zhang, B., Wang, X., Zhu, A., Ma, K., Lv, Y., Wang, et al. (2015). Enzyme-initiated reversible addition–fragmentation chain transfer polymerization.* Macromolecules, *48(21), 7792–7802. Copyright (2015) American Chemical Society.*

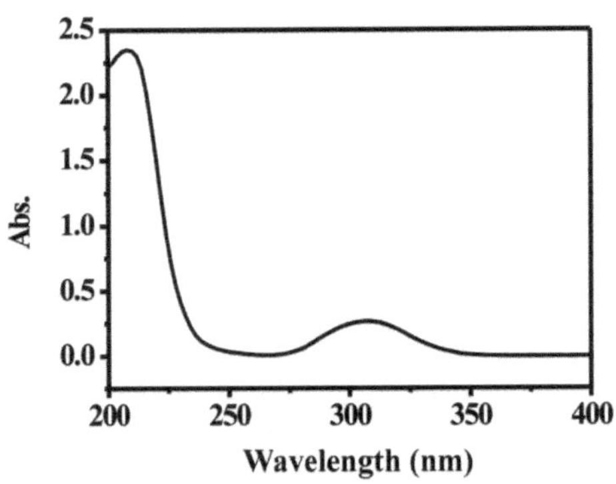

Fig. 6 UV–Vis absorption spectrum of $PDMA_{32}$ in water synthesized by HRP-initiated RAFT polymerization. *Reprinted with permission from Zhang, B., Wang, X., Zhu, A., Ma, K., Lv, Y., Wang, et al. (2015). Enzyme-initiated reversible addition–fragmentation chain transfer polymerization.* Macromolecules, *48(21), 7792–7802. Copyright (2015) American Chemical Society.*

And the UV–vis absorption spectrum of the corresponding CTA provides the reference for quantification of the thiocarbonyl end groups retained in polymers synthesized by RAFT (Skrabania, Miasnikova, Bivigou-Koumba, Zehm, & Laschewsky, 2011).

2.6 Monomers and CTA selection

This synthesis protocol was also applied to other monomers, including more activated monomers (MAMs) and less activated monomers (LAMs). Suitable

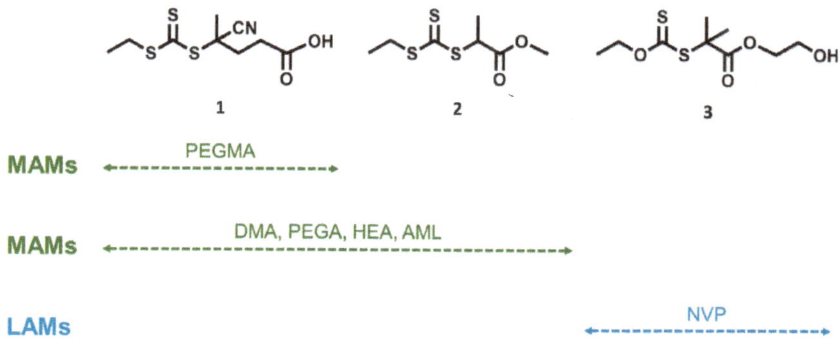

Fig. 7 Exemplary CTAs for RAFT polymerization of more activated monomers (MAMs) and less activated monomers (LAMs). Abbreviations of monomers used for HRP-initiated RAFT polymerization: PEG(M)A = polyethylene glycol methyl ether (meth)acrylate, DMA = *N,N*-dimethylacrylamide, HEA = 2-hydroxyethylacrylate, NVP = *N*-vinylpyrolidone and AML = 4-acryloylmorpholine.

CTAs should be chosen to endow RAFT control for each monomer (Fig. 7) (Keddie, Moad, Rizzardo, & Thang, 2012; Perrier, 2017). Excellent RAFT control was achieved for MAMs including DMA, 2-hydroxyethylacrylate (HEA), polyethylene glycol methyl ether (meth)acrylate (PEG(M)A) as well as for a LAM *N*-vinylpyrolidone (NVP) (Zhang et al., 2015). This suggests the excellent versatility of this HRP/H_2O_2/ACAC ternary initiating system.

2.7 Heterogeneous dispersion polymerization

Heterogeneous dispersion polymerization was also conducted using this HRP-initiated RAFT system. A hydrophilic polymer bearing a RAFT end-group was used as a macromolecular chain transfer agent (macro-CTA), instead of a CTA in the abovementioned protocol. HRP-initiated RAFT-PISA via dispersion polymerization of MEA using PDMA and PPEGMA as stabilizers for at 30 °C was demonstrated (Zhang et al., 2015). PDMA-PMEA and PPEGMA-PMEA block copolymers with predetermined MW and low dispersity were obtained. The morphology and size of resulting nanoparticles were characterized by transmission electron microscopy (TEM) and dynamic light scattering (DLS).

2.7.1 TEM characterization

The obtained nanoparticle dispersions were diluted to ~0.5% w/v solids. The diluted dispersions were dropped on carbon-coated copper grids followed by drying under vacuum at 40 °C for 12 h. The number-average

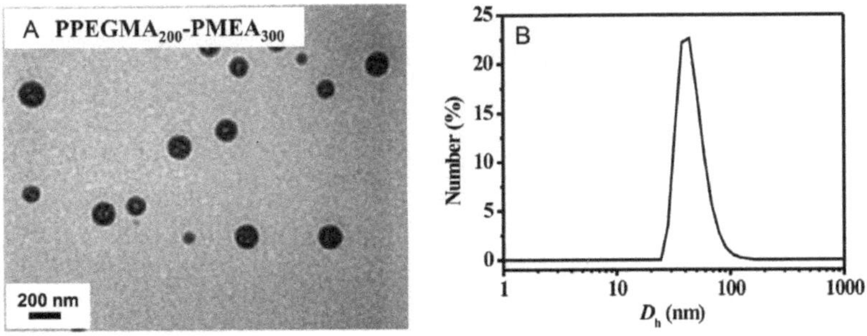

Fig. 8 HRP-initiated RAFT dispersion polymerization of MEA using $PPEGMA_{200}$ macro-CTA: [macro-CTA]:[HRP]:[H_2O_2]:[ACAC] = 1:0.00054:0.2:4, 10% *w/v* solids, PBS (pH 7), 30 °C. (A) TEM micrograph and (B) DLS result of $PPEGMA_{200}$-b-$PMEA_{300}$ nanoparticles. *Adapted with permission from Zhang, B., Wang, X., Zhu, A., Ma, K., Lv, Y., Wang, et al. (2015). Enzyme-initiated reversible addition–fragmentation chain transfer polymerization.* Macromolecules, *48(21), 7792–7802. Copyright (2015) American Chemical Society.*

diameter (D_n) can be determined by measurement of particles in the TEM image (Xiao, Sun, & Du, 2017; Zhang et al., 2017). For example, $PPEGMA_{200}$-$PMEA_{300}$ nanospheres (126 ± 31 nm) were observed (Fig. 8A).

2.7.2 DLS analysis

For example, the obtained $PPEGMA_{200}$-$PMEA_{300}$ dispersions were diluted to ~0.1% w/v solids for DLS test with a Malvern ZS90 sizer. The hydrodynamic diameter (D_h) (80 nm) and polydispersity (PDI) were provided by the Malvern zetasizer software (Fig. 8B).

3. Protocols for RAFT polymerization using P2Ox-HRP cascade catalysis

In this section, we mainly focus on the synthesis of multiblock copolymers and UHMW polymers via RAFT polymerization mediated by P2Ox-HRP cascade catalysis.

3.1 Decablock copolymer synthesis

The synthesis of decablock $PDMA_{150}$-$(PDMA_{50})_9$ is taken as a representative example to illustrate the general protocol for the synthesis of multiblock copolymers via P2Ox-HRP catalyzed RAFT polymerization in open vessels.

3.1.1 Materials

Phosphate buffer solution (PBS, pH 7, 20 mM).
Horseradish peroxidase (HRP, ≥100 U/mg).
Pyranose 2-oxidase (P2Ox, 1229 U/mL).
D-(+)-glucose (also termed as glucose).
Acetylacetone (ACAC).
N,N-dimethylacrylamide (DMA, passed through a Al_2O_3 column to remove the inhibitor prior to polymerization).
4-Cyano-4-(ethylsulfanylthiocarbonyl)sulfanylpentanoic acid (CTA1 in Fig. 7, synthesized as previously reported (Thang et al., 1999))
BHT used as a quenching agent.

3.1.2 Polymerization procedure

The detailed feed ratio and characterization data is shown in Table 1.

(1) First block

The first block was synthesized under the following conditions. [CTA1]:[DMA]:[HRP]:[ACAC] = 1:150:0.00054:2, [P2Ox] = 2.7 U/mL, [glucose] = 80 mM, monomer concentration [DMA] = 10% w/v.

- **(i)** CTA1 (0.0087 g, 0.033 mmol), DMA (0.4913 g, 4.956 mmol), ACAC (6.6 mg, 0.066 mmol), glucose (0.0721 g, 0.40 mmol) and PBS (5 mL) were added to a vessel under stirring with a speed of 100 rpm at 30 °C.
- **(ii)** P2Ox (11 μL, 13.5 U) was then added to deoxygenate the above solution for 10 min.
- **(iii)** HRP stock solution (26 μL, 0.0178 μmol, 0.77 mg) was added to initiate the polymerization in an open vessel for 1.5 h.

(2) Subsequent blocking cycles

- **(i)** DMA (0.1637 g, 1.651 mmol), ACAC (0.66 mg, 0.0066 mmol), glucose (0.0144 g, 0.08 mmol), P2Ox (2.2 μL, 2.7 U) were dissolved in PBS (1 mL) and then transferred immediately into the polymerization vial with stirring (100 ppm).
- **(ii)** HRP stock solution (22 μL, 0.0036 μmol, 0.154 mg) was added to initiate the polymerization in an open vessel, which was then kept stirring for 1 h.

(3) Sampling and quenching

At the end of each block synthesis, an aliquot was sampled and quenched by addition of a small amount of BHT for 1H NMR and GPC analysis.

Table 1 Summary for the synthesis of $PDMA_{150}$-$(PDMA_{50})_9$ decablock polymer in an open vessel.

Cycle	1	2	3	4	5	6	7	8	9	10
Monomer	DMA	DMA	DMA	DMA	DMA	DMA	DMA	DMA	DMA	DMA
$DP_{targeted}$	150	50	50	50	50	50	50	50	50	50
$m_{CTA\ added}$ (mg)	8.7									
$m_{monomer\ added}$ (mg)	491.3	163.7	163.7	163.7	163.7	163.7	163.7	163.7	163.7	163.7
$m_{glucose\ added}$ (mg)	72.1	14.4	14.4	14.4	14.4	14.4	14.4	14.4	14.4	14.4
$m_{ACAC\ added}$ (mg)	6.6	0.66	0.66	0.66	0.66	0.66	0.66	0.66	0.66	0.66
$V_{P2Ox\ added}$ (μL)	11	2.2	2.2	2.2	2.2	2.2	2.2	2.2	2.2	2.2
$m_{HRP\ added}$ (mg)	0.77	0.154	0.154	0.154	0.154	0.154	0.154	0.154	0.154	0.154
$V_{PBS\ added}$ (mL)	5	1	1	1	1	1	1	1	1	1
[M] (%*w*/*v*)	10.0	11.1	11.8	12.4	12.8	13.2	13.5	13.7	13.9	14.1
Time (h)	1.5	1	1	1	1	1	1	1	1	1
Conv.(%)[a]	95	92	94	95	97	99	98	98	97	96
$M_{n,th}$ (kg/mol)[b]	14.1	18.7	23.3	28.0	32.8	37.7	42.6	47.5	52.3	57.0
$M_{n,GPC}$ (kg/mol)[c]	13.6	18.1	23.7	29.3	36.2	42.1	48.1	55.2	60.3	62.1
Đ[c]	1.21	1.24	1.27	1.31	1.28	1.30	1.32	1.34	1.38	1.46

[a]Monomer conversion determined by ^1H NMR in D_2O.
[b]Molecular weights calculated by conversion.
[c]Molecular parameters determined by GPC (DMF, PMMA standard).
Reprinted with permission Liu, Z., Lv, Y. & An, Z. (2017). Enzymatic cascade catalysis for the synthesis of multiblock and ultrahigh-molecular-weight polymers with oxygen tolerance. *Angewandte Chemie International Edition*, *56*(44), 13852–13856. Copyright 2017, Wiley.

3.1.3 Decablock copolymer analysis

Well-defined decablock (co)polymers of acrylamides, DMA and/or 4-acryloylmorpholine (AML) with MWs of 68–97 kg/mol were successfully prepared by applying this RAFT polymerization mediated by P2Ox-HRP cascade catalysis via iterative polymerizations with near-quantitative conversion for each step, in open vessels (Liu et al., 2017). For example, the ^{1}H NMR spectra in Fig. 9 indicates near-quantitative monomer conversion (>95%) for each block within 60–90 min. The symmetric GPC traces in Fig. 9 with negligible tailing at low MWs show a gradual increase in MW and narrow dispersities (Đ < 1.5), thus suggesting good RAFT control and high blocking efficiencies. Thanks to the high polymerization rate and the one-pot synthetic procedure, such high-order block copolymers can be obtained within as short as 10.5 h.

3.1.4 Suitable DPs for the first block

It should be noted that a first block with a DP of 100 or 150 was typically chosen in this work. The DP of the first block and subsequent blocks can certainly be flexible. However, when targeting low DPs (50–100), the polymerization rate was lower than that for targeting larger DPs (200–500) (Liu et al., 2017; Zhang et al., 2015). This is presumably attributed to inhibited enzyme activity by the higher concentrations of CTA at lower target DPs. High monomer conversion could also be achieved if longer polymerization times and/or higher concentrations of ACAC were used.

3.2 UHMW polymer synthesis

Herein, an exemplary synthesis of PDMA targeting DP 10000 is given to illustrate the general procedure for the synthesis of UHMW polymers via P2Ox-HRP catalyzed RAFT polymerization in sealed vessels without prior degassing.

3.2.1 Polymerization procedure

Compared with the synthesis of low-molecular-weight homopolymers, the synthesis of UHMW polymers calls for more thorough deoxygenation, so glucose concentration was increased from 80 to 100 mM, and the molar ratio of [CTA1]:[HRP]:[ACAC] was set at 1:0.00216:16, [P2Ox] = 2.7 U/mL, [DMA] = 30% w/v.

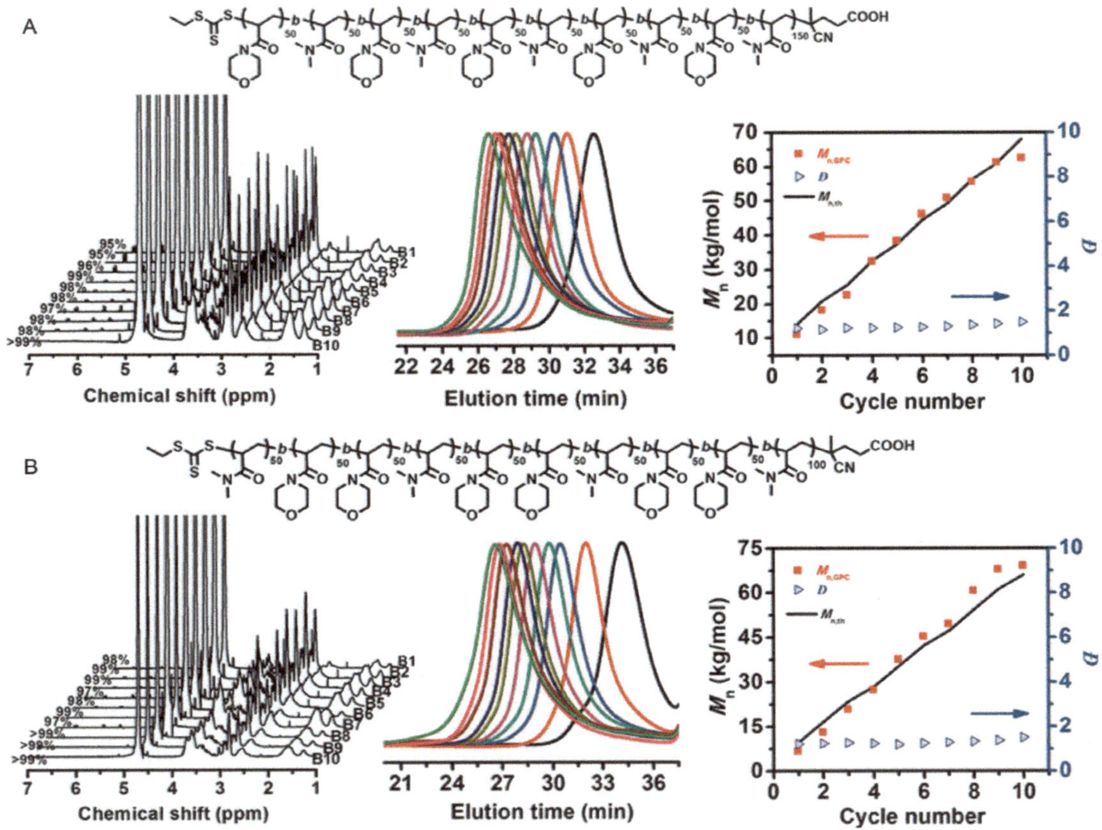

Fig. 9 ^{1}H NMR spectra (D_2O), GPC traces (RI-GPC, DMF, PMMA standards), evolution of MW and dispersity with blocking cycle for the synthesis in PBS (pH 7) of decablock copolymers, $PDMA_{150}$-($PAML_{50}$-$PDMA_{50}$)$_4$-$PAML_{50}$ (A) and $PDMA_{100}$-($PAML_{50}$-$PAML_{50}$-$PDMA_{50}$)$_3$ (B), via RAFT polymerization mediated by P2Ox-HRP in open vessels. *Reprinted with permission Liu, Z., Lv, Y. & An, Z. (2017). Enzymatic cascade catalysis for the synthesis of multiblock and ultrahigh-molecular-weight polymers with oxygen tolerance.* Angewandte Chemie International Edition, *56(44), 13852–13856. Copyright 2017, Wiley.*

(1) Given the poor solubility of CTA1 in PBS, CTA1 stock solution was prepared beforehand using DMSO.
(2) A CTA1 stock solution (24 μL, 1.518 μmol, 0.3998 mg), DMA (1.4988 g, 0.0151 mol), ACAC (0.0024 g, 24.17 μmol), glucose (0.0901 g, 0.5 mmol), and PBS (5 mL) were added to a vessel at a stirring of 100 rpm at 30 °C.
(3) A P2Ox stock solution (11 μL, 13.5 U) was added to deoxygenate the solution for 10 min.
(4) HRP (28 μL, 0.0033 μmol, 0.1426 mg) was added to the reaction solution, followed by immediate sealing of the polymerization vial.
(5) After polymerization for 5 h, a small amount of BHT was added to quench the polymerization.

3.2.2 UHMW PDMAs analysis

UHMW PDMAs (up to 2.3×10^6 g mol^{-1}) were readily prepared via RAFT synthesis mediated by P2Ox-HRP cascade catalysis in closed vessels without extra degassing (Liu et al., 2017). ^{1}H NMR analysis suggested that high conversions (>90%) were achieved in 5 h. Kinetic study in Fig. 10A for a polymerization of DMA with a target DP of 10,000 in a sealed vessel without prior degassing indicates a well-controlled pseudo-first-order process. Monomodal GPC traces in Fig. 10B show relatively low dispersities (Đ < 1.4) and the measured MWs determined by triple-detection GPC were close to theoretical values.

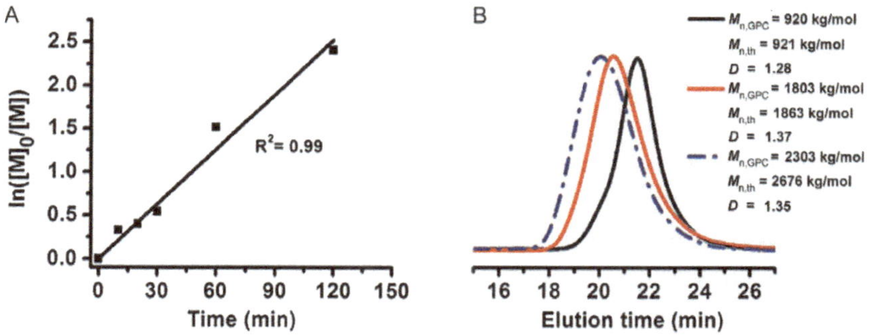

Fig. 10 (A) Kinetic plot for polymerization a targeting DP = 10,000 in a sealed vessel without prior degassing: [CTA]/[HRP]/[ACAC] = 1:0.00216:16, [glucose] = 100 mM, [P2Ox] = 2.7 U/mL, [M] = 30% w/v, PBS (pH 7), 30 °C, prepolymerization incubation for 10 min. (B) GPC traces (triple-detection GPC, DMF) for polymerizations targeting DP = 10,000, 20,000, and 30,000. *Reproduced with permission Liu, Z., Lv, Y. & An, Z. (2017). Enzymatic cascade catalysis for the synthesis of multiblock and ultrahigh-molecular-weight polymers with oxygen tolerance.* Angewandte Chemie International Edition, *56(44), 13852–13856. Copyright 2017, Wiley.*

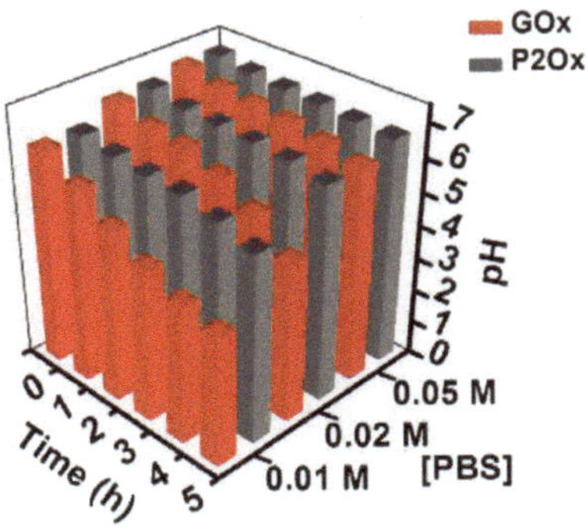

Fig. 11 pH trends of GOx and P2Ox deoxygenated solutions in PBS (0.01–0.05 M, pH 7). [GOx] or [P2Ox] = 2.7 U/mL, [CTA] = 0.005 M, [glucose] = 100 mM, 30 °C. *Reprinted with permission Liu, Z., Lv, Y. & An, Z. (2017). Enzymatic cascade catalysis for the synthesis of multiblock and ultrahigh-molecular-weight polymers with oxygen tolerance.* Angewandte Chemie International Edition, *56(44), 13852–13856. Copyright 2017, Wiley.*

3.3 pH trends during deoxygenation

As we mentioned in Section 1.1, pH is of great importance for HRP-initiated RAFT polymerization. RAFT syntheses initiated by HRP were conducted in buffer solution at pH 7 to render high initiation efficiency as well as good RAFT control. Lower or even no conversion was achieved at lower pH (3–4), presumably due to decreased enzyme activity and/or poor stability of HRP in this condition (Zhang et al., 2015). When cooperating with oxidase deoxygenation, pH also affects enzyme activity, deoxygenation efficiency, and thus rate of polymerization and polymerization control. The pH trends of GOx and P2Ox deoxygenated buffer solutions of different phosphate concentrations were investigated (Fig. 11). The D-glucono-δ-lactone generated in GOx deoxygenation is susceptible to hydrolysis into gluconic acid, which can gradually decrease the pH of the solution. Therefore, buffer solutions with a moderately high concentration (~50 mM) are preferred to maintain the pH value during the time of polymerization. Compared with GOx, the 2-keto-D-glucose produced in P2Ox deoxygenation is fairly resistant to hydrolysis, so there is less pH fluctuation even in buffer solution with a low concentration (~10 mM).

4. Conclusions

Enzyme-initiated RAFT polymerization has been developed as an efficient and versatile strategy for controlled synthesis of polymers under mild conditions. The HRP/H_2O_2/ACAC ternary initiating system exhibits excellent compatibility with the RAFT process, which has been demonstrated in the polymerization of diverse monomer families with good RAFT control under both homogeneous and heterogeneous conditions. Furthermore, oxygen-tolerant RAFT with high efficiency and excellent control has also been developed by using GOx or P2Ox oxidase deoxygenation instead of external degassing. Remarkably, efficient synthesis of precise multiblock copolymers (up to 10 blocks) and UHMW polymers has been achieved via facile RAFT assisted by P2Ox-HRP cascade catalysis directly conducted in open or sealed vessels without additional deoxygenation operation. The enzymatic catalysis is highly efficient and only requires low concentrations (~μM level generally or even lower under extreme conditions) of enzymes to be used.

Enzyme-initiated RAFT polymerization has shown promises for sustainable synthesis of precision polymers. It is anticipated that functional polymers with complex composition and architecture can be prepared via enzyme-catalyzed RAFT for potential applications, and this calls for an expansion of enzymatic polymerization systems and especially for further efforts on catalytic mechanism exploration. Moreover, continuous production via robust oxygen-tolerant RAFT mediated by enzyme catalysis incorporating enzyme immobilization and stabilization techniques is highly desired for large-scale production of well-defined polymers.

Acknowledgments

The authors thank the financial support provided by the National Natural Science Foundation of China (21871175).

References

Anastasaki, A., Nikolaou, V., Zhang, Q., Burns, J., Samanta, S. R., Waldron, C., et al. (2014). Copper(II)/tertiary amine synergy in Photoinduced living radical polymerization: Accelerated synthesis of ω-functional and α,ω-Heterofunctional poly(acrylates). *Journal of the American Chemical Society*, *136*(3), 1141–1149.

Canning, S. L., Smith, G. N., & Armes, S. P. (2016). A critical appraisal of RAFT-mediated polymerization-induced self-assembly. *Macromolecules*, *49*(6), 1985–2001.

Chapman, R., Gormley, A. J., Herpoldt, K.-L., & Stevens, M. M. (2014). Highly controlled open vessel RAFT polymerizations by enzyme degassing. *Macromolecules*, *47*(24), 8541–8547.

Chapman, R., Gormley, A. J., Stenzel, M. H., & Stevens, M. M. (2016). Combinatorial low-volume synthesis of well-defined polymers by enzyme degassing. *Angewandte Chemie International Edition*, *55*(14), 4500–4503.

Chen, M., Zhong, M., & Johnson, J. A. (2016). Light-controlled radical polymerization: Mechanisms, methods, and applications. *Chemical Reviews*, *116*(17), 10167–10211.

Chiefari, J., Chong, Y. K., Ercole, F., Krstina, J., Jeffery, J., Le, T. P. T., et al. (1998). Living free-radical polymerization by reversible addition – fragmentation chain transfer: The RAFT process. *Macromolecules*, *31*(16), 5559–5562.

Dadashi-Silab, S., Doran, S., & Yagci, Y. (2016). Photoinduced electron transfer reactions for macromolecular syntheses. *Chemical Reviews*, *116*(17), 10212–10275.

Danielson, A. P., Van Kuren, D. B., Lucius, M. E., Makaroff, K., Williams, C., Page, R. C., et al. (2016). Well-defined macromolecules using horseradish peroxidase as a RAFT initiase. *Macromolecular Rapid Communications*, *37*(4), 362–367.

Emery, O., Lalot, T., Brigodiot, M., & Maréchal, E. (1997). Free-radical polymerization of acrylamide by horseradish peroxidase-mediated initiation. *Journal of Polymer Science Part A: Polymer Chemistry*, *35*(15), 3331–3333.

Enciso, A. E., Fu, L., Lathwal, S., Olszewski, M., Wang, Z., Das, S. R., et al. (2018). Biocatalytic "oxygen-fueled" atom transfer radical polymerization. *Angewandte Chemie International Edition*, *57*(49), 16157–16161.

Enciso, A. E., Fu, L., Russell, A. J., & Matyjaszewski, K. (2018). A breathing atom-transfer radical polymerization: Fully oxygen-tolerant polymerization inspired by aerobic respiration of cells. *Angewandte Chemie International Edition*, *57*(4), 933–936.

Fodor, C., Gajewska, B., Rifaie-Graham, O., Apebende, E. A., Pollard, J., & Bruns, N. (2016). Laccase-catalyzed controlled radical polymerization of N-vinylimidazole. *Polymer Chemistry*, *7*(43), 6617–6625.

Giffhorn, F. (2000). Fungal pyranose oxidases: Occurrence, properties and biotechnical applications in carbohydrate chemistry. *Applied Microbiology and Biotechnology*, *54*(6), 727–740.

Guo, H. Z., & An, Z. S. (2018). Enzyme catalysis for RAFT polymerization and functionalization: A sustainable strategy for synthesis of precision polymers. *Acta Polymerica Sinica*, *10*, 1253–1261.

Jennifer, L. D. L. H., Zhang, X., Chaduc, I., Brunel, F., Lansalot, M., & D'Agosto, F. (2016). The effect of Hydrophile topology in RAFT-mediated polymerization-induced self-assembly. *Angewandte Chemie International Edition*, *55*(11), 3739–3743.

Keddie, D. J., Moad, G., Rizzardo, E., & Thang, S. H. (2012). RAFT agent design and synthesis. *Macromolecules*, *45*(13), 5321–5342.

Kobayashi, S., & Makino, A. (2009). Enzymatic polymer synthesis: An opportunity for green polymer chemistry. *Chemical Reviews*, *109*(11), 5288–5353.

Konkolewicz, D., Schröder, K., Buback, J., Bernhard, S., & Matyjaszewski, K. (2012). Visible light and sunlight Photoinduced ATRP with ppm of Cu catalyst. *ACS Macro Letters*, *1*(10), 1219–1223.

Leibfarth, F. A., Mattson, K. M., Fors, B. P., Collins, H. A., & Hawker, C. J. (2013). External regulation of controlled polymerizations. *Angewandte Chemie International Edition*, *52*(1), 199–210.

Lin, Y., Wu, L., Huang, Y., Ren, J., & Qu, X. (2015). Positional assembly of hemin and gold nanoparticles in graphene–mesoporous silica nanohybrids for tandem catalysis. *Chemical Science*, *6*(2), 1272–1276.

Liu, Z., Lv, Y., & An, Z. (2017). Enzymatic cascade catalysis for the synthesis of multiblock and ultrahigh-molecular-weight polymers with oxygen tolerance. *Angewandte Chemie International Edition, 56*(44), 13852–13856.

Liu, Z., Lv, Y., Zhu, A., & An, Z. (2018). One-enzyme triple catalysis: Employing the promiscuity of horseradish peroxidase for synthesis and functionalization of well-defined polymers. *ACS Macro Letters*, 7(1), 1–6.

Lv, Y., Liu, Z., Zhu, A., & An, Z. (2017). Glucose oxidase deoxygenation – redox initiation for RAFT polymerization in air. *Journal of Polymer Science Part A: Polymer Chemistry, 55*(1), 164–174.

Matyjaszewski, K., Coca, S., Gaynor, S. G., Wei, M., & Woodworth, B. E. (1998). Controlled radical polymerization in the presence of oxygen. *Macromolecules, 31*(17), 5967–5969.

McCormick, C. L., & Lowe, A. B. (2004). Aqueous RAFT polymerization: Recent developments in synthesis of functional water-soluble (co)polymers with controlled structures. *Accounts of Chemical Research, 37*(5), 312–325.

Moad, G., Rizzardo, E., & Thang, S. H. (2008). Toward living radical polymerization. *Accounts of Chemical Research, 41*(9), 1133–1142.

Ng, Y.-H., di Lena, F., & Chai, C. L. L. (2011a). Metalloenzymatic radical polymerization using alkyl halides as initiators. *Polymer Chemistry, 2*(3), 589–594.

Ng, Y.-H., di Lena, F., & Chai, C. L. L. (2011b). PolyPEGA with predetermined molecular weights from enzyme-mediated radical polymerization in water. *Chemical Communications, 47*(22), 6464–6466.

O'malley, J. J., & Ulmer, R. W. (1973). Thermal stability of glucose oxidase and its admixtures with synthetic polymers. *Biotechnology and Bioengineering, 15*(5), 917–925.

Perrier, S. (2017). 50th anniversary perspective: RAFT polymerization—A user guide. *Macromolecules, 50*(19), 7433–7447.

Reetz, M. T. (2013). Biocatalysis in organic chemistry and biotechnology: Past, present, and future. *Journal of the American Chemical Society, 135*(34), 12480–12496.

Reyhani, A., Nothling, M. D., Ranji-Burachaloo, H., McKenzie, T. G., Fu, Q., Tan, S., et al. (2018). Blood-catalyzed RAFT polymerization. *Angewandte Chemie International Edition, 57*(32), 10288–10292.

Rodriguez, K. J., Gajewska, B., Pollard, J., Pellizzoni, M. M., Fodor, C., & Bruns, N. (2018). Repurposing biocatalysts to control radical polymerizations. *ACS Macro Letters*, 7(9), 1111–1119.

Rosen, B. M., & Percec, V. (2009). Single-electron transfer and single-electron transfer degenerative chain transfer living radical polymerization. *Chemical Reviews, 109*(11), 5069–5119.

Schneiderman, D. K., Ting, J. M., Purchel, A. A., Miranda, R., Tirrell, M. V., Reineke, T. M., et al. (2018). Open-to-air RAFT polymerization in complex solvents: From whisky to fermentation broth. *ACS Macro Letters*, 7(4), 406–411.

Shanmugam, S., Xu, J., & Boyer, C. (2015). Exploiting metalloporphyrins for selective living radical polymerization tunable over visible wavelengths. *Journal of the American Chemical Society, 137*(28), 9174–9185.

Shen, L., Lu, Q., Zhu, A., Lv, X., & An, Z. (2017). Photocontrolled RAFT polymerization mediated by a supramolecular catalyst. *ACS Macro Letters, 6*(6), 625–631.

Shenoy, R., Tibbitt, M. W., Anseth, K. S., & Bowman, C. N. (2013). Formation of core–shell particles by interfacial radical polymerization initiated by a glucose oxidase-mediated redox system. *Chemistry of Materials, 25*(5), 761–767.

Sigg, S. J., Seidi, F., Renggli, K., Silva, T. B., Kali, G., & Bruns, N. (2011). Horseradish peroxidase as a catalyst for atom transfer radical polymerization. *Macromolecular Rapid Communications, 32*(21), 1710–1715.

Silva, T. B., Spulber, M., Kocik, M. K., Seidi, F., Charan, H., Rother, M., et al. (2013). Hemoglobin and red blood cells catalyze atom transfer radical polymerization. *Biomacromolecules, 14*(8), 2703–2712.

Skrabania, K., Miasnikova, A., Bivigou-Koumba, A. M., Zehm, D., & Laschewsky, A. (2011). Examining the UV-vis absorption of RAFT chain transfer agents and their use for polymer analysis. *Polymer Chemistry*, *2*(9), 2074–2083.

Straathof, A. J. J. (2014). Transformation of biomass into commodity chemicals using enzymes or cells. *Chemical Reviews*, *114*(3), 1871–1908.

Tan, J., Liu, D., Bai, Y., Huang, C., Li, X., He, J., et al. (2017). Enzyme-assisted photo-initiated polymerization-induced self-assembly: An oxygen-tolerant method for preparing block copolymer Nano-objects in open vessels and multiwell plates. *Macromolecules*, *50*(15), 5798–5806.

Tan, J., Xu, Q., Li, X., He, J., Zhang, Y., Dai, X., et al. (2018). Enzyme-PISA: An efficient method for preparing well-defined polymer nano-objects under mild conditions. *Macromolecular Rapid Communications*, *39*(9), e1700871.

Teixeira, D., Lalot, T., Brigodiot, M., & Maréchal, E. (1999). β-Diketones as key compounds in free-radical polymerization by enzyme-mediated initiation. *Macromolecules*, *32*(1), 70–72.

Thang, S. H., Chong, Y. K., Mayadunne, R. T. A., Moad, G., & Rizzardo, E. (1999). A novel synthesis of functional dithioesters, dithiocarbamates, xanthates and trithiocarbonates. *Tetrahedron Letters*, *40*(12), 2435–2438.

Torrelo, G., Hanefeld, U., & Hollmann, F. (2015). Biocatalysis. *Catalysis Letters*, *145*(1), 309–345.

van Oers, M. C. M., Rutjes, F., & van Hest, J. C. M. (2014). Cascade reactions in nanoreactors. *Current Opinion in Biotechnology*, *28*, 10–16.

Veitch, N. C. (2004). Horseradish peroxidase: A modern view of a classic enzyme. *Phytochemistry*, *65*(3), 249–259.

Vriezema, D. M., Garcia, P. M. L., Sancho Oltra, N., Hatzakis, N. S., Kuiper, S. M., Nolte, R. J. M., et al. (2007). Positional assembly of enzymes in polymersome nanoreactors for cascade reactions. *Angewandte Chemie International Edition*, *46*(39), 7378–7382.

Wang, X., & An, Z. (2019). New insights into RAFT dispersion polymerization-induced self-assembly: From monomer library, morphological control, and stability to driving forces. *Macromolecular Rapid Communications*, *40*(2), e1800325.

Wang, Y., Fu, L., & Matyjaszewski, K. (2018). Enzyme-deoxygenated low parts per million atom transfer radical polymerization in Miniemulsion and ab initio emulsion. *ACS Macro Letters*, 7(11), 1317–1321.

Wang, J., Rivero, M., Muñoz Bonilla, A., Sanchez-Marcos, J., Xue, W., Chen, G., et al. (2016). Natural RAFT polymerization: Recyclable-catalyst-aided, opened-to-air, and sunlight-photolyzed RAFT polymerizations. *ACS Macro Letters*, *5*(11), 1278–1282.

Wang, X., Shen, L., & An, Z. (2018). Dispersion polymerization in environmentally benign solvents via reversible deactivation radical polymerization. *Progress in Polymer Science*, *83*, 1–27.

Warren, N. J., & Armes, S. P. (2014). Polymerization-induced self-assembly of block copolymer nano-objects via RAFT aqueous dispersion polymerization. *Journal of the American Chemical Society*, *136*(29), 10174–10185.

Xiao, Y., Sun, H., & Du, J. (2017). Sugar-breathing glycopolymersomes for regulating glucose level. *Journal of the American Chemical Society*, *139*(22), 7640–7647.

Xu, J., Jung, K., Atme, A., Shanmugam, S., & Boyer, C. (2014). A robust and versatile Photoinduced living polymerization of conjugated and unconjugated monomers and its oxygen tolerance. *Journal of the American Chemical Society*, *136*(14), 5508–5519.

Xu, J., Jung, K., & Boyer, C. (2014). Oxygen tolerance study of photoinduced electron transfer–reversible addition–fragmentation chain transfer (PET-RAFT) polymerization mediated by Ru(bpy)3Cl2. *Macromolecules*, *47*(13), 4217–4229.

Xu, J., Shanmugam, S., Duong, H. T., & Boyer, C. (2015). Organo-photocatalysts for photoinduced electron transfer-reversible addition–fragmentation chain transfer (PET-RAFT) polymerization. *Polymer Chemistry*, *6*(31), 5615–5624.

Yeow, J., & Boyer, C. (2017). Photoinitiated polymerization-induced self-assembly (photo-PISA): New insights and opportunities. *Advanced science (Weinheim, Baden-Württemberg, Germany)*, *4*(7), 1700137.

Zavada, S. R., McHardy, N. R., & Scott, T. F. (2014). Oxygen-mediated enzymatic polymerization of thiol–ene hydrogels. *Journal of Materials Chemistry B*, *2*(17), 2598–2605.

Zhang, W., D'Agosto, F., Boyron, O., Rieger, J., & Charleux, B. (2012). Toward a better understanding of the parameters that lead to the formation of nonspherical polystyrene particles via RAFT-mediated one-pot aqueous emulsion polymerization. *Macromolecules*, *45*(10), 4075–4084.

Zhang, L., Lu, Q., Lv, X., Shen, L., Zhang, B., & An, Z. (2017). In situ cross-linking as a platform for the synthesis of triblock copolymer vesicles with diverse surface chemistry and enhanced stability via RAFT dispersion polymerization. *Macromolecules*, *50*(5), 2165–2174.

Zhang, H., Trout, W. S., Liu, S., Andrade, G. A., Hudson, D. A., Scinto, S. L., et al. (2016). Rapid bioorthogonal chemistry turn-on through enzymatic or long wavelength photocatalytic activation of Tetrazine ligation. *Journal of the American Chemical Society*, *138*(18), 5978–5983.

Zhang, B., Wang, X., Zhu, A., Ma, K., Lv, Y., Wang, X., et al. (2015). Enzyme-initiated reversible addition–fragmentation chain transfer polymerization. *Macromolecules*, *48*(21), 7792–7802.

CHAPTER TWELVE

Radical addition polymerization: Enzymatic template-free synthesis of conjugated polymers and their nanostructure fabrication

Jorge Romero-García*, Antonio Ledezma-Pérez, Manuel Martínez-Cartagena, Carmen Alvarado-Canché, Paola Jiménez-Cárdenas, Arxel De-León, Carlos Gallardo-Vega
Centro de Investigación en Química Aplicada, Saltillo, Coah., México
*Corresponding author: e-mail address: jorge.romero@ciqa.edu.mx

Contents

Abstract

Conjugated polymers are attractive for many applications due to their unique properties. Their molecular structure can easily be tuned, making them suitable for an enormous number of specific applications. Conjugated polymers have the potential to achieve electrical properties similar to those of noncrystalline inorganic semiconductors; however, their chemical structure is much more complex and somewhat resembles that of biomacromolecules. The molecular conformation and interactions of conjugated polymers play an important role in their functionality. The use of enzymes has emerged as a highly valuable alternative method to synthesize these polymers and is very useful

Methods in Enzymology, Volume 627
ISSN 0076-6879
https://doi.org/10.1016/bs.mie.2019.08.009

in the fabrication of their nanostructures. Here, we present established strategies for the synthesis of conjugated polymers in template-free systems that do not interfere with the preparation of their nanostructures. These strategies are based on the use of peroxidases (class III; EC 1.11.1.7, donor: hydrogen peroxide oxidoreductase), which are enzymes that have the ability to catalyze the oxidation of a number of compounds (including aromatics such as aniline, pyrrole, thiophene and some of their derivatives), in the presence of hydrogen peroxide, to obtain conjugated polymers.

1. Introduction

In the last two decades, enzymes have acquired great relevance in polymer synthesis because of their special catalytic characteristics (Jiang & Loos, 2016). In addition to being effective, enzymatically catalyzed polymerizations are often less harmful to the environment than traditional chemical polymerizations, (Shoda, Uyama, Kadokawa, Kimura, & Kobayashi, 2016). Moreover, enzymatic reactions are carried out under mild conditions (temperature and pressure) and have high stereoselectivity and regioselectivity that avoid laborious purification steps to improve the quality of the end products (Percec, 2013). These selectivities also allow the introduction of side- or end-chain functional groups (Puskas, Sen, & Kasper, 2008).

Examples of enzymatic polymerization with different architectures have been described for a broad range of polymer families including the following: polyesters (Pellis, Hanson, Comerford, Clark, & Farmer, 2019), polyamides (Lima, dos Anjos, Orozco, & Porto, 2019), vinyl polymers (Uzawa, Kobayashi, Fukushima, Taniguchi, & Nakahira, 2013), among many others. However, unlike the conventional synthesis of polymers, the use of enzymes still presents economic limitations for extended application. An interesting strategy for enzymatic polymerization is the synthesis of high performance and technologically relevant polymers, such as electronically conducting or semiconducting conjugated polymers. A conjugated polymer (CP) is a macromolecule where the main –C–C– backbone has alternating sigma and π-bonds. The continuous overlapping of π-orbitals generates high electronic conjugation, resulting in the formation of so called "molecular wires." Such intermixing of molecular orbitals generates a semiconductor band structure with valence and conduction bands, or, alternatively, CP also possess conductivity due to mobile protons or other ions (Hoeben, Jonkheijm, Meijer, & Schenning, 2005). The possibility to modulate the chemical structure of the CP through multiple pathways allows the design of materials with properties tailored for specific applications. Optical and optoelectronic properties are of particular interest. The fluorescent, electrical and electrochemical properties of CPs show promise for several applications, such as biosensing (Pan, Gonuguntla, Li, and Trau (2017), chemosensing (Liu et al., 2018), bioimaging, 3D cell scaffolds (Alegret, Dominguez-Alfaro, & Mecerreyes, 2019) and electronic devices (Bao, Braun, Wang, Liu, & Fahlman, 2019). The idea of using peroxidases in the synthesis of CP polymers may have emerge from the fact that metals, especially transition metal cations in the presence of H_2O_2 have high performance as a catalyst in their synthesis (Della Pina, Falletta, & Rossi, 2011). Peroxidases belong to the group of oxireductase metaloenzymes that contain one or more transition metal cations in their catalytic center. They are the most investigated enzymes for the polymerization of conjugated polymers. A detail information about the history of the subject can be find elsewhere (Higashimura, 2019; Uyama, 2019). Several peroxidases have been reported for this purpose; the most used are horseradish peroxidase (HRP) and the peroxidase of the shell of the soybean seeds (SBP). Although, HRP has been extensively studied for decades, both SBP and HRP have many similarities, with Fe at the active center, and are often referred to as classical plant peroxidases (Ryan, Carolan, & Ó'Fágáin, 2006). The catalytic cycle of these peroxidases is initiated with their oxidation by hydrogen peroxide, going from the native state (PN) through two catalytically active forms, compound I (PNi) and compound II (Pii), before returning to the native form. Each of these active forms of the enzyme oxidizes a total of two aromatic molecules to produce free radicals (Job & Dunford, 1976). The generated radicals diffuse into the reaction medium and combine to form dimers, trimers, further oligomers and eventually polymers through a nonenzymatic reaction, implying step growth by addition polymerization followed by propagation and termination sequences, similar to that occurring in classical chemical oxidation polymerization (Fig. 1). A large number of monomers and their derivatives have been used in the enzymatic or biocatalytic synthesis of polymers conjugated by means of peroxidases. A large part of these polymerizations have been carried out with the help of molecules or macromolecules that act as templates for the linear orientation (head-to-tail coupling) during the growth of the CP (Leppänen et al., 2013; Liu et al., 1999). However, usually these template molecules are polyanions or polycations that reach a high degree of complexation within the CP, which makes it challenging to obtain the required polymers for specific applications, such as fibers (Wang, Romero, Mates, Zhu, & Winokur, 2000) or other type of nanostructures (Arizmendi et al., 2006).

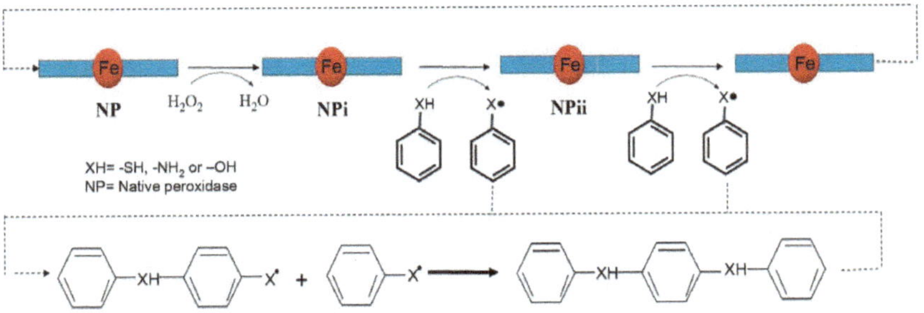

Fig. 1 Schematic showing the peroxidase cycle for catalysis in the template-free synthesis of conjugated polymers by radical addition polymerization.

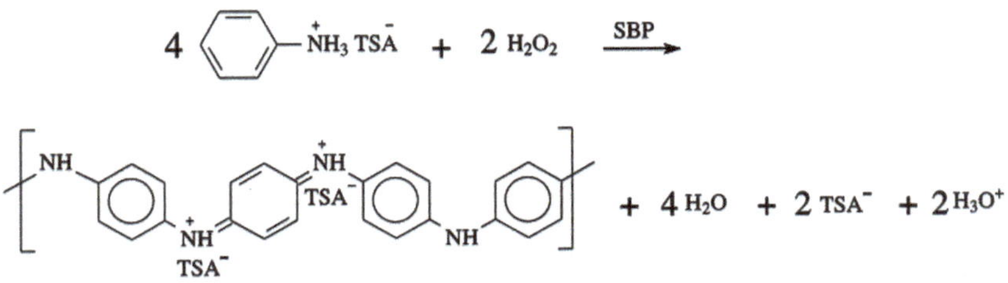

Fig. 2 Schematic representing the route of PANi polymerization catalyzed by SBP.

2. Radical addition polymerization: Enzymatic template-free synthesis of conjugated polymers

In this chapter, we present a detailed protocols for the synthesis of polyaniline (PANi) (Fig. 2) by using HRP and SBP peroxidases as catalyst through a radical addition polymerization (RAP), as described in the next sections. Throughout this chapter, the PANi is presented as a model for the enzymatic template-free synthesis of conjugated polymers, but it is worth mentioning that the model can be adapted for the synthesis of other conjugated polymers.

2.1 Purification of chemicals

2.1.1 Precautions

For all syntheses, it is necessary follow all the safety rules and use protection equipment designed to work in classical chemical laboratories.

2.1.2 Equipment

- Distillation apparatus

2.1.3 Chemicals

- Aniline (Química Dinámica, México)

2.1.4 Protocol

1. Distill aniline under reduced pressure over stannous chloride and potassium hydroxide. Collect the middle fraction in a round-bottom flask, closed with a septum. Store it at −28 °C prior to use.

2.2 Polymerization and polymer purification procedures

2.2.1 Polyaniline synthesis

2.2.1.1 Equipment

- Water bath with temperature control
- Jacked round-bottom flask
- Centrifuge (17,000 rpm)
- pH meter

2.2.1.2 Chemicals

- Distilled aniline (Química Dinámica, México) prepared as described in Section 2.1
- Ammonium hydroxide (Química Dinámica, México)
- Hydrogen peroxide (30%, Sigma-Aldrich)
- *N*-Methyl-2-pyrrolidone (Sigma-Aldrich)
- Hydrochloric acid (Sigma-Aldrich)
- 4-Toluenesulfonic acid (Merck)
- Soybean peroxidase (SBP 64 U/mg, Sigma)
- Ammonium persulfate (Sigma-Aldrich)
- Centrifuge (Beckman Coulter, Allegra™ 25R)

2.2.1.3 Protocol

Polyaniline is synthesized in a 250 mL jacketed round-bottom flask connected to a water bath with temperature control. Polyaniline produced via chemical synthesis is used as the reference material.

2.2.1.4 Chemical route

The polymer is obtained according a previously reported procedure with small modifications (Wei & Hsueh, 1989).

1. Carry out preparation and reaction under nitrogen atmosphere.
2. Use a 250 mL jacketed round-bottom flask, equipped with a stirring magnetic bar. Keep the recipient <5 °C by recirculating water through the jacketed round-bottom flask, using the water bath with temperature control.
3. Dissolve 1 mL of distilled aniline (0.107 mol) in 60 mL of 1 M HCl.

4. Add dropwise over 15 min a solution of ammonium persulfate, prepared by dissolving 2.82 g (0.125 mol) $(NH_4)_2S_2O_8$ in 30 mL of 1 M HCl. Keep the reaction under medium agitation.
5. After 2 h, centrifuge (10,000 rpm, 10 min), wash and centrifuge 4 times with 10 mL of 1 M HCl.
6. Transfer the precipitate into a breaker containing 40 mL of 1 M HCl.
7. Keep the stirring the mixture for 4 h. Follow with centrifugation (10,000 rpm, 10 min).
8. Freeze-dry overnight. Around 9 g of HCl-doped polyaniline salt is obtained (76% yield).
9. To convert the HCl-doped polyaniline salt to the base form, place 5 g of polymer in an excess amount of NH_4OH (50 ml, 0.1 N). Maintain the suspension at room temperature for 3 h.
10. Centrifuge (10,000 rpm, 10 min). Wash with 50 mL of distilled water and then centrifuge (10,000 rpm, 10 min). Freeze-dry overnight. A blue powder of the base form of polyaniline is obtained.

2.2.1.5 Enzymatic route

The enzymatic polymerization of aniline is carried out in distilled water with soybean peroxidase as a catalyst (Cruz-Silva et al., 2005).

1. Carry out preparation and reaction under nitrogen atmosphere.
2. Use a 250 mL jacketed round-bottom flask, equipped with a stirring magnetic bar. Keep the recipient <5 °C by recirculating water through the jacketed round-bottom flask, using the water bath with temperature control.
3. Dissolve 1.83 g (0.014 mol) of distilled aniline, in 60 mL of distilled water.
4. Adjust the initial pH to 3 with a 25 wt% of 4-toluenesulfonic acid solution.
5. Add 0.5 mg/mL of soybean peroxidase, follow by stepwise addition of H_2O_2 (80 μL each 12 min) to reach a 1:1 M ratio with aniline.
6. Keep the reaction under continuous agitation and purged with nitrogen for 8 h at <3 °C.
7. Collect the precipitate of the reaction by filtration and wash successively with methanol, and an aqueous solution of 4-toluenesulfonic acid (5% w/v). Finally, freeze-dry the product.

2.3 Analysis of the polyaniline

In order to compare the main structural features of the polymers obtained from chemical and enzymatic synthesis, the polymers are characterized by UV–vis and FT-IR spectroscopy, as well as WAXS analysis and the determination of electric conductivity.

2.3.1 UV–vis spectroscopy

UV–visible analysis is performed in order to characterize the polyaniline with regards to the chemical structure determined for a wide range of oxidation states. However, emeraldine is the only material that can be doped to the highly conductive state and has equal proportions of amine (–NH–) and imine (=N–) sites. This half oxidized PANi (emeraldine base) is doped in the presence of acids, becomes protonated, and transforms into the conductive PANi (emeraldine salt) form. That process originates the formation of polaronic or bipolaronic structure (Canales, Torras, Fabregat, Meneguzzi, & Alemań, 2014). These polyaniline forms differ in their light absorption properties.

The UV–vis electronic absorption spectra of PANI solutions are recorded on a Shimadzu UV 2401 using *N*-methyl-2-pyrrolidinone as a solvent. Fig. 3 shows the spectra for chemically and enzymatically synthesized PANi. In both spectra two bands are observed: one at 330nm (band I), associated with the π-π* transition of the aromatic rings, and the second one at 639nm (band II), assigned to a benzenoid to quinoid excitonic transition (Kulkarni, Viswanatah, Marimuthu, & Seth, 2004).

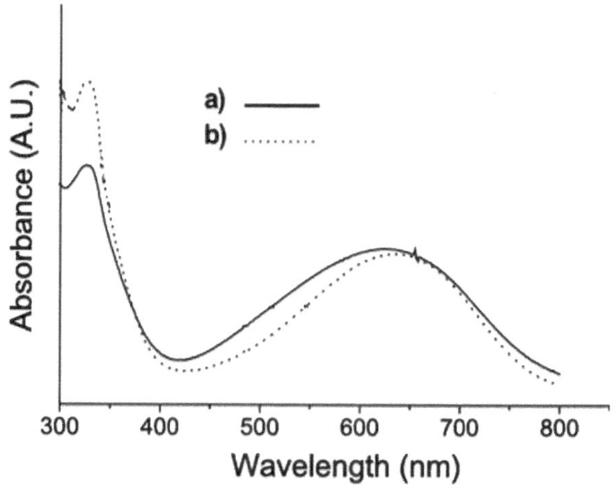

Fig. 3 UV–vis absorption spectra of PANi: (a) chemically and (b) enzymatically synthesized. *Adapted with permission from Cruz-Silva, R., Romero-García, J., Angulo-Sánchez, L. J., Ledezma-Pérez, A., Arias-Marín, E., Moggio, I. & Flores-Loyola, E. (2005). Template-free enzymatic synthesis of electrically conducting polyaniline using soybean peroxidase.* European Polymer Journal, 41, *1129–1135. Copyright 2005 Elsevier.*

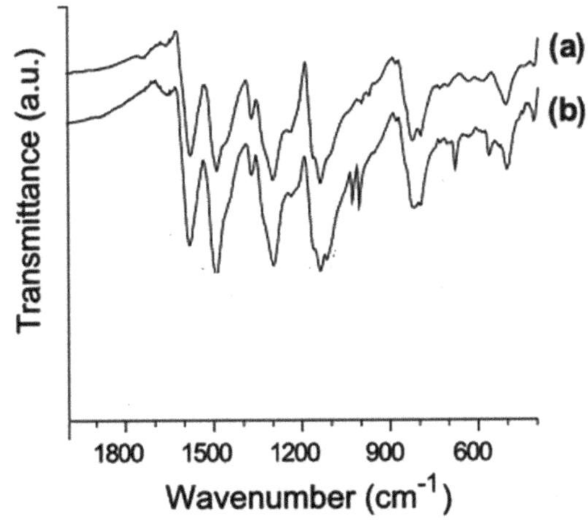

Fig. 4 FT-IR spectra of PANi: (a) chemically and (b) enzymatically synthesized. *Adapted with permission from Cruz-Silva, R., Romero-García, J., Angulo-Sánchez, L. J., Ledezma-Pérez, A., Arias-Marín, E., Moggio, I. & Flores-Loyola, E. (2005). Template-free enzymatic synthesis of electrically conducting polyaniline using soybean peroxidase.* European Polymer Journal, 41, *1129–1135. Copyright 2005 Elsevier.*

2.3.2 FT-IR spectroscopy

FT-IR analysis is another fundamental tool because it reveals the type of groups present, as well the way in which the aniline monomer rings are coupled to generate structures with different levels of oxidation. Fourier transformed infrared (FT-IR) spectra are measured in KBr pellets on a Nicolet Magna 550 FT-IR spectrophotometer. Fig. 4 shows the FT-IR absorption spectra of PANI synthesized by the chemical and enzymatic routes. Both spectra are quite similar where the typical bands are observed, including the ring stretching of the quinoid diamine (Q) and the benzenoide diamine (B) units at 1598 and 1500 cm^{-1}, respectively. The peak at 1375 cm^{-1}, characteristic of a standard PANi base, is assigned to C–N stretching of Q-B units, and the peak at 1305 cm^{-1} is due to C–N stretching in the B units (Kang, Neoh, & Tan, 1998). The peaks located at 1170 and 828 cm^{-1} are associated with in-plane and out of plane bending of the aromatic C–H. The two peaks located at 1010 and 1030 cm^{-1}, as well as the peak at 696 cm^{-1}, suggest the presence of S=O groups attached to the aromatic rings (Zheng, Angelopoulos, Epstein, & MacDiarmid, 1997) remaining from the TSA traces in enzymatically synthesized PANI, which were not possible to eliminate after dedoping treatment.

2.3.3 Wide angle X-ray diffraction

The influence of main chain alignment in the emeraldine salt form drives the enhancement of specific structural properties. It is known that in a single

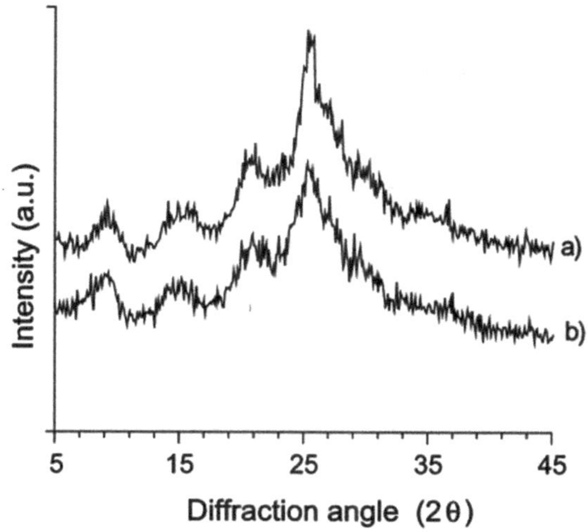

Fig. 5 WAXD patterns of PANi: (a) chemically and (b) enzymatically synthesized. *Adapted with permission from Cruz-Silva, R., Romero-García, J., Angulo-Sánchez, L. J., Ledezma-Pérez, A., Arias-Marín, E., Moggio, I. & Flores-Loyola, E. (2005). Template-free enzymatic synthesis of electrically conducting polyaniline using soybean peroxidase.* European Polymer Journal, 41, *1129–1135. Copyright 2005 Elsevier.*

isolated polymer chain, conduction only occurs along the polymer chain axis. By contrast, in the tridimensional polymer structural arrangement, transport of charge carriers also occurs between neighboring chains by the hopping mechanisms (Pouget, Jozefowiczt, Epstein, Tang, & MacDiarmid, 1991). It is well known that emeraldine in its basic form is practically amorphous, whereas in the salt form it has a well-established semicrystalline structure (Liao, Li, & Xu, 2019).

Wide angle X-ray diffraction patterns (WAXD) are collected on a Siemens D-5000 X-ray diffractometer with a CuKa radiation source, (intensity 25 mA, acceleration voltage 35 kV).

The data are collected in the 2θ mode with a scan rate of 0.3°/min at room temperature. Fig. 5 shows WAXD diffraction for polyaniline synthesized both chemically and enzymatically. The diffraction peaks occur at 9.3°, 15.3°, 2 1° and 25.5°, representing the crystalline phase corresponding to the characteristic emeraldine salt form, known as ES-I (Pouget et al., 1991).

2.3.4 Electrical conductivity determination

The conducting form of polyaniline, which is the emeraldine salt (ES), is obtained upon p-doping of the emeraldine base (EB). The doping process consists of protonation of the imine nitrogen atoms in acidic conditions to produce a highly conducting state without changing the number of π-electrons in the polymer structure (Avlyanova, Min, MacDiarmida, & Epstein, 1995). The electrical conductivity of PANi is associated with the doping of counter ions. In contrast to other conjugated polymers, the electrical conductivity of PANI is affected by doping and protonation of the polymer segments. Doping is a process that allows the polymer to return to its original state with little or no damage to the backbone. The doped form of PANi consists of a conjugated polymer chain backbone and a delocalized π-system, and is the only state with electrical conductivity (Cruz-Silva et al., 2005).

Electrical conductivity measurements are done according to the two-probe technique. Electrical conductivity of PANI synthesized by chemical or enzymatic oxidation is 5.4 and 2.4 S/cm, respectively.

2.3.4.1 Equipment

- Digital multimeter from Ex-tech Instruments
- Stainless cylinder (5 mm diameter and 7.5 mm height)
- Manual press

2.3.4.2 Protocol

1. Fill the stainless cylinder with polyaniline powders.
2. With the manual press compress the polyaniline powders.
3. Cover the bottom and top surfaces of the pellets with a thin layer of conductive silver paint.
4. Measure resistance with the digital multimeter.
5. Convert resistance to electrical conductivity using Ohm's Law.

3. Nanostructure fabrication by enzymatic template-free synthesis of conjugated polymers

PANi has received great attention because of its potential for many technological applications, such as sensors, transparent conductors, electrochromic devices, supercapacitors, rechargeable batteries, flexible electrodes, smart windows, fuel cells, solar cells, sensors, anticorrosion coatings, conducting fibers, actuators, toxic metal recovery, electromagnetic interference shielding, and antistatic charge dissipation among others (Wang, Tran, & Kaner, 2011). Such applications are feasible due to its most relevant properties: it is the only conducting polymer whose electronic structure (and, thus, its electrical conductivity) can be controlled in a reversible manner, it shows good environmental stability, and it is easy to synthesize in large quantities starting from a relatively cheap monomer (Cruz-Silva, Roman, & Romero, 2011). However, the main restrictions that prevent widespread use of this polymer is material that does not melt and is difficult to dissolve it in most nontoxic common solvents (Cao, Smith, & Heeger, 1992), or under other severe functionalization conditions to process the polymer (Wei, Wang, Long, Bobeczko, & Epstein, 1996). In the last two decades, alternatives have emerged for the processing of PANi. One of it is the preparation of polyaniline in colloidal form. A great number of reports have been published on the fabrication of PANi dispersions containing particles with different sizes and morphologies. Usually, the preparation is done by chemical polymerization of aniline in an aqueous acid solution with suitable stabilizers, for example, water soluble polymers (Bongiovanni, Yslas, Rivarola, & Barbero, 2018). However, very few reports exist of colloidal dispersions of PANi for the preparation of colloidal dispersions to obtain nanostructures using enzymes (Stejskal, Trchová, Bober, & Humpolíček, 2015).

3.1 Preparation of colloidal dispersions nanoparticles of polyaniline by enzymatic polymerization

In the next section, we describe some protocols for the preparation of PANi dispersions by a radical addition polymerization using water soluble polymers as steric stabilizers. The enzymatic polymerization of aniline is carry out in distilled water by the use of HRP as a catalyst in the presence of poly(vinyl alcohol) as a steric stabilizer (Cruz-Silva et al., 2006).

3.1.1 Equipment

- Three neck 50 mL reactor
- Magnetic stirring plate
- Water ice bath
- Centrifuge (Beckman Coulter, Allegra™ 25R; 17,000 rpm)

3.1.2 Chemicals

- Distilled aniline (Química Dinámica, México) prepared as described in Section 2.1.
- Hydrogen peroxide (30 wt% solution, Sigma-Aldrich).
- HRP (Type II, 240 U/mg, RZ = 1.9, Sigma-Aldrich).
- *p*-Toluenesulfonic acid monohydrate (TSA, +99%, Merck)

- Poly(vinyl alcohol) (PVAL, hydrolyzed 99C%, Mw 89,000–98,000, Sigma-Aldrich).
- (±)-Camphor-10-sulfonic acid (β) (CSA, +99%, Sigma-Aldrich).

3.1.3 Protocol

1. Carry out preparation and reaction under nitrogen atmosphere.
2. Use a three neck 50 mL reactor, equipped with stirring magnetic bar. Keep the recipient <5 °C in a water ice/bath.
3. Add 18 mL of a PVAL solution (5.6 wt%), 0.002 M of aniline, and 0.002 M of doping acid (CSA or TSA).
4. Keep the reaction mixture under vigorous magnetic stirring in the water/ice bath.
5. Add a 2 mL of a freshly prepared HRP solution in distilled water (2.5 mg/mL) to the reaction mixture.
6. Initiate the reaction by adding 26 mL of hydrogen peroxide (30 wt%) in step additions every 3 min until an equimolar hydrogen peroxide/aniline ratio is achieved (eight additions).
7. After 2.5 h of reaction time, collect the dark green PANI dispersion by centrifugation (10,000 rpm, 10 min).

3.2 Analysis of colloidal dispersions of polyaniline containing nanoparticles

In order to compare the main structural features of the PANi nanoparticles obtained by enzymatic synthesis, the materials are characterized by UV–vis and FT-IR spectroscopy, as well as WAXS analysis, and the determination of electric conductivity as previously described in Section 2.3. Further characterization data and methods of the PANi nanoparticles are described below.

3.2.1 UV-Vis spectroscopy

The UV–vis electronic absorption spectra of PANI solutions are recorded on a Shimadzu UV in acidic medium, using TSA as doping agent. Fig. 6 shows the presence of a peaks at 400 and 800 nm, which is evidence of the formation of a polaron (Stejskala et al., 1999).

3.2.2 SEM morphology of dry colloidal particles of PANi

The morphology of the PANi particles is determined by means of scanning electronic microscopy (SEM), using a TOPCOM SM-510 instrument. The samples were previously covered with gold–palladium. Fig. 7 shows a representative SEM image of dry colloidal PANi particles.

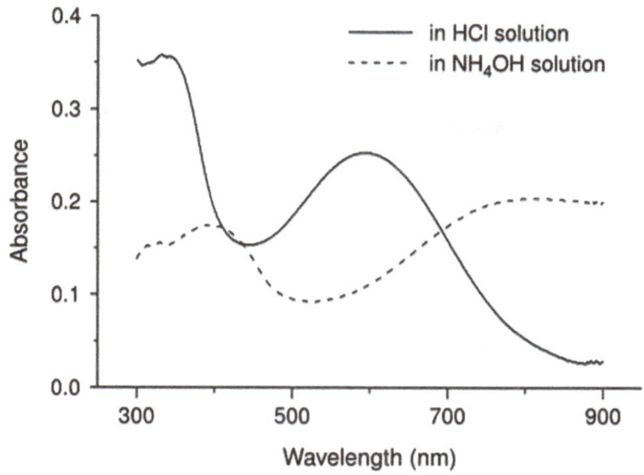

Fig. 6 UV–visible spectra of PANi dispersion particles using HRP as catalyst and p-toluenesulfonic acid (TSA) as doping agent. *Adapted with permission from Cruz-Silva, R., Ruiz-Flores, C., Arizmendi, L., Romero-García J., Arias-Marin E., et al., (2006). Enzymatic synthesis of colloidal polyaniline particles.* Polymer, *47, 1563–1568. Copyright 2006 Elsevier.*

Fig. 7 SEM images of de-doped PANi particles synthesized using HRP as catalyst, PVAL as steric stabilizer and TSA as doping acid during synthesis. The white bar on the images corresponds to 1 μm. *Adapted with permission from Cruz-Silva, R., Ruiz-Flores, C., Arizmendi, L., Romero-García J., Arias-Marin E., et al., (2006). Enzymatic synthesis of colloidal polyaniline particles.* Polymer, *47, 1563–1568. Copyright 2006 Elsevier.*

When TSA is used as doping agent, the particles have spherical shape and a narrow size distribution with an average diameter of 240 nm (Fig. 8).

In contrast, when the synthesis of the colloidal dispersion of PANi is carried out in the presence of HCl, chain-shaped structures are present

Fig. 8 SEM images of de-doped PANi particles synthesized using HRP as catalyst, PVAL as steric stabilizer and HCl as doping acid during synthesis. The white bar on the images corresponds to 1 μm. *Adapted with permission from Cruz-Silva, R., Ruiz-Flores, C., Arizmendi, L., Romero-García J., Arias-Marin E., et al., (2006). Enzymatic synthesis of colloidal polyaniline particles.* Polymer, *47, 1563–1568. Copyright 2006 Elsevier.*

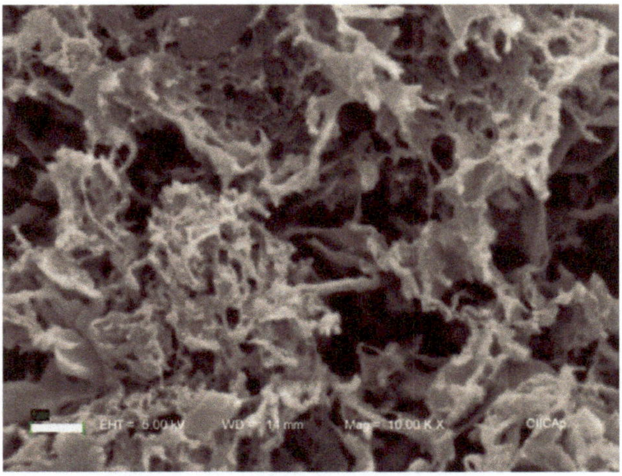

Fig. 9 SEM images of dedoped PANi particles synthesized using HRP as catalyst, PVAL as steric stabilizer and CSA as doping acid during synthesis. The white bar on the images corresponds to 1 μm. *Adapted with permission from Cruz-Silva, R., Ruiz-Flores, C., Arizmendi, L., Romero-García J., Arias-Marin E., et al., (2006). Enzymatic synthesis of colloidal polyaniline particles.* Polymer, *47, 1563–1568. Copyright 2006 Elsevier.*

(Fig. 8). When CSA is used as a doping agent during the synthesis of the colloidal dispersion of PANi, flake-shaped or needle-like structures occur and the particles show low colloidal stability (Fig. 9).

4. Conclusions

The use of peroxidases (i.e., HRP and SBP and many other peroxidases) is effective catalyst in the enzymatic template-free synthesis of conjugated polymers through radical addition polymerization. This technique is an environmentally friendly process because the only byproduct of the reaction is water. Peroxidases carry out double oxidation one-electron with H_2O_2 to form two water molecules. During the catalytic cycle two molecules of monomer (substrate) are oxidized to give two radical molecules ($2MH \rightarrow 2M^{\cdot} + 2H^{+} + 2e^{-}$), followed by radical coupling with each other ($2M^{\cdot} \longrightarrow M-M$) and subsequently continuing with the radical addition polymerization process. Furthermore, with this technique is possible to match the optical, electrical and electronic properties of conjugated polymers synthesized by chemical or electrochemical routes.

The enzymatic template-free synthesis of conjugated polymers has relevant advantages over molecular template-assisted enzymatic polymerization using electrolytes and other macromolecules, such as sulfonated polystyrene, DNA or poly(ethylene glycol, among others (Kim, Kumar, Bruno, & Samuelson, 2006; Kim, Uyama, & Kobayashi, 2003; Ma, Zhang, Zhang, & He, 2004). With this technique, water soluble and well-defined conjugated polymers can be formed, but strongly complexed with the template. This complexing makes it difficult to obtain bulk conjugated polymers that are required in special applications. Such is the case of fibers, nanostructures, sensors and biosensors, as well as other type of electronic devices.

Acknowledgments

This project was funded by CONACYT under project CB-2016-287954 MEXICO.

References

Alegret, N., Dominguez-Alfaro, A., & Mecerreyes, D. (2019). 3D scaffolds based on conductive polymers for biomedical applications. *Biomacromolecules*, *20*, 73–89.

Arizmendi, L., Romero-Garcia, J., Arias-Marin, E., Moggio, I., Castillon, F. F., & Farias, M. H. (2006). Enzymatic synthesis of colloidal polyaniline particles. *Polymer*, *47*, 1563–1568.

Avlyanova, K. J., Min, Y., MacDiarmida, G. A., & Epstein, J. A. (1995). Polyaniline: Conformational changes induced in solution by variation of solvent and doping level. *Synthetic Metals*, *72*, 65–71.

Bao, Q., Braun, S., Wang, C., Liu, X., & Fahlman, M. (2019). Interfaces of (ultra)thin polymer films in organic electronics. *Advances Materials Interfaces*, *6*, 1800897.

Bongiovanni, A. S., Yslas, I. Y., Rivarola, R. C., & Barbero, A. C. (2018). Synthesis of polyaniline (PANI) and functionalized polyaniline (F-PANI) nanoparticles with controlled size by solvent displacement method. *Nanotechnology*, *29*, 125604.

Canales, M., Torras, J., Fabregat, G., Meneguzzi, A., & Alemań, C. (2014). Polyaniline emeraldine salt in the amorphous solid state: Polaron versus bipolaron. *The Journal of Physical Chemistry B*, *118*, 11552–11562.

Cao, Y., Smith, P., & Heeger, J. A. (1992). Counter-ion induced processibility of conducting polyaniline and of conducting polyblends of polyaniline in bulk polymers. *Synthetic Metals*, *48*, 91–97.

Cruz-Silva, R., Roman, P., & Romero, J. (2011). Enzymatic synthesis of polyaniline and other electrically conductive polymers. In K. Loos (Ed.), *Biocatalysis in polymer chemistry* (pp. 187–210). Weinheim: Wiley-VCH GmbH & Co. KGaA.

Cruz-Silva, R., Romero-García, J., Angulo-Sánchez, L. J., Ledezma-Pérez, A., Arias-Marín, E., Moggio, I., et al. (2005). Template-free enzymatic synthesis of electrically conducting polyaniline using soybean peroxidase. *European Polymer Journal*, *41*, 1129–1135.

Cruz-Silva, R., Ruiz-Flores, C., Arizmendi, L., Romero-García, J., Arias-Marin, E., et al. (2006). Enzymatic synthesis of colloidal polyaniline particles. *Polymer*, *47*, 1563–1568.

Della Pina, C., Falletta, E., & Rossi, M. (2011). Conductive materials by metal catalyzed polymerization. *Catalysis Today*, *160*, 11–27.

Higashimura, H. (2019). Synthesis of poly(aromatic)s II: Enzyme-model complexes as catalyst. In S. Kobayashi, H. Uyama, & J.-I. Kodokawa (Eds.), *Enzymatic polymerization towards green polymer chemistry* (pp. 267–305). Singapore: Springer Nature Singapure Pte Ltd.

Hoeben, M. J. F., Jonkheijm, P., Meijer, W. E., & Schenning, J. H. P. A. (2005). About supramolecular assemblies of π-conjugated systems. *Chemical Reviews*, *105*(4), 1491–1546.

Jiang, Y., & Loos, K. (2016). Enzymatic synthesis of biobased polyesters and polyamides. *Polymers*, *8*, 243.

Job, D., & Dunford, H. B. (1976). Substituent effect on the oxidation of phenols and aromatic amines by horseradish peroxidase compound 1. *European Journal of Biochemistry*, *66*, 607–614.

Kang, T. E., Neoh, G. K., & Tan, L. K. (1998). Polyaniline: A polymer with many interesting intrinsic redox states. *Progress in Polymer Science*, *23*, 211–324.

Kim, S. C., Kumar, J., Bruno, F. F., & Samuelson, L. A. (2006). Synthesis and properties of self-doped polyaniline with polycationic templates via biocatalysis. *Journal of Macromolecular Science, Part A*, *43*, 2007–2018.

Kim, Y. J., Uyama, H., & Kobayashi, S. (2003). Peroxidase-catalyzed polymerization of phenol with poly(ethylene glycol) by template polymerization of phenol in water. *Macromolecules*, *36*, 5058–5060.

Kulkarni, M. V., Viswanatah, A. K., Marimuthu, R., & Seth, T. (2004). Synthesis and characterization of polyaniline doped with organic acids. *Journal of Polymer Science Part A: Polymer Chemistry*, *42*, 2043–2049.

Leppänen, A.-S., Xu, C., Liu, J., Wang, X., Pesonen, M., & Willför, S. (2013). Anionic polysaccharides as templates for the synthesis of conducting polyaniline and as structural matrix for conducting biocomposites. *Macromolecules Rapid Communications*, *34*, 1056–1061.

Liao, G., Li, Q., & Xu, Z. (2019). The chemical modification of polyaniline with enhanced properties: A review. *Progress in Organic Coatings*, *126*, 35–43.

Lima, N. R., dos Anjos, S. C., Orozco, M. V. E., & Porto, M. L. A. (2019). Versatility of Candida Antarctica lipase in the amide bond formation applied in organic synthesis and biotechnological processes. *Molecular Catalysis*, *446*, 75–105.

Liu, W., Cholli, L. A., Nagarajan, R., Kumar, J., Tripathy, S., Bruno, F. F., et al. (1999). *Journal of the American Chemical Society*, *121*, 11345–11355.

Liu, Y., Wang, Y.-M., Zhu, W.-Y., Zhang, H.-C., Tang, H., & Jiang, H.-J. (2018). Conjugated polymer nanoparticles-based fluorescent biosensor for ultra-sensitive detection of hydroquinone. *Analytica Chemical Acta, 1012*, 60–65.

Ma, Y., Zhang, J., Zhang, G., & He, H. (2004). Polyaniline nanowires on Si surfaces fabricated with DNA templates. *Journal of the American Chemical Society, 126*, 7097–7101.

Pan, H. M., Gonuguntla, S., Li, S., & Trau, D. (2017). Conjugated polymers for biosensor devices. In P. Ducheyne, D. W. Grainger, K. E. Healy, D. W. Hutmacher, & C. J. Kirkpatrick (Eds.), *Comprehensive biomaterials II: Vol. 3* (pp. 716–754). Oxford: Elsevier.

Pellis, A., Hanson, A. P., Comerford, W. J., Clark, H. J., & Farmer, J. T. (2019). Enzymatic synthesis of unsaturated polyesters: Functionalization and reversibility of the aza-Michael addition of pendants. *Polymer Chemistry, 10*, 843–851.

Percec, V. (2013). Advances in polymer science. In *Hierarchical macromolecular structures: 60 Years after the Staudinger Nobel Prize*: Vol. 261-262. New York: Springer. Cham.

Pouget, P. J., Jozefowiczt, E. M., Epstein, J. A., Tang, X., & MacDiarmid, G. A. (1991). X-ray structure of polyaniline. *Macromolecules, 24*, 779–789.

Puskas, J., Sen, Y. S., & Kasper, R. J. (2008). Green polymer chemistry: Telechelic poly (ethylene glycol)s via enzymatic catalysis. *Journal of Polymer Science: Part A: Polymer Chemistry, 46*, 3024–3028.

Ryan, B. J., Carolan, N., & Ó'Fágáin, C. (2006). Horseradish and soybean peroxidases: Comparable tools for alternative niches? *Trends Biotechnology, 24*, 355–363.

Shoda, S., Uyama, H., Kadokawa, J., Kimura, S., & Kobayashi, S. (2016). Enzymes as green catalysts for precision macromolecular synthesis. *Chemical Reviews, 116*, 2307–2413.

Stejskal, J., Trchová, M., Bober, P., & Humpolíček, P. (2015). Conducting polymers: Polyaniline. In *Encyclopedia of polymer science and technology*. New York: John Wiley & Sons, Inc.

Stejskala, J., Spirkova, M., Riede, A., Helmstedt, M., Mokrevac, P., & Prok, J. (1999). Polyaniline dispersions 8. The control of particle morphology. *Polymer, 40*, 2487–2492.

Uyama, H. (2019). Synthesis of poly(aromatic)s I: Oxireductase as catalyst. In S. Kobayashi, H. Uyama, & J.-I. Kodokawa (Eds.), *Enzymatic polymerization towards green polymer chemistry* (pp. 267–305). Singapure: Springer Nature Singapure Pte Ltd.

Uzawa, S., Kobayashi, A., Fukushima, H., Taniguchi, T., & Nakahira, T. (2013). Enzymatic emulsifier-free emulsion polymerization to prepare polystyrene particles using horseradish peroxidase as a catalyst. *Polymer Journal, 45*, 354–358.

Wang, H. L., Romero, R. J., Mates, B. R., Zhu, Y., & Winokur, M. J. (2000). Effect of processing conditions on the properties of high molecular weight conductive polyaniline fiber. *Journal Polymer Science Part B: Polymer Physics, 38*, 194–204.

Wang, Y., Tran, H. D., & Kaner, R. B. (2011). Applications of oligomers for nanostructured conducting polymers. *Macromolecular Rapid Communications, 2011*(32), 35–49.

Wei, Y., & Hsueh, K. F. (1989). Thermal analysis of chemically synthesized polyaniline and effects of thermal aging on conductivity. *Journal of Science Part A: Polymer Chemistry, 49*, 153–164.

Wei, X.-L., Wang, Z. Y., Long, M. S., Bobeczko, C., & Epstein, J. A. (1996). Synthesis and physical properties of highly sulfonated polyaniline. *Journal of the American Chemical Society, 118*, 2545–2555.

Zheng, W., Angelopoulos, M., Epstein, J. A., & MacDiarmid, G. (1997). Experimental evidence for hydrogen bonding in polyaniline: Mechanism of aggregate formation and dependency on oxidation state. *Macromolecules, 30*, 2953–2955.

Further reading

Grimaud, F., Faucard, P., Tarquis, L., Pizzut-Serin, S., Roblin, P., Morel, S., et al. (2018). Enzymatic synthesis of polysaccharide-based copolymers. *Green Chemistry, 20*, 4012–4022.

CHAPTER THIRTEEN

Surface functionalization of polyester

Felice Quartinello[a], Georg M. Guebitz[a,b,*], Doris Ribitsch[a,b]
[a]Institute for Environmental Biotechnology, IFA Tulln, University of Natural Resources and Life Sciences, Vienna, Austria
[b]ACIB—Austrian Centre of Industrial Biotechnology, Graz, Austria
*Corresponding author: e-mail address: guebitz@boku.ac.at

Contents

Abstract

Surface functionalization such as hydrophilization is an important step in the polymer manufacturing process and a key requirement for application of polyesters. Conventional methods like chemical or plasma treatment are often toxic, expensive and adversely affect the mechanical properties of the polymer. Enzymes have proven to be an attractive alternative for surface hydrolysis and functionalization of synthetic polymers since they work under mild and environmental friendly process conditions while preserving the mechanical properties. This chapter deals with the enzymatic surface treatment of polyesters and in particular with current methods for the analysis of polymer hydrolysis and of changes of surface properties.

Methods in Enzymology, Volume 627
ISSN 0076-6879
https://doi.org/10.1016/bs.mie.2019.08.007

1. Introduction

Polyester is a common term used to describe a wide range of synthetic and semi-synthetic materials used in a broad range of applications (Wei & Zimmermann, 2017). Due to an increasing demand worldwide, the production of plastic materials which are produced from fossil fuels has grown with the fastest rate compared to any other group of materials. The main reason of their strong position in market is mainly due to their versatile performances combined with the highly competitive price (Kim & Song, 2006). Furthermore, physical and mechanical properties of polyesters can be modified or added with addition of reinforcing fillers, colors, foaming compound, flame retardant, plasticizers, antimicrobial agents etc. It is well known that the polyester surface is quite inert and hydrophobic, therefore an activation step is needed. Beside the conventional functionalization process, chemical treatments at high temperatures are currently very common. Here, alkaline treatment is carried out in the presence of 4–20% alkaline solution (NaOH or KOH) (Abdelaal, Sobahi, & Makki, 2008; Azfarniam & Norouzi, 2016; Manalo, Wani, Azwa, & Karunasena, 2015) which can also occur in presence of phase catalyst like titanium (Rahman & East, 2009; Sever, Sarikanat, Erkan, Erdo, & Erden, 2012). Besides, acid pre-treatment can be involved using very high concentration of mineral acid such as sulfuric acid (Yoshioka, Sato, & Okuwaki, 1994). These chemical treatments represent a costly operation process and are very toxic since different neutralization and purification steps are necessary. Moreover, chemical functionalization is characterized by a mechanism defined as "all or nothing" (Guebitz & Cavaco-Paulo, 2008), where the adsorption of the highly concentrated alkaline or acid solution does not affect only the polymer surface but also the bulk structure, resulting into pitting corrosion and reduces tensile strength. An alternative method for surface modification is plasma treatment. This method is rather expensive and difficult to control, which in the end can lead to "ageing" of the polymer, a possible result of the rearrangement and different orientation of the polar groups (Chan & Ko, n.d.; Guruvenket, Rao, Komath, & Raichur, 2004). The increased industrialization of biology gave far-reaching benefits from different points of view: driving toward a more environmentally friendly economic growth, sustainable energy together benefits for future generations (Acero et al., 2011; Araújo, Casal, & Cavaco-Paulo, 2008), which can be also available in polyester surface functionalization. The main driving force behind using "green" approaches is given by inspiration from natural processes.

Table 1 List of enzymes applied for polyester functionalization.

Organism	Class	Substrate	Reference
Thermobifida cellulosilytica	Cutinase (EC3.1.1.74)	PET, PLA, PEF	Pellis et al. (2016)) and Weinberger et al. (2017)
Pelosinus fermentans	Lipase (EC3.1.1.3)	PBAT	Biundo et al. (2016)
Clostridium botulinum	Cutinase (EC3.1.1.74)	PBAT	Perz et al. (2016)
Humicola insolens	Cutinase (EC3.1.1.74)	PET	Ronkvist, Xie, Lu, and Gross (2009)

It is well known that a large variety of organisms (including bacteria and fungi, Table 1) show the capability to use their enzymes to degrade natural polyesters. Nowadays, these "natural tools" are applied in a wide range of industrial applications, and also in polyester treatment. Enzymatic functionalization has the advantages to be performed at mild operation conditions with high specificity/selectivity. Moreover, enzyme adsorption occurs on the surface of the polymer, which avoids reduction of mechanical properties (Guebitz & Cavaco-Paulo, 2008). Meanwhile, the polymer industry also had to deal with major changes toward the synthesis/production of more sustainable and bio-based and therefore easier functionalizable polymers. This is the reason behind the synthesis of analogous bio-based polymers such as Poly(butylene adipate-co-butylene terephthalate) (PBAT), biodegradable ones such as PolyLactic Acid and biobased polymers like Poly(ethylene 2,5-furandicarboxylate).

1.1 Poly(ethylene terephthalate) (PET)

PET is currently the most produced polyester worldwide, with applications as synthetic textile fiber, films, medical devices, and packaging material. It is synthesized from ethylene glycol (EG) and terephthalic acid (TA) (Pellis, Gamerith, et al., 2016) (Fig. 1). The lack of reactive groups on PET for further application has been improved via different approaches. The enzymatic functionalization showed the potential to overcome such limitations and allows generation of more hydrophilic reactive groups at the surface without affecting the bulk properties of the polymer. Several polyesterases are reported to be active in the hydrolysis of PET. Two high homologous cutinases from *Thermobifida cellulosilytica* (Gamerith et al., 2017; Herrero Acero et al., 2011; Vecchiato et al., 2017) showed high hydrolytic efficiency.

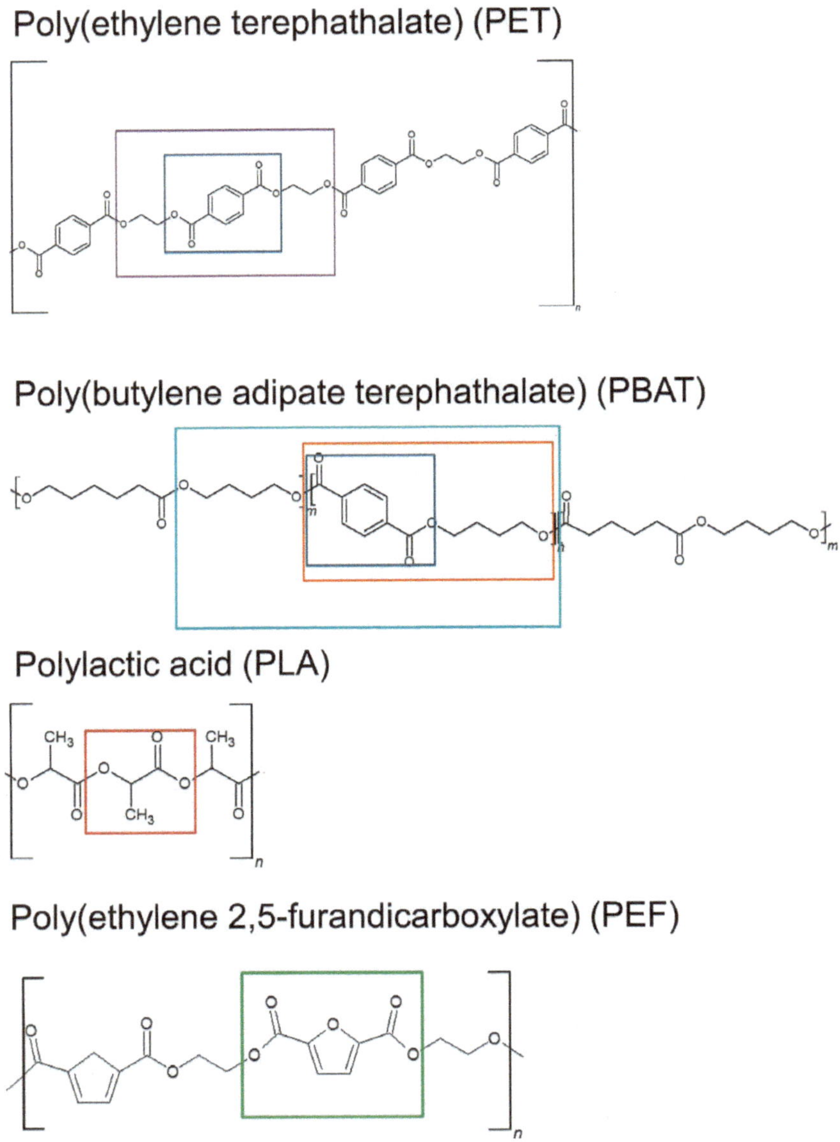

Fig. 1 Monomers and oligomers from polyesters hydrolysis: Ta (blue), MHET (yellow) and BHET (purple) from PET; Ta (blue), BTa (orange) and BTaB (cyan) from PBAT; La (red) from PLA and FDCA (green) from PEF.

Different PET-substrates are incubated with cutinases from *Thermobifida fusca* (Matsumura et al., 2008; Then et al., 2015), *Fusarium solani* (Kwon, Suk, Ho, Keun, & Kwang, 2009), *Pseudomonas mendocina* (Hsieh & Cram, 2014) and *Humicola insolens* cutinases (HiC) (Quartinello et al., 2017;

Ronkvist et al., 2009; Roth et al., 2014), reporting to be active on PET. Among cutinases, HiC showed high temperature stability at 70°C, which corresponds to an optimal hydrolysis temperature, being close to the T_g of PET. At such temperature, polyester chains in the amorphous area have high mobility which leads to higher accessibility of the enzyme to the ester bonds. The surface hydrolysis of PET with polyesterase improved antistatic behavior, wetting and dying abilities, which find perfect application in textile manufacture (Vertommen, Nierstrasz, Van Der Veer, & Warmoeskerken, 2005).

1.2 Poly(butylene adipate-co-butylene terephthalate) (PBAT)

PBAT is a biodegradable random copolymer composed of adipic acid, 1,4-butanediol and terephthalic acid. Adipic acid and 1,4-butanediol are polymerized into their polyester dimer, DMT and 1,4-butanediol are also forming dimers. In a second step, the two dimers are transesterified to create PBAT (Fig. 1). It is a random copolymer, the ratio of the dimers defines the dispersity of the polymer chain lengths. PBAT is generally marketed as a fully biodegradable alternative to low-density polyethylene, having many similar properties including flexibility and resilience allowing many applications such as plastic bags and wraps. PBAT is produced at industrial scale and used as a component of polymer blends e.g. in combination with starch. The biological decomposability of PBAT under composting conditions is demonstrated by different research groups (Kijchavengkul et al., 2010; Lucas et al., 2008). Hydrolysis of PBAT is not only demonstrated by cutinases from aerobic *Humicola insolens* and *Thermobifida cellulosilytica* (Perz et al., 2016) but also by enzymes from anaerobic *Clostridium botulinum* (Perz, Baumschlager, et al., 2016), *Pelosinus fermentans* (Biundo et al., 2016) and *Pseudomonas pseudoalcaligenes* (Wallace et al., 2017).

1.3 PolyLactic acid (PLA)

PLA is a thermoplastic, biocompatible and biodegradable polymer, which can be obtained in fermentation processes from degraded waste material (corn, sugarcane, or sugar beet pulp) (Fig. 1). Due to its biocompatibility, PLA is used for biomedical applications taking advantage of the fact that it can be absorbed by humans (Pellis et al., 2016). It is also widely used as a packaging material, as feedstock material in desktop fused filament fabrication 3D printers, exploiting its ability to be biodegradable. On the other hand, this polymer shows limited application due to low toughness at room temperature and low melt strength. Therefore, copolymerization with plastizicers or blending with other material is necessary, which requires prior activation. Different lipases from *Aspergillus niger*, *Candida cylindracea*, and *Candida rugosa* are tested on PLA and optimum surface hydrolysis conditions are reported, enabling the preservation of their mechanical or structural properties (Lee & Song, 2011). Cutinases have also been investigated for their ability to hydrolyze PLA films and/or fibers gaining attention for applications in polymer surface functionalization (Pellis et al., 2015; Vecchiato et al., 2017). Ribitsch et al. (Ribitsch et al., 2012) reported that cutinase from *Thermobifida halotolerans* hydrolyzes PLA films, leading to the release of lactic acid monomers. The resulting polar groups on the PLA surface cause a decreasing of the hydrophobicity of the polymer.

1.4 Poly(ethylene 2,5-furandicarboxylate) (PEF)

Polyethylene 2,5-furandicarboxylate is an attractive fully bioderived polymer that can be produced by polycondensation of 2,5-furandicarboxylic acid (FDCA) and ethylene glycol (Fig. 1). As aromatic polyester containing ethylene glycol, it exhibits great potential to replace polyethylene terephthalate (PET). Compared to PET, this polyester offers different benefits: superior glass transition and lower melting point, recyclable and competitive costs at industrial scale. PEF exhibits mostly similar physical and mechanical properties to PET, identifying PEF as potential substitute of PET in some application. A life-cycle assessment of bioderived PEF against fossil-fuel based PET demonstrated the reduction of greenhouse emission of up to 55% (Burgess et al., 2014). Enzymatic hydrolysis of ester bonds on the surfaces of PEF was demonstrated for films with different crystallinities via incubation with cutinases from *Thermobifida cellulosilytica* and from *Humicola insolens* (Weinberger et al., 2017), where around 20% of weight lost was shown within 72 h.

1.5 Analysis of enzymatic effects on polyester surfaces

Partial hydrolysis of ester bonds on polyester surface or fibers has been detected via different analytical methods. Enzymatically released monomers or small oligomers can be quantified by UV detection at 245 nm (Ribitsch et al., 2011) or 260 nm (Weinberger et al., 2017), following separation by reversed-phase high performance liquid chromatography (HPLC), or refractive index detector (HPLC-RI) in case of sugar related compounds. Furthermore, released higher oligomers can be monitored via Gel permeation chromatography (GPC) (Pellis et al., 2015). GPC considers dispersity (Ð) as well as molecular weight. Smaller analytes enter into the pores more easily and spend therefore more time in those pores leading to increased retention times. Conversely, larger analytes spend less time in the pores and are quickly eluted. Oligomers are, as well, characterized by time-of-flight mass spectrometry (TOF-MS) (Jackson & Simonsick, 1997), which determines the mass to charge ratio of ions via time of flight measurements. Ions are accelerated by an electric field. This acceleration results in an ion with the same kinetic energy as any other ion that has the same charge. For instance, the speed of the ions depends on the mass-to-charge ratio (heavier ions of the same charge reach lower speeds, although ions with higher charge will also increase in velocity).

Changes in water permeability are also relevant regarding the degree of functionalization of the synthetic polymers (Alisch-Mark, Herrmann, & Zimmermann, 2006; Fischer-Colbrie et al., 2004; Gamerith et al., 2017; Mueller, 2006). The hydraulic permeability of fabrics or membranes is a test which makes it possible to determine the polymer surface functionalization. In the same way, the Water Contact Angle (WCA) is a direct index of the presence of surface functional groups. The contact angle is the angle (Vecchiato, Ahrens, et al., 2017), conventionally measured through the liquid, where a liquid–vapor interface meets a solid surface. It quantifies the wettability of a solid surface by a liquid via the Young equation (Eq. 1) (Li, Huang, Chen, & Lai, 2017).

$$\gamma_{SG} - \gamma_{SL} - \gamma_{LG} \cos\theta_C = 0 \tag{1}$$

γ_{SG} is the interfacial energy, γ_{SL} is the solid–liquid interfacial energy, γ_{LG} is surface tension, θ_C is the equilibrium contact angle. A given system of solid, liquid, and vapor at a given temperature and pressure has a unique equilibrium contact angle.

The functional groups obtained from the enzymatic hydrolysis have been coupled with reactive compounds. The improved uptake of Toluidine Blue O ($C_{15}H_{16}N_3S$, cationic dye) (Vecchiato, Ahrens, et al., 2017) is a reflection of the carboxylic groups generated from ester bonds cleavage, which is quantified by colorimetric measurement. In a similar way, esterification with fluorescent alkyl bromide 2(bromomethyl)naphthalene (DMF) (Donelli et al., 2009; Herrero Acero et al., 2011) facilitates determination of reactive groups on enzyme treated surfaces (Fig. 2).

Fourier Transformed Infrared spectroscopy (FTIR) (Donelli, Freddi, Nierstrasz, & Taddei, 2010; Quartinello et al., 2017) is another method that determines the chemical changes in enzyme-treated surfaces in the micron range of the polymer. The transmittance and reflectance of the infrared rays

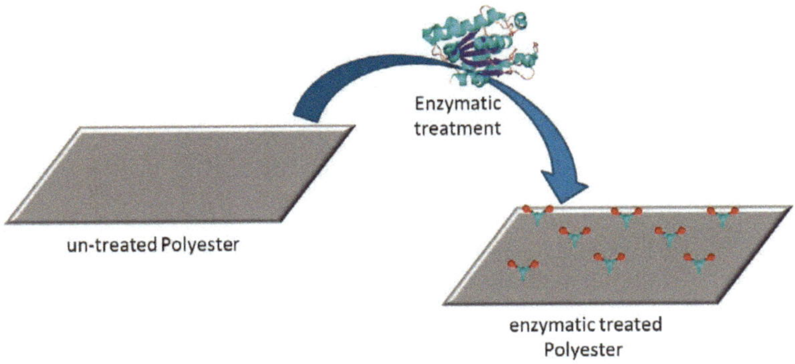

Fig. 2 Coupling of reactive dyes and fluorescence compounds for surface functional groups determination.

at different frequencies result in a certain spectral pattern related to the molecular bonding of the polymer. Due to the cleavage of ester bonds via enzymatic treatment, their corresponding bands resulted rearranged or even more decreased in comparison with the untreated polymer. Using attenuated total reflectance FTIR, more accurate differences can be detected, as changes in crystallinity/amorphous regions from polyesters (Donelli et al., 2009; Ronkvist et al., 2009). The results indicate that the amorphous region decreased after the enzymatic process. Another spectroscopic method is X-ray photoelectron spectroscopy (XPS), which is able to define the chemical composition of specific functional groups on the top of 10nm of the polymer surface (Brueckner, Eberl, Heumann, Rabe, & Guebitz, 2008). As expected, after treatment the number of –OH and –COOH groups are increased and vice versa the amount of C—C bonds is decreased (Feuerhack, Kisner, Pezzin, & Zimmermann, 2009).

Surface topography of enzyme treated samples has been characterized using Scanning Electron Microscopy (SEM). SEM reveals cracks, scaling and increased roughness of the surface. In high resolution atomic force microscopy (AFM), a special cantilever with a small tip is exploited to scan the sample surface. The deflection of such cantilever is measured via a laser beam, which reflects into a photodetector leading the generation of the surface map. As well for SEM, AFM is a method to measure the roughness of the surface after the enzymatic treatment (Weinberger et al., 2017).

2. Protocols

General laboratory safety procedures and personal protective materials (lab coat, gloves and safety glasses) should be used during the experiments.

2.1 Enzymatic treatment of polyesters

2.1.1 Standard esterase activity assay

Equipment

- Enzymes
- Buffers
- *p*-Nitrophenyl butyrate (*p*-NPB)
- Dimethyl sulfoxide (DMSO)
- 96 Well-plate microtiter
- Plate reader (Tecan INFINITE M200)

Method

Determination of esterase activity is performed with *p*-NPB as a standard substrate at 30°C. The assay mixture is prepared by mixing 40 μL substrate stock solution (86 μL *p*-NPB dissolved in 1000 μL DMSO) with 1 mL of buffer pH 7. Then, 200 μL of the assay mixture are incubated with 20 μL enzyme solution in a 96-well microtiter plate. Esterase activity is determined by following the increase of absorbance of *p*-nitrophenol at 405 nm ($\varepsilon_{405\ nm} = 9.36\ mMol^{-1}\ cm^{-1}$) in a TECAN plate reader (Fig. 3). For blank reaction, the assay is performed using 20 μL buffer instead of enzyme. Activity is calculated in Units (U), where 1 U is defined as the amount of enzyme required to hydrolyze 1 μmol of substrate per minute under the given assay conditions.

2.1.2 Enzymatic treatment of polyester

Equipment

- Enzymes
- Substrates (film and powder or fabric)
    - PET
    - PBAT
    - PLA
    - PEF
- Triton X-100

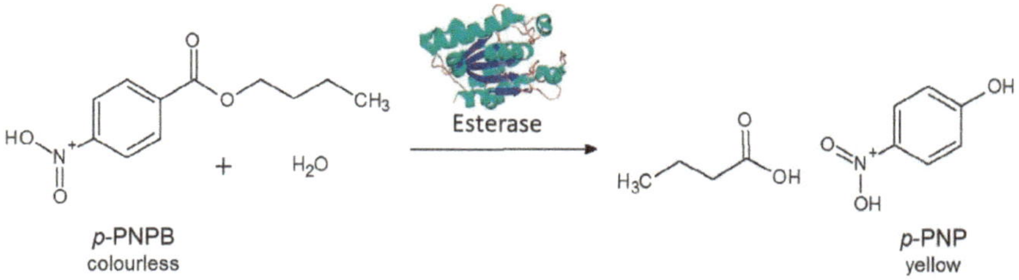

Fig. 3 Model substrate reaction.

- Na_2CO_3
- Ultrapure (MQ)-water
- Potassium phosphate buffer
- Tris(hydroxymethyl)aminomethane hydrochloride (Tris) buffer
- Orbital shaker (IKA KS 4000 ic control, Germany)

Method

Respective polyester films are cut into 2cm × 2cm pieces and washed three times in order to remove impurities. The washing steps are performed in the order: Triton X-100 ($5\,gL^{-1}$), Na_2CO_3 (100mM), and MQ-water. In each step the film is incubated for 30min at 50°C and dried with a constant air flow for 5min at 21°C. Then, films are incubated in buffer pH 7 with the enzyme (final volume 2mL with 5μM of enzyme) and 150rpm as described in Table 2.

At certain time points, samples are withdrawn from the reaction for further analysis of degradation products. After incubation, films are washed as described before in order to completely remove the protein from the surface. Blank reactions containing film in buffer without enzyme are performed under the same conditions. All experiments are performed in triplicates (Fig. 4).

2.1.3 Enzyme adsorption and mass loss detection by QCM-D

Equipment

- QCM-D E4 system (QSense, Sweden)
- QCM-D sensor (QSX 301, Microvacuum)
- Spin coater (WS-650MZ-23NPP, Laurell Technologies)

Method

QCM-D E4 system with four flow cells is a useful method to monitor enzymatic hydrolysis of polyester thin films coated onto QCM-D sensors. QCM-D is a piezoacoustic resonator technique that measures changes in resonance frequencies Δf_i (Hz) and energy dissipations ΔD_i of a piezoquartz crystal embedded into the QCM-D sensor. During polymer hydrolysis, the fundamental tone ($i=1$) as well as the six oscillation overtones ($i=$odd numbers between 3 and 13) are continuously measured. Measured changes of Δf_i values are converted into changes in the adlayer mass on the sensor surface, Δ_m (ng/cm^2), using the Sauerbrey equation (Eq. 2)

$$\Delta m = C * \frac{-\Delta fi}{i} \tag{2}$$

Table 2 List of enzymes applied for polyester functionalization.

Polymer	Enzymes	Conditions	Reference
PET	ThC_1	50°C Potassium phosphate buffer (100 mM, pH 7)	Quartinello et al. (2017) and Vecchiato, Ahrens, et al. (2017)
	HiC	65°C Potassium phosphate buffer (100 mM, pH 7)	
PBAT	PfL1	50°C Sodium phosphate buffer (100 mM, pH 7)	Biundo et al. (2016) and Biundo, Ribitsch, Steinkellner, Gruber, and Guebitz (2017)
	Cbotu_EstB	50°C Tris–HCl (100 mM, pH 7)	
PLA	ThC_1	37°C Potassium phosphate buffer (100 mM, pH 7)	Ortner et al. (2017) and Pellis et al. (2017)
	HiC	37°C Potassium phosphate buffer (100 mM, pH 7)	
PEF	ThC_1	50°C Potassium phosphate buffer (100 mM, pH 7)	Weinberger et al. (2017)
	HiC	65°C Potassium phosphate buffer (100 mM, pH 7)	

in which C (=17.7 $ng \cdot cm^{-2}\ Hz^{-1}$) is the sensor-specific mass sensitivity constant. The fifth overtone (i = 5) is used for plots and calculations. Preparation of polyester films is done by dissolving the polymer in chloroform to a final concentration of 0.5% (w/w). Then, 40 μL of the polyester solution are transferred onto a QCM-D sensor mounted on a spin coater, followed by spinning the sensor for 1 min at 4000 rpm with an acceleration of 1500 $rpm\,s^{-1}$. For calculation of the mass of each coated film, the decrease in the resonance frequencies of the sensor before and after the coating step is measured on the QCM-D in air at 20°C. Each polyester-coated QCM-D sensor is pre-equilibrated to the solution of hydrolysis experiments by immersing the sensor in a solution for 14 h at 25°C. The sensor is mounted into the flow cell of the QCM-D instrument which runs the experimental

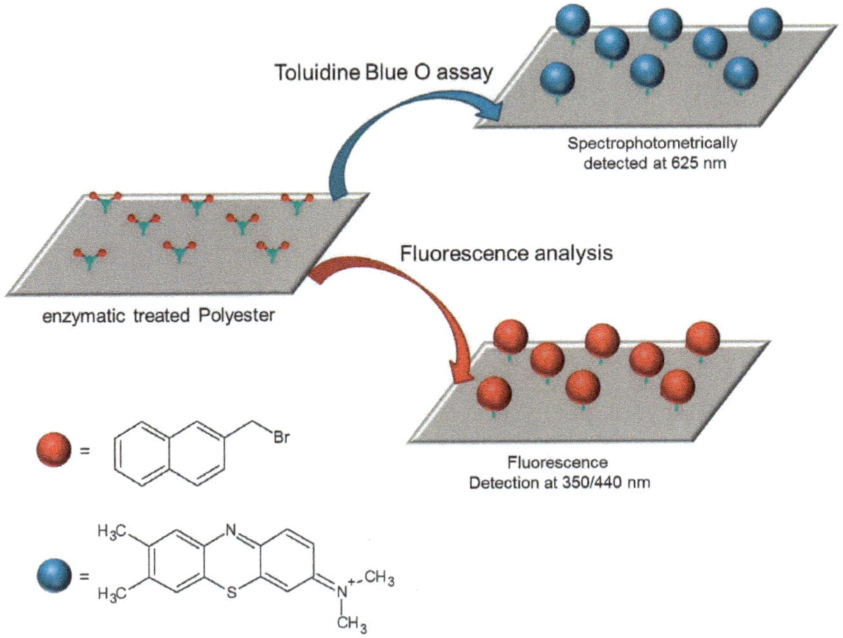

Fig. 4 Enzymatic surface functionalization.

solution over the sensor at a constant volumetric flow rate (20 μL min^{-1}) and a constant temperature. Upon reaching constant Δf_i and ΔD_i values, the solutions are changed to ones that contain one of the polyesterase while all other solution parameters remain the same. Hydrolysis of the polyesters results in increasing resonance frequencies and is followed until constant Δf_i and ΔD_i values are reached. Sensor from the flow cells are washed by dipping into MQ water and dried in a N_2 stream. In order to quantify the fraction of the coated polyester mass which has been removed by enzymatic hydrolysis, the resonance frequency of each sensor dried in air at 20°C after hydrolysis is compared with the untreated sample.

2.2 Detection of release products by chromatography

Equipment

- Terepthalic acid (Ta)
- Bis(hydroxyl terephtalate) (BHET)
- Mono (4-hydroxylbutyl terephthalate) (BTa)
- Bis(4-hydroxybutyl terephthalate) (BTaB)
- Lactic acid (LA)
- 2,5-furandicarboxylic acid (FDCA)

- Methanol
- MQ-water
- Formic acid
- 2% of $K_4[Fe(CN)_6] \cdot 3H_2O$
- 2% of $ZnSO_4 \cdot 7H_2O$
- 0.45 μm filters (PTFE/PA)
- HPLC vials
- HPLC (Agilent Technologies, 1260 Infinity, Palo Alto, CA, USA)
- Reversed phase column C18 (YMC 30, 250 × 4.6 mm ID, S-5 μm) (Agilent Technologies, 1260 Infinity, Palo Alto, CA, USA)
- Photodiode array detector (Agilent Technologies, 1290 Infinity II, Vienna, Austria)
- Agilent Technologies HPLC System (Agilent Technologies1260 Infinity)
- 17,369 6.0 mm ID* 40 mm L HHR-H, 5 μm Guard column (Agilent Technologies1260 Infinity)
- 18,055 7.8 mm ID*300 mm LGMHHR-N, 5 μm TSK gel liquid chromatography column (Tosoh Bioscience, Tessendro, Belgium)
- Tetrahydrofuran (THF)
- Agilent Technologies G1362A refractive index detector
- Polystyrene Calibration standards (400-2,000,000 Da)
- Filter paper (595/12, Whatman GmbH, Dassel, Germany)
- Reversed-phase C18 rapid resolution column (Waters Xterra, Milford, MA, USA) of 3.0 × 15 mm and 3.5 μm particle diameter
- Time-of-flight mass spectrometer (6230 TOF LC/MS, Agilent Technologies)
- Electrospray interface (Agilent Technologies, 1290 Infinity II, Vienna, Austria)
- Column ION-300 (Transgenomic, Inc.)
- Centrifuge (Hettich MIKRO 200 R, Tuttlingen, Germany)

2.2.1 High-performance liquid chromatography (HPLC)

Method

After enzymatic treatment of polyester powder or films, the enzyme is precipitated using ice cold MeOH. Samples are centrifuged at 14.000 rpm and 0°C for 15 min. Obtained supernatants are filtered through 0.45 μm PTFE or PA filters and filled in HPLC vials.

Degradation products released from PET, PBAT and PEF during enzymatic hydrolysis are analyzed by reversed phase HPLC (C18 column) using a H_2O/MeOH/HCOOH gradient. The flow rate is set to 0.85 mL min^{-1} and the column is kept constant at 40°C. The injection volume is 10 μL. Analytes are detected by a photodiode array detector at a wavelength of 245 nm (for PET and PBAT release products) and 260 nm (for PEF).

For PET functionalization studies, calibration curves for the main release products Ta and BHET are prepared. In case of PBAT degradation studies, calibration curves for Ta, BTa and BTaB are produced. PEF hydrolysis is quantified based on a standard FDCA calibration curve.

Following to enzymatic incubation, PLA degradation products are clarified by the Carrez-precipitation method. Therefore, 20 μL of 6.2% $K_4[Fe(CN)_6]\cdot 3H_2O$ solution are added to the samples, vortexed and incubated for 1 min. Then, 20 μL of 2% $ZnSO_4\cdot 7H_2O$ solution are added to the mixture. After vortexing and incubation for 5 min, samples are centrifuged (30 min, 14,000 rpm, 25°C), supernatants are filtered through a 0.45 μm filter membrane directly into glass vials for HPLC-RI. Flow is set to 0.325 mL min^{-1}, 0.01 N H_2SO_4 is used as mobile phase. Temperature is set to 45°C, the injection volume to 40 μL.

2.2.2 Gel permeation chromatography (GPC)

Method

Polyesters are dissolved in THF (addition of 250 ppm BHT as inhibitor) and filtered through filter paper. Gel permeation chromatography is performed using THF as eluent with the flow of 1 mL min^{-1} at 30°C. Molecular weight of the products is analyzed by polystyrene calibration standards.

2.2.3 Liquid chromatography/time of flight mass spectrometry (LC–MS/TOF)

Method

Samples are diluted in MeOH and filtered through 0.20 μm PTFE filters. As mobile phase A, mQ H_2O with 0.1% formic acid is used, the mobile phase B consists of MeOH. The flow rate is set to 1.0 mL min^{-1} and the column is maintained at a temperature of 30°C. Injection volume is 1.0 μL. Analytes are detected by a photodiode array detector at the wavelength of 254 nm. HPLC is connected to a time-of-flight mass spectrometer equipped with an electrospray interface under the following operating parameters: capillary 3500 V, nebulizer 20 psig, drying gas 10 L min^{-1}, gas temperature 325°C, fragmentor 200 V, skimmer 65 V,

OCT 1 RF Vpp 750 V. The mass axis is calibrated using the mixture provided by the manufacturer over the m/z 50–3000 range. A second orthogonal sprayer with a solution is used as a continuous calibration using the following reference masses: 121.0508 and 922.0097 m/z. Spectra are recorded over the 100–3000 m/z range at a scan rate of 2 spectra.

2.3 Surface analysis

Equipment

- Enzymatically treated polyester
- Triton X-100
- Na_2CO_3
- MQ-water
- Toluidine Blue O (TBO)
- Tris/HCl buffer (100 mM, pH 8.6)
- Sodium lauryl sulfate (SDS)
- *N,N*-dimethylformamide (DMF)
- 2-(bromomethyl)naphthalene
- 96 well-plate reader
- FS920 fluorescence spectrometer (Edinburgh Instruments, Livingston, UK) equipped with a Hamamatsu R928P red-sensitive photomultiplier (wavelength range from 200 to 850 nm).
- Drop Shape Analysis System DSA 100 (Kruss GmbH, Hamburg, Germany)
- Multiplate reader (Varioskan, Thermo)

2.3.1 Hydraulic permeability

Hydraulic permeability is measured by a dead-end setup using a 25 mL Amicon cell and a membrane diameter of 25 mm (Millipore). The cell is connected to a water reservoir which is under nitrogen pressure enabling the adjustment of trans-membrane pressure Δp, which is set to a maximum of 0.3 bar. The permeate is collected in a beaker for a certain time and the amount of water is determined gravimetrically. The permeate flux J is calculated according to Eq. (3), with the permeate volume V in a particular measuring time Δt and the effective membrane (filter) area A.

$$J = \frac{V}{A^* \Delta t} \tag{3}$$

Results are reported as water permeability P, calculated according to Eq. (4).

$$P = \frac{V}{\Delta p} \tag{4}$$

Three polyester membrane samples for each state are analyzed; average values are reported.

2.3.2 Hydrophobicity by water contact angle

Method

Hydrophobicity of enzymatically treated polyester film is quantified via Water Contact Angle (WCA) measurements. Polymer films are measured after enzymatic treatments. The proteins are washed from the surface using three consecutive washing as described above. Afterward, films are rinsed with MQ-water and dried. A sample incubated with buffer (100 mM, pH 7) under the same conditions as above is used as a blank. Polymeric films are analyzed with the Drop Shape Analysis System DSA 100 using ddH_2O as test liquid with a drop size of 5 μL and a deposition speed of 60 U min^{-1}. The water contact angle is measured after 1 s from the deposition of the drop, and the data are analyzed using a video recording method. Each treatment is performed in triplicates for both samples in order to allow a homogeneous distribution of the water on the sample. The data are obtained from the average of nine measurements.

The surface free energy (SFE) is determined using the video recording method after the incubation of polyester with enzyme for 24 and 72 h. A blank incubated only with buffer is used as the control. Analysis is done by three different liquid solutions (water, glycerol, olive oil). Owens-Wendt-Rabel-Kaelble (OWRK) method is applied as a standard approach for calculating the surface free energy of polymeric films from water contact angle. Briefly, as interaction between two phases (liquid and polymer surface in this case) can be split in different components (polar, dispersive, and hydrogen bonding) and the OWRK model is able to discriminate polar and dispersive energy fraction of the surface energy (Janssen, De Palma, Verlaak, Heremans, & Dehaen, 2006).

2.3.3 Determination of carboxylic groups on the surface

Surface modification is quantified by determination of carboxylic groups formed on the polyester surface after enzymatic treatment. Surface carboxylic groups (SC) are determined using the Toluidine Blue O (TBO) method as reported by Rödiger et al. with some modifications (Vecchiato, Ahrens, et al., 2017).

Therefore, samples and blanks are incubated in 0.1% TBO/100 mM Tris HCl pH 8.6 for 15 min at 50°C and 130 rpm (6 mL). Then, samples are washed five times in 100 mM Tris HCl pH 8.6 for 5 min until the washing solution is clear. Samples are transferred to 20 mL vials to desorb the TBO bound to the surface carboxylic groups, by washing with 20% SDS for 30 min, at 50°C and 130 rpm. The released TBO is quantified spectrophotometrically by measuring the absorbance at 625 nm and 23°C. From each solution, 200 μL are transferred into a well of a 96- well plate reader. SC are calculated according to the following Eq. 5:

$$SC = (A * V)/(A_s * d * \varepsilon) \tag{5}$$

A is the absorption at 625 nm. V is the volume of desorption solution [L], A_s the area of the sample surface [mm^{-2}], d the light path way [cm] and ε the extinction coefficient of TBO [54,800 $L\,mol^{-1}\,cm^{-1}$]. SC is the surface density carboxylic groups [$nmol\,mm^{-2}$].

2.3.4 Fluorescence analysis

Films of 50*10 mm size are alkylated in 20 mL of DMF containing 2-(bromomethyl) naphthalene (0.05 M) and potassium fluoride (0.02 M). Samples are incubated for 3 h at room temperature under shaking conditions at 100 rpm. Thereafter films are washed with DMF in order to remove unreacted reagents and rinsed with distilled water. Photoluminiscence emission and excitation spectra are measured using a fluorescence spectrometer. Emission and excitation spectra of surface modified polymers are quantified with excitation and emission of 350 and 440 nm, under the angle of 45°. Fluorescence emission values are directly correlated with the amount of coupled 2-(bromomethyl naphthalene), which is giving information of the amount of carboxylic groups generated after the enzymatic treatment.

2.4 Spectroscopy measurements

2.4.1 Fourier transform-infrared spectroscopy (FTIR)

Measurements are performed on a NEXUS Thermo Nicolet FTIR spectrometer employing an Attenuated Total Reflectance (ATR) accessory mod Smart Performer. All spectra are obtained with a Ge crystal cell Spectra are normalized before data processing (Ortner et al., 2017; Quartinello et al., 2017; Vecchiato, Ahrens, et al., 2017).

2.4.2 X-ray photoelectron spectroscopy (XPS)

XPS is performed on an Multiprobe UHV-surface-analysis system (Omicron Nanotechnology) equipped with a DAR400 X-ray source, an Al anode and a quartz-crystal monochromator XM 500 using an X-ray excitation energy of 1486.70 eV (Al Ka1-line). The monochromated X-ray line width is 0.3 eV. The analyzed surface area is approximately 1 mm^2.

2.5 Microscopy

2.5.1 Scanning electron microscopy (SEM)

The surface morphology of enzymatically hydrolyzed films is observed after 48 h using SEM. A control sample consisting of a film without enzymatic treatment is also measured. All SEM images are acquired by collecting secondary electrons on a Hitachi 3030TM (Metrohm INULA GmbH, Vienna, Austria) working at Energy Dispersive X-ray Spectrom (EDX) acceleration voltage.

2.5.2 Atomic force microscopy (AFM)

Atomic force microscopy (AFM) is used to obtain high resolution 3D images of polymer films. Samples are mounted on metallic plates using epoxy glue and subsequently characterized using an MFP-3D AFM (Asylum Research, Oxford Instruments, Santa Barbara, CA, USA). Measurements are carried out in the air at room temperature working in dynamic mode. Cantilevers, characterized by a resonant frequency of 320 kHz and force constant of 40 nN/nm (NSG30, NT-MDT, Moscow, RU), are used working at low oscillation amplitudes with a half free-amplitude set point. Images are acquired at 256 × 256 pixels at a one line/s scan speed. All AFM data are analyzed using Gwyddion open-source SPM analysis software (Pellis et al., 2015) in particular three-dimensional reconstructions are used to evaluate surface roughness. Surface roughness is computed as the root-mean-square (RMS) value of the height irregularities of AFM images.

Acknowledgments

This project received funding from the European Union's Horizon 2020 research and innovation program under the grant agreement 641942 (Resyntex project) and Austrian Centre of Industrial Biotechnology (ACIB), which is funding by the Austrian BMWFW, BMVIT, SFG, Standortagentur Tirol, Government of Lower Austria and Business Agency Vienna through the Austrina FFG-COMET-Funding Program.

References

Abdelaal, M. Y., Sobahi, T. R., & Makki, M. S. I. (2008). Chemical degradation of poly(ethylene terephthalate). *International Journal of Polymeric Materials*, *57*, 73–80. https://doi.org/10.1080/00914030701329080.

Acero, E. H., Ribitsch, D., Steinkellner, G., Gruber, K., Greimel, K., Eiteljoerg, I., et al. (2011). Enzymatic surface hydrolysis of PET: Effect of structural diversity on kinetic properties of cutinases from thermobifida. *Macromolecules*, *44*, 4632–4640. https://doi.org/10.1021/ma200949p.

Alisch-Mark, M., Herrmann, A., & Zimmermann, W. (2006). Increase of the hydrophilicity of polyethylene terephthalate fibres by hydrolases from Thermomonospora fusca and Fusarium solani f. sp. pisi. *Biotechnology Letters*, *28*, 681–685. https://doi.org/10.1007/s10529-006-9041-7.

Araújo, R., Casal, M., & Cavaco-Paulo, A. (2008). Application of enzymes for textile fibres processing. *Biocatalysis and Biotransformation*, *26*, 332–349. https://doi.org/10.1080/10242420802390457.

Azfarniam, L., & Norouzi, M. (2016). Multifunctional polyester fabric using a multicomponent treatment. *Fibers and Polymers*, *17*, 298–304. https://doi.org/10.1007/s12221-016-5579-8.

Biundo, A., Hromic, A., Pavkov-Keller, T., Gruber, K., Quartinello, F., Haernvall, K., et al. (2016). Characterization of a poly(butylene adipate-co-terephthalate)-hydrolyzing lipase from Pelosinus fermentans. *Applied Microbiology and Biotechnology*, *100*, 1753–1764. https://doi.org/10.1007/s00253-015-7031-1.

Biundo, A., Ribitsch, D., Steinkellner, G., Gruber, K., & Guebitz, G. M. (2017). Polyester hydrolysis is enhanced by a truncated esterase: Less is more. *Biotechnology Journal*, *12*, 1–5. https://doi.org/10.1002/biot.201600450.

Brueckner, T., Eberl, A., Heumann, S., Rabe, M., & Guebitz, G. M. (2008). Enzymatic and chemical hydrolysis of poly (ethylene terephthalate) fabrics. *Journal of Polymer Science, Part A: Polymer Chemistry*, *46*, 6435–6443. https://doi.org/10.1002/pola.22952.

Burgess, S. K., Leisen, J. E., Kraftschik, B. E., Mubarak, C. R., Kriegel, R. M., & Koros, W. J. (2014). Chain mobility, thermal, and mechanical properties of poly(ethylene furanoate) compared to poly(ethylene terephthalate). *Macromolecules*, *47*, 1383–1391. https://doi.org/10.1021/ma5000199.

C. Chan, T. Ko, Polymer surface modification by plasmas and photons, (n.d.).

Donelli, I., Freddi, G., Nierstrasz, V. A., & Taddei, P. (2010). Surface structure and properties of poly-(ethylene terephthalate) hydrolyzed by alkali and cutinase. *Polymer Degradation and Stability*, *95*, 1542–1550. https://doi.org/10.1016/j.polymdegradstab.2010.06.011.

Donelli, I., Taddei, P., Smet, P. F., Poelman, D., Nierstrasz, V. A., & Freddi, G. (2009). Enzymatic surface modification and functionalization of PET: A water contact angle, FTIR, and fluorescence spectroscopy study. *Biotechnology and Bioengineering*, *103*, 845–856. https://doi.org/10.1002/bit.22316.

Feuerhack, A., Kisner, A., Pezzin, S. H., & Zimmermann, W. (2009). Biocatalytic surface modification of knitted fabrics made of poly (ethylene terephthalate) with hydrolytic enzymes from Thermobifida fusca KW3b. *2422*, 357–364. https://doi.org/10.1080/10242420802360948.

Fischer-Colbrie, G., Heumann, S., Liebminger, S., Almansa, E., Cavaco-Paulo, A., & Guebitz, G. M. (2004). New enzymes with potential for PET surface modification. *Biocatalysis and Biotransformation*, *22*, 341–346. https://doi.org/10.1080/10242420400024565.

Gamerith, C., Gajda, M., Ortner, A., Acero, E. H., Guebitz, G. M., & Ulbricht, M. (2017). Enzymatic hydrolysis of poly(ethyleneterephthalate) used for and analysed by pore modification of track-etched membranes. *New Biotechnology*, *39*, 42–50. https://doi.org/10.1016/j.nbt.2017.06.007.

Guebitz, G. M., & Cavaco-Paulo, A. (2008). Enzymes go big: Surface hydrolysis and functionalisation of synthetic polymers. *Trends in Biotechnology, 26*, 32–38. https://doi.org/10.1016/j.tibtech.2007.10.003.

Guruvenket, S., Rao, G. M., Komath, M., & Raichur, A. M. (2004). Plasma surface modification of polystyrene and polyethylene. *Applied Surface Scienc, 236*, 278–284. https://doi.org/10.1016/j.apsusc.2004.04.033.

Hsieh, Y., & Cram, L. A. (2014). Enzymatic hydrolysis to improve wetting and absorbency of polyester fabrics. *Textile Research Journal, 68*. https://doi.org/10.1177/004051759806800501.

Jackson, C. A., & Simonsick, W. J. (1997). Application of mass spectrometry to the characterization of polymers. *Current Opinion in Solid State and Materials Science, 2*, 661–667.

Janssen, D., De Palma, R., Verlaak, S., Heremans, P., & Dehaen, W. (2006). Static solvent contact angle measurements , surface free energy and wettability determination of various self-assembled monolayers on silicon dioxide. *Thin Solid Films, 515*, 1433–1438. https://doi.org/10.1016/j.tsf.2006.04.006.

Kijchavengkul, T., Auras, R., Rubino, M., Selke, S., Ngouajio, M., & Fernandez, R. T. (2010). Biodegradation and hydrolysis rate of aliphatic aromatic polyester. *Polymer Degradation and Stability, 95*, 2641–2647. https://doi.org/10.1016/j.polymdegradstab.2010.07.018.

Kim, H. R., & Song, W. S. (2006). Lipase treatment of polyester fabrics. *Fibers and Polymers, 7*, 339–343.

Kwon, M., Suk, H., Ho, T., Keun, B., & Kwang, J. (2009). High-level expression and characterization of Fusarium solani cutinase in Pichia pastoris. *Protein Expression and Purification, 68*, 104–109. https://doi.org/10.1016/j.pep.2009.06.021.

Lee, S. H., & Song, W. S. (2011). Enzymatic hydrolysis of polylactic acid fiber. *Applied Biochemistry and Biotechnology, 164*, 89–102. https://doi.org/10.1007/s12010-010-9117-7.

Li, S., Huang, J., Chen, Z., & Lai, Y. (2017). A review on special wettability textiles: Theoretical models, fabrication technologies and multifunctional applications. *Journal of Materials Chemistry A, 5*, 31–55. https://doi.org/10.1039/c6ta07984a.

Lucas, N., Bienaime, C., Belloy, C., Queneudec, M., Silvestre, F., & Nava-saucedo, J. (2008). Chemosphere polymer biodegradation: Mechanisms and estimation techniques. *Chemosphere, 73*, 429–442. https://doi.org/10.1016/j.chemosphere.2008.06.064.

Manalo, A. C., Wani, E., Azwa, N., & Karunasena, W. (2015). Effects of alkali treatment and elevated temperature on the mechanical properties of bamboo fi bre e polyester composites. *Composites: Part B, 80*, 73–83. https://doi.org/10.1016/j.compositesb.2015.05.033.

Matsumura, S., Hlil, A. R., Lepiller, C., Gaudet, J., Guay, D., Shi, Z., et al. (2008). Ionomers for proton exchange membrane fuel cells with sulfonic acid groups on the end-groups: Novel branched poly(ether-ketone)s. *American Chemical Society Polymeric Preprints (American Chemical Society, Division of Polymer Chemistry), 49*, 511–512. https://doi.org/10.1002/pola.

Mueller, R. (2006). Biological degradation of synthetic polyesters—Enzymes as potential catalysts for polyester recycling. *Process Biochemistry, 41*, 2124–2128. https://doi.org/10.1016/j.procbio.2006.05.018.

Ortner, A., Pellis, A., Gamerith, C., Yebra, A. O., Scaini, D., Kaluzna, I., et al. (2017). Superhydrophobic functionalization of cutinase activated poly(lactic acid) surfaces. *Green Chemistry, 19*, 816–822. https://doi.org/10.1039/c6gc03150a.

Pellis, A., Acero, E. H., Weber, H., Obersriebnig, M., Breinbauer, R., Srebotnik, E., et al. (2015). Biocatalyzed approach for the surface functionalization of poly(L-lactic acid) films using hydrolytic enzymes. *Biotechnology Journal, 10*, 1739–1749. https://doi.org/10.1002/biot.201500074.

Pellis, A., Gamerith, C., Ghazaryan, G., Ortner, A., Herrero Acero, E., & Guebitz, G. M. (2016). Ultrasound-enhanced enzymatic hydrolysis of poly(ethylene terephthalate). *Bioresource Technology, 218*, 1298–1302. https://doi.org/10.1016/j.biortech.2016.07.106.

Pellis, A., Herrero Acero, E., Ferrario, V., Ribitsch, D., Guebitz, G. M., & Gardossi, L. (2016). The closure of the cycle: Enzymatic synthesis and functionalization of bio-based polyesters. *Trends in Biotechnology*, *34*, 316–328. https://doi.org/10.1016/j.tibtech.2015.12.009.

Pellis, A., Silvestrini, L., Scaini, D., Coburn, J. M., Gardossi, L., Kaplan, D. L., et al. (2017). Enzyme-catalyzed functionalization of poly (L-lactic acid) for drug delivery applications. *Process Biochemistry*, *59 (*, 77–83. https://doi.org/10.1016/j.procbio.2016.10.014.

Perz, V., Baumschlager, A., Bleymaier, K., Zitzenbacher, S., Hromic, A., Steinkellner, G., et al. (2016). Hydrolysis of synthetic polyesters by clostridium botulinum esterases. *Biotechnology and Bioengineering*, *113*, 1024–1034. https://doi.org/10.1002/bit.25874.

Perz, V., Bleymaier, K., Sinkel, C., Kueper, U., Bonnekessel, M., Ribitsch, D., et al. (2016). Substrate specificities of cutinases on aliphatic-aromatic polyesters and on their model substrates. *New Biotechnology*, *33*, 295–304. https://doi.org/10.1016/j.nbt.2015.11.004.

Quartinello, F., Vajnhandl, S., Volmajer Valh, J., Farmer, T. J., Vončina, B., Lobnik, A., et al. (2017). Synergistic chemo-enzymatic hydrolysis of poly(ethylene terephthalate) from textile waste. *Microbial Biotechnology*, *10*, 1376–1383. https://doi.org/10.1111/1751-7915.12734.

Rahman, M., & East, G. C. (2009). Titanium dioxide particle-induced alkaline degradation of poly(ethylene terephthalate): Application to medical textiles. *Textile Research Journal*, *79*, 728–736. https://doi.org/10.1177/0040517507081916.

Ribitsch, D., Acero, E. H., Greimel, K., Dellacher, A., Zitzenbacher, S., Marold, A., et al. (2012). A new esterase from Thermobifida halotolerans hydrolyses polyethylene terephthalate (PET) and polylactic acid (PLA). *Polymers (Basel)*, *4*, 617–629. https://doi.org/10.3390/polym4010617.

Ribitsch, D., Heumann, S., Trotscha, E., Herrero Acero, E., Greimel, K., Leber, R., et al. (2011). Hydrolysis of polyethyleneterephthalate by p-nitrobenzylesterase from Bacillus subtilis. *Biotechnology Progress*, *27*, 951–960. https://doi.org/10.1002/btpr.610.

Ronkvist, Å. M., Xie, W., Lu, W., & Gross, R. A. (2009). Cutinase-Catalyzed hydrolysis of poly(ethylene terephthalate). *Macromolecules*, *42*, 5128–5138. https://doi.org/10.1021/ma9005318.

Roth, C., Wei, R., Oeser, T., Then, J., Föllner, C., Zimmermann, W., et al. (2014). Structural and functional studies on a thermostable polyethylene terephthalate degrading hydrolase from Thermobifida fusca. *Applied Microbiology and Biotechnology*, *98*, 7815–7823. https://doi.org/10.1007/s00253-014-5672-0.

Sever, K., Sarikanat, M., Erkan, G., Erdo, Ü. H., & Erden, S. (2012). Surface treatments of jute fabric: The influence of surface characteristics on jute fabrics and mechanical properties of jute/polyester composites. *Industrial Crops and Products*, *35*, 22–30. https://doi.org/10.1016/j.indcrop.2011.05.020.

Then, J., Wei, R., Oeser, T., Barth, M., Belis??rio-Ferrari, M. R., Schmidt, J., et al. (2015). Ca^{2+} and Mg^{2+} binding site engineering increases the degradation of polyethylene terephthalate films by polyester hydrolases from Thermobifida fusca. *Biotechnology Journal*, *10*, 592–598. https://doi.org/10.1002/biot.201400620.

Vecchiato, S., Ahrens, J., Pellis, A., Scaini, D., Mueller, B., Herrero Acero, E., et al. (2017). Enzymatic functionalization of HMLS-polyethylene terephthalate fabrics improves the adhesion to rubber. *ACS Sustainable Chemistry & Engineering*, *5*, 6456–6465. https://doi.org/10.1021/acssuschemeng.7b00475.

Vecchiato, S., Skopek, L., Jankova, S., Pellis, A., Ipsmiller, W., Aldrian, A., et al. (2017). Enzymatic recycling of high-value phosphor flame-retardant pigment and glucose from rayon fibers. *ACS Sustainable Chemistry & Engineering*, *6*, 2386–2394. acssuschemeng.7b03840. https://doi.org/10.1021/acssuschemeng.7b03840.

Vertommen, M. A. M. E., Nierstrasz, V. A., Van Der Veer, M., & Warmoeskerken, M. M. C. G. (2005). Enzymatic surface modification of poly(ethylene terephthalate). *Journal of Biotechnology*, *120*, 376–386. https://doi.org/10.1016/j.jbiotec.2005.06.015.

Wallace, P. W., Haernvall, K., Ribitsch, D., Zitzenbacher, S., Schittmayer, M., Steinkellner, G., et al. (2017). PpEst is a novel PBAT degrading polyesterase identified by proteomic screening of Pseudomonas pseudoalcaligenes. *Applied Microbiology and Biotechnology*, *101*, 2291–2303. https://doi.org/10.1007/s00253-016-7992-8.

Wei, R., & Zimmermann, W. (2017). Microbial enzymes for the recycling of recalcitrant petroleum-based plastics: How far are we? *Microbial Biotechnology*, *10*, 1308–1322. https://doi.org/10.1111/1751-7915.12710.

Weinberger, S., Haernvall, K., Scaini, D., Ghazaryan, G., Zumstein, M. T., Sander, M., et al. (2017). Enzymatic surface hydrolysis of poly(ethylene furanoate) thin films of various crystallinities. *Green Chemistry*, *19*, 5381–5384. https://doi.org/10.1039/C7GC02905E.

Yoshioka, T., Sato, T., & Okuwaki, A. (1994). Hydrolysis of waste PET by sulfuric acid at 150°C for a chemical recycling. *Journal of Applied Polymer Science*, *52*, 1353–1355. https://doi.org/10.1002/app.1994.070520919.

FSC
www.fsc.org
MIX
Papier | Fördert
gute Waldnutzung
FSC® C083411

Kolibri 360 GmbH
Ferdinand-Jühlke-Straße 7
99095 Erfurt, Deutschland
produktsicherheit@kolibri360.de